WORLDS OF FLO

Worlds of Flow

*A History of Hydrodynamics from
the Bernoullis to Prandtl*

OLIVIER DARRIGOL

OXFORD

UNIVERSITY PRESS

OXFORD

UNIVERSITY PRESS

Great Clarendon Street, Oxford OX2 6DP

Oxford University Press is a department of the University of Oxford.
It furthers the University's objective of excellence in research, scholarship,
and education by publishing worldwide in

Oxford New York

Auckland Cape Town Dar es Salaam Hong Kong Karachi
Kuala Lumpur Madrid Melbourne Mexico City Nairobi
New Delhi Shanghai Taipei Toronto

With offices in

Argentina Austria Brazil Chile Czech Republic France Greece
Guatemala Hungary Italy Japan Poland Portugal Singapore
South Korea Switzerland Thailand Turkey Ukraine Vietnam

Oxford is a registered trade mark of Oxford University Press
in the UK and in certain other countries

Published in the United States
by Oxford University Press Inc., New York

British Library Cataloguing in Publication Data

(Data available)

Library of Congress Cataloguing in Publication Data

Darrigol, Olivier.
Worlds of flow : a history of hydrodynamics from the Bernoullis to
    Prandtl / Oliver Darrigol.
      p. cm.
Includes bibliographical references and index.
ISBN 0-19-856843-6 (alk. paper)
1. Hydrodynamics–History. I. Title.
QC151.D25 2005
532'.5'09–dc22
200501963

Typeset by SPI Publisher Services, Pondicherry, India.
Printed in the UK
on acid-free paper by the MPG Books Group

ISBN 978–0–19–856843–8 (Hbk.)   978–0–19–955911–4 (Pbk.)

10 9 8 7 6 5 4 3 2 1

PREFACE

> Now I think hydrodynamics is to be the root of all physical science, and is at present second to none in the beauty of its mathematics.[1] (William Thomson, December 1857)

In Victorian Britain, William Thomson wanted hydrodynamics to reign over all the physical sciences, from the intimate constitution of ether and matter to the beautiful pattern of ship waves. Although today's scientists tend to ascribe a less dominant role to this theory, it still inspires many human endeavors and it challenges the skills of a large community of experts. Many of the problems of modern hydrodynamics have roots in Thomson's times. Despite tremendous improvements in methods, the basic equations and the basic questions have been more stable in this field than in any other domain of physics. These circumstances probably explain the interest that students of hydrodynamics often manifest in the history of their subject.

The present book aims at satisfying this curiosity, in a way that only requires an elementary knowledge of modern fluid mechanics. Through its focus on the concrete, worldly circumstances of theoretical progress, it should appeal to a great variety of users of hydrodynamics. By attending to the technicalities of concept formation and evolution, it may help budding specialists understand the foundations of their own field. By exploring deep genetic connections with other domains of knowledge, it should also interest historians of science. In addition, this history answers a few questions recently raised in the philosophy of science, regarding the failures of purely deductive conceptions of physical theory and the role of asymptotic theory and singular approximations.

Hydrodynamics may be defined as the art of subjecting flow to the general principles of dynamics. This book tells how, in the eighteenth century, a small elite of Swiss and French geometers inaugurated this theory; how nineteenth-century engineers, physicists, and mathematicians widely expanded its concepts, methods, and purposes; and how its concrete applications nevertheless failed until, in the first third of the twentieth century, it evolved into a reliable guide for the engineers of flow.

There were many natural and artificial worlds of flow to which hydrodynamics was meant to be relevant: the hydraulics of conduits, rivers, and canals; the physiology of blood and sap circulation; aspects of navigation including tides, waves, and ship resistance; acoustic phenomena; the damping of the seconds pendulum in air; atmospheric motion; bird flight, ballooning, and aviation; British theories of the ether as a perfect liquid; and hydrodynamic analogies for electromagnetic phenomena. This book offers glimpses into each of these worlds, with an emphasis on aspects of fluid mechanics in which compressibility and thermal effects play no role (hydrodynamics in the narrow sense).

---

[1]Thomson to Stokes, 20 Dec. 1857, *ST*.

There were also many socio-professional worlds of flow, which may be arranged into two categories: one comprising the academic elite of mathematics, natural philosophy, and astronomy; and another including the practical men of hydraulics, naval architecture, construction engineering, meteorology, and instrument making. Although the founders of hydrodynamics belonged to the first category, they hoped to serve the world of engineers as part of their pursuit of the enlightenment ideal of a rational unification of knowledge. The new hydrodynamics largely frustrated this desire. It is indeed a cliché of the history of fluid mechanics that until the twentieth century at least two separate disciplines of flow existed—hydrodynamics and hydraulics—which implied utterly different methods and professional identities, and which evolved independently of each other. Whereas hydro-dynamicists applied advanced mathematics to flows rarely encountered by engineers, hydraulicians used simple empirical or semi-empirical formulas that defied deeper theory.[2]

D'Alembert's paradox of 1768, according to which the steady motion of a perfect liquid exerts no force on a fully-immersed solid, quickly became the emblem of the split between the ideal and the practical worlds of flow. As may be retrospectively judged, the cause of this misfit was not any error in the application of the laws of mechanics. The correct equations of fluid motion, namely Euler's and Navier's, were indeed known at early stages of the history of hydrodynamics. The real difficulty lay in analyzing the consequences of these equations. The results and solutions that could be derived from them contradicted common observation—indeed they forbade the soaring of birds and made water rush at unreal velocities in channels or conduits. In 1786, the prominent hydraulician Charles Bossut recorded this impotence of the new fluid mechanics:

> These great geometers [d'Alembert and Euler] seem to have exhausted the resources that can be drawn from analysis to determine the motion of fluids: their formulas are so complex, by the nature of things, that we may only regard them as geometric truths, and not as symbols fit to paint the sensible image of the actual and physical motion of a fluid.

More than a century later, the prominent hydrodynamicist Wilhelm Wien similarly lamented:[3]

> In hydrodynamics … the real processes differ so much from the theoretical conclu-sions that engineers have had to develop their own approach to hydrodynamics questions, usually called hydraulics. In this approach, however, both foundations and conclusions lack rigor to such an extent that most results remain confined to empirical formulas of very limited validity.

For a long time, knowledge of the fundamental equations of the theory proved utterly insufficient in practice—a philosophically interesting situation on which more will be said in the conclusions in Chapter 8. Yet there was no lack of attempts, all through the nineteenth century, to make hydrodynamics more relevant to the practical problems of flow. Despite their frequent failure, these attempts were the source of almost every conceptual innovation in the science of flow of those times. For this reason this book is largely a history of the transgression of the borders between ideal and practical worlds of

---

[2]In this book, I indulge in the convenient neologisms 'hydrodynamicist' and 'hydraulician'.

[3]Bossut [1786], vol. 1 p. xv; Wien [1900] p. V.

flow, and of the several subcultures that favored them. Polytechnique-trained French engineers such as Navier and Saint-Venant formed one of these subcultures. In Victorian Britain, natural philosophers such as Stokes, Thomson, and Rayleigh, and theory-oriented engineers or engineering professors such as Russell, Rankine, Froude, and Reynolds formed two others. In Germany, the quintessential polymath Hermann Helmholtz by himself defined still another subculture of mediation between the ideal and the practical.

By elevating themselves above and yet struggling to act in the many practical worlds of flow, investigators like Saint-Venant, Stokes, or Helmholtz blurred the borders between these worlds. In their view, the retardation of water in hydraulic conduits, the friction of water on ship hulls, the subduction of winds near the ground, or the steering of airships, all belonged to the same category of phenomena. Any insight into one of them was transferable to the others. They trusted that ultimately these phenomena would be covered by the same mathematical theory. This synthetic perspective and the belief in theoretical transfer and cross-fertilization contrasted with the cultural isolation of the average practitioner. Whereas, for example, a Rankine or a Froude found it natural to apply pipe retardation formulas to the skin resistance of ships, most naval architects and hydraulicians ignored one another's fields and thus sometimes arrived at similar laws without knowing it.

The opposition between ideal and real worlds of flow, the fertility of attempts to reconcile them, and the bridging of different practical worlds, are the organizing themes of the following narrative. Chapter 1 recounts the eighteenth-century emergence of an ideal world of flow under the rule of partial differential equations by tracing the efforts of Daniel and Johann Bernoulli, Jean le Rond d'Alembert, Leonhard Euler, and Joseph Louis Lagrange. The creation of this theory was intimately bound with the formulation of the general dynamics of connected systems, with an extension of the concept of pressure, and with the elaboration of partial differential calculus. The importance of these broader innovations led nineteenth-century physicists to place great value in hydrodynamic theory; it also led to great dismay as they recognized the absurd consequences of this elegant theory.

As d'Alembert admitted, the paradox of vanishing resistance threatened to confine the new hydrodynamics to the realm of pure abstraction. There was nonetheless one sort of fluid motion, namely water waves, for which hydrodynamicists from Lagrange to Boussinesq harvested results that proved important to tide prediction, ship resistance, and ship rolling. Chapter 2 is devoted to these important advances. Although the mathematicians Laplace, Lagrange, Poisson, and Cauchy here obtained significant results, the more 'physical', application-oriented approaches to water waves came from members of the above-mentioned subcultures of mediation, namely Airy, Russell, Stokes, Thomson, and Rayleigh in Britain, and Saint-Venant and Boussinesq in France. These investigators successfully explained a great variety of observed wave behaviors. In the 1870s, Boussinesq and Rayleigh even managed to explain Russell's 'great solitary wave,' which had long perplexed wave theorists.

The analysis of other kinds of flow involved greater difficulties. In 1843, the founding father of British hydrodynamics, George Gabriel Stokes, imagined three possible ways nature could have chosen to escape d'Alembert's paradox: fluid friction, the formation of surfaces of finite slip of fluid over fluid, and instability leading to turbulence in the wake of the immersed solid. Much of the history of nineteenth-century hydrodynamics can be seen as a successive exploration of these three areas of research, to which Stokes himself largely contributed.

Chapter 3 of this book is devoted to the first option, namely, the introduction of viscosity. Navier inaugurated this approach in 1822, by analogy with the molecular theory of elasticity that he also invented. Although the relevant equation, now called the Navier–Stokes equation, was rediscovered four or five times, it failed to explain the hydraulic retardation for which it was intended, and only succeeded in the cases of pendulum oscillations and capillary flow. Some fifty years elapsed before physicists commonly agreed that this failure was only superficial and adopted the Navier–Stokes equation as the general foundation of hydrodynamics.

According to Helmholtz, viscosity alone could not be held responsible for the drastic difference between real flows and the ideal flows described by French mathematicians. For slightly viscous fluids such as air and water, the main defect of earlier theories was rather the assumption that the velocity of the flow derived from a potential. As is recounted in Chapter 4, in 1858 Helmholtz showed that, in the lack of a velocity potential, vortices existed in the fluid and obeyed simple laws of conservation in the incompressible, inviscid case. Ten years later, he argued that unstable vortex sheets, equivalent to surfaces of finite slip of fluid over fluid, were formed at the edges of solid walls. He thus explained the tendency of water and air to form coherent jets when projected into a quiet mass of the same fluid, as well as the convoluted decay of these jets. This idea of discontinuous fluid motion wonderfully bridged different worlds of flow: originally meant to solve a paradox of organ pipes, it turned out to provide the dead-water solution of d'Alembert's paradox, an explanation of the observed velocity of trade winds, some clues about the formation of water waves under wind, and even an anticipation of the meteorological front theory.

Helmholtz's considerations involved two special kinds of instability: the growth of discontinuity surfaces at the edges of solid walls; and the growth and spiral unrolling of any small bump on a surface of discontinuity. Chapter 5 is devoted to these and other flow instabilities contemplated by nineteenth-century hydrodynamicists. Owing to the difficulties inherent in any mathematical investigation of these questions, opinions diverged on whether some basic forms of perfect-liquid motion were stable or not. Stokes tended to favor instability because he believed he could thus recover slightly-viscous fluid behavior within the perfect-liquid picture. Thomson tended to favor stability because he hoped to construct permanent molecules out of vortex rings in a perfect liquid. Beyond this playful controversy, in the 1880s Rayleigh and Reynolds made decisive progress on the problem of the stability of parallel flow within both viscous and non-viscous fluids.

By their very nature, proofs of instability provide a negative kind of information, namely, certain fluid motions that seem to result from the fundamental equations never occur in nature because they are utterly unstable. Although the way a perturbation grows may sometimes indicate features of the final motion, the primary source of knowledge of this motion was by necessity experimental. Since the beginning of the nineteenth century, both hydraulicians and hydrodynamicists were aware of the turbulent character of the flow occurring in hydraulic conduits and in open channels. As is recounted in Chapter 6, the unpredictable, confused character of turbulent flow did not scare off every nineteenth-century theorist. Saint-Venant and his disciple Boussinesq sought to describe hydraulic flow through large-scale averaging and effective viscosity. Similarly, Reynolds later conceived a kinetic theory of turbulent momentum transport.

By the end of the century, a variety of mediations between ideal and real worlds of flow had led to the new concepts of viscous stress, vortex motion, discontinuity surface, instability, and turbulence. However, none of these conceptual innovations fully achieved the intended mastery of real flows. Further progress resulted from the extension and orchestration of nineteenth-century concepts of flow, with a focus on rendering the high-Reynolds-number flows most frequently encountered in natural and technical worlds. A much more efficient kind of fluid mechanics thus emerged at the beginning of the twentieth century, based on the boundary-layer and wing theories developed by Ludwig Prandtl and his disciples. These synthetic achievements, and anticipations by Rankine and Froude in the context of ship design, form the subject of the seventh and final chapter of this book. In the conclusions in Chapter 8, I examine the mechanisms of theory evolution through application, their neglect in Kuhnian philosophy, their pervasiveness in the history of major physical theories, and the special form they take in the case of fluid mechanics.

By emphasizing the sort of conceptual innovations that are induced by challenges from the natural and technical worlds, the present history of hydrodynamics leaves aside a few abundant developments of a more formal or mathematical nature.[4] Although this selective approach conveniently reduces the amount of relevant sources, the preparation of this book required much original research. The historians' interest in hydrodynamics indeed seems to have been inversely proportional to its historical importance. The two main reasons for this neglect have been the technical difficulty of the subject, and the focus of historians of the nineteenth and twentieth centuries on entirely new theories such as electrodynamics, thermodynamics, relativity, and quantum theory. There are only two global surveys of the history of hydrodynamics, written by modern leaders in this field. The first, by Hunter Rouse and Simon Ince, is a useful series of short biographies of leading hydraulicians and hydrodynamicists from antiquity to the present. The second, by Gregori Tokaty, is a historical sketch of the main conceptual advances in fluid mechanics.[5]

A few particular aspects of the history of hydrodynamics have received more detailed attention. To the late Clifford Truesdell we owe a penetrating account of Euler's foundational contributions, which must however be balanced with Gérard Grimberg's more exact assessment of d'Alembert's role. On water waves and on open-channel flow, we have Saint-Venant's schematic but technically competent histories, as well as recent contributions by Robin Bullough, Alex Craik, and John Miles. The profiles of polytechnique-trained engineers, who play essential parts in this story, are well described in Bruno Belhoste's and Antoine Picon's works. On Saint-Venant's own fluid mechanics, Chiara

---

[4]The best-known nineteenth-century treatises on hydrodynamics, those of Basset [1888] and Lamb [1879, 1895], had large chapters on the Lagrangian treatment of the motion of solids immersed in a perfect liquid, the stability of vortex systems, or German mathematical theorems of potential flow, none of which receives much attention in this book. Basset and Lamb also had chapters on the effects of fluid compressibility, including sound waves and shock waves, which I have completely left aside. My incursion into early twentieth-century hydrodynamics is even more limited: I have exclusively attended to the Göttingen school, although the Cambridge school harvested results of fundamental import to the broader development of hydrodynamics.

[5]Rouse and Ince [1957]; Tokaty [1971]. On the more recent history of fluid mechanics, cf. Lighthill [1995].

Melucci has written an informative dissertation. On tides, there is the clearly-written book by the modern expert David Edgar Cartwright. The monumental biography of Kelvin by Crosbie Smith and Norton Wise is a rich source on Victorian science and Kelvin's interest in flow analogies and navigation. On ship hydrodynamics, there is the excellent (unfortunately unpublished) dissertation by Thomas Wright. On aerodynamics and the theory of flight, John Anderson's recent, easily readable history stands out. In a freshly published book, John Ackroyd, Brian Axcell, and Anatoly Ruban give competently commented translations of a few historical papers. Information on Prandtl's school, Theodore von Kármán, aerodynamics, and applied mathematics in the early twentieth century are found in two rich books by Kármán himself and by Paul Hanle, and in a short insightful study by Giovanni Battimelli. I have myself published a few articles on some aspects of the history of hydrodynamics. Much of their content is included in this book, with the kind permission of the University of California Press, the Société Mathématique de France, and Springer Verlag.[6]

Part of the research for this book was done in Paris, within the Rehseis research team of CNRS and Paris VII. It is a pleasure to thank the director of that team, Karine Chemla, for intellectual stimulation and institutional support, and my friends Martha-Cecilia Bustamante, Nadine de Courtenay, and Edward Jurkowitz for fruitful discussions. Long, pleasant stays at the Max Planck Institut für Wissenschaftsgeschichte in Berlin, at the Dibner Institute in Cambridge, MA, at Harvard University's History of Science department, and at UC–Berkeley's OHST have greatly eased my compilation of bibliographical and archival materials, as well as my subsequent reflections. I am very grateful to Jürgen Renn, Jed Buchwald, Peter Galison, and Cathryn Carson, who welcomed me in these institutions and offered useful suggestions for my work. I am also indebted to a few hydrodynamicists for their warm support and for helpful criticism: Alex Craik, Marie Farge, Elizabeth Guazzelli, Etienne Guyon, and John Hinch (also Saint-Venant, in one of my dreams).

It is not easy, and perhaps not desirable, to reduce the pleasant gurgling of a mountain creek or the great surfs of the Silver Coast to the dry symbolism of a few fundamental equations. Neither is it easy to predict tides, pipe retardation, or ship resistance from first principles. Over the two centuries spanned in this book, a few valiant men gradually mastered these more technical problems and even gained some insights into the more poetical kinds of flow. I hope the story of their efforts will convince my reader that the life of hydrodynamics has been—and will be—as beautiful as some of the flows it purports to explain.

[6]Truesdell [1954]; Grimberg [1998]; Saint-Venant [1887c], [1888]; Bullough [1988]; Craik [2004]; Belhoste [1994]; Picon [1992]; Melucci [1996]; Cartwright [1999]; Smith and Wise [1989]; Wright [1983]; Anderson [1997]; Ackroyd, Axcell, and Ruban [2001] (of which I became aware only after writing Chapter 7); Kármán [1954]; Hanle [1982]; Battimelli [1984]; Darrigol [1998], [2002a], [2002b], [2002c], [2003]. Soon after the hardback edition of the present book came out, Michael Eckert published his excellent *The dawn of fluid mechanics*, Eckert [2005], which mainly covers the rise of applied fluid mechanics in the first half of the twentieth century, with emphasis on Prandtl's school and its contribution to aeronautic industry.

CONTENTS

Conventions and notation   **xiii**

1   The dynamical equations   **1**
    1.1   Daniel Bernoulli's *Hydrodynamica*   4
    1.2   Johann Bernoulli's *Hydraulica*   9
    1.3   D'Alembert's fluid dynamics   11
    1.4   Euler's equations   23
    1.5   Lagrange's analysis   26

2   Water waves   **31**
    2.1   French mathematicians   32
    2.2   Scott Russell, the naval engineer   47
    2.3   Tides and waves   56
    2.4   Finite waves   69
    2.5   The principle of interference   85

3   Viscosity   **101**
    3.1   Mathematicians' versus engineers' fluids   103
    3.2   Navier: molecular mechanics of solids and fluids   109
    3.3   Cauchy: stress and strain   119
    3.4   Poisson: the rigors of discontinuity   122
    3.5   Saint-Venant: slides and shears   126
    3.6   Stokes: the pendulum   135
    3.7   The Hagen–Poiseuille law   140

4   Vortices   **145**
    4.1   Sound the organ   146
    4.2   Vortex motion   148
    4.3   Vortex sheets   159
    4.4   Foehn, cyclones, and storms   166
    4.5   Trade winds   172
    4.6   Wave formation   178

5   Instability   **183**
    5.1   Divergent flows   184
    5.2   Discontinuous flow   188
    5.3   Vortex atoms   190
    5.4   The Thomson–Stokes debate   197
    5.5   Parallel flow   208

6   Turbulence   **219**
    6.1   Hydraulic phenomenology   221
    6.2   Saint-Venant on tumultuous waters   229
    6.3   Boussinesq on open channels   233
    6.4   The turbulent ether   239
    6.5   Reynolds's criterion   243

7   Drag and lift   **264**
    7.1   Tentative theories   265
    7.2   Ship resistance   273
    7.3   Boundary layers   283
    7.4   Wing theory   302

8   Conclusion   **323**

Appendix A Modern discussion of d'Alembert's paradox   326

Bibliography   **329**
    Bibliographic abbreviations   329
    Bibliography of primary literature   330
    Bibliography of secondary literature   344

Index   **350**

CONVENTIONS AND NOTATION

For the ease of the reader, uniform notation is used throughout this book, which of course implies some departure from original notation. To a large extent, this liberty amounts to a permutation of letters. Less innocuously, Cartesian-component equations are rendered as vector equations (although no hydrodynamicist used the vector notation before Prandtl and Sommerfeld), and Daniel Bernoulli's statements of proportions are translated into equations with dimensional proportionality constants. This is not to deny the importance of investigating how Johann Bernoulli, Euler, and others gradually reached the modern concept of a physics equation, why Prandtl promoted the vector notation, and how these changes affected theory. Such enquiries would, however, exceed the scope of the present book.

$d\mathbf{l}$, $d\mathbf{S}$, $d\tau$	elements of length, surface, and volume, respectively
g	acceleration of gravity
h	height
$\mathbf{k}$	wave vector
P	pressure
$\mathbf{r}$	position vector
t	time
u, v	x- and y-components of the velocity, respectively
$\mathbf{v}$	fluid velocity
Δ	Laplacian operator
ε	effective, large-scale viscosity (including eddy viscosity)
μ	viscosity parameter
ν	kinematic viscosity μ/ρ
ρ	fluid density
σ	deformation of a water surface, sometimes pulsation
τ	shear stress
τ_{ij}	stress tensor
φ	velocity potential ($\mathbf{v} = \nabla\varphi$)
ψ	stream function for two-dimensional, incompressible flow ($d\psi = -v\,dx + u\,dy$)
$\boldsymbol{\omega}$	vorticity $\nabla \times \mathbf{v}$
ω	either vorticity value or pulsation, according to the context
∇	gradient operator
$\nabla \times \mathbf{v}$	curl of the vector field $\mathbf{v}$
$\nabla \cdot \mathbf{v}$	divergence of the vector field $\mathbf{v}$

- Citations are in the author–date format and refer to one of the two bibliographies (primary and secondary). Page numbers refer to the last-mentioned source in the bibliographical item. Abbreviations are listed at the start of the bibliography.
- Translations are my own, unless the source is itself a translation.

1

THE DYNAMICAL EQUATIONS

> Admittedly, as useful a matter as the motion of fluid and related sciences has always been an object of thought. Yet until this day neither our knowledge of pure mathematics nor our command of the mathematical principles of nature have permitted a successful treatment.[1] (Daniel Bernoulli, September 1734)

Modern derivations of the fundamental equations for non-viscous fluids have an air of evidence. The fluid is divided into volume elements, and the acceleration of a volume element is equated to a force divided by a mass. The force on the element $d\tau$ is the sum of an external action $\mathbf{f}d\tau$ (e.g. gravity) and of the resultant $-(\nabla P)d\tau$ of the pressures exerted on the surface of the element by the surrounding fluid. If $\mathbf{v}(\mathbf{r}, t)$ denotes the velocity of the fluid at the point $\mathbf{r}$ and time t, and $\mathbf{r}(t)$ is the position of the element at time t, then the acceleration of the element is the time derivative of $\mathbf{v}[\mathbf{r}(t), t]$, that is, $\partial\mathbf{v}/\partial t + (\mathbf{v} \cdot \nabla)\mathbf{v}$. The mass of the element is the product of its density ρ and its volume $d\tau$. Hence Euler's equation follows:

$$\rho\left(\frac{\partial\mathbf{v}}{\partial t} + (\mathbf{v} \cdot \nabla)\mathbf{v}\right) = \mathbf{f} - \nabla P. \tag{1.1}$$

The conservation of the mass of a fluid element during its motion further gives the 'continuity equation':

$$\nabla \cdot \rho\mathbf{v} + \frac{\partial\rho}{\partial t} = 0, \tag{1.2}$$

so named because it assumes that the fluid remains continuous and does not burst into droplets during its motion.

A closer look at this derivation shows that it relies on concepts and idealizations that are by no means obvious. Firstly, it assumes that the action of the surrounding fluid on a given fluid portion can be represented by a normal pressure on its surface, whereas molecular intuition suggests a more complex distribution of the forces between inner and outer molecules. In fact, in real fluids the pressure is only normal in the case of rest; viscosity implies a tangential component of the pressure. In order to justify the existence and properties of internal pressure without appealing to the molecular picture, one could examine experience. Unfortunately, experience only informs us of the pressure exerted on the surface of immersed objects, not of the pressure of the fluid on itself. The latter notion requires an unwarranted idealization.

Another difficulty lurks in the application of Newton's second law to fluid elements. If the fluid is thought of as an assembly of molecules, this application does not immediately

---

[1]D. Bernoulli to Shoepflin, in *Mercure Suisse* (Sept. 1734) pp. 42–50; also in Bernoulli [2002] pp. 87–90.

follow from the validity of the law for individual molecules, because a volume element is not necessarily made of the same molecules during its evolution in time. If the fluid is instead regarded as a true continuum, then an unwarranted extension of the law to infinitesimal elements of mass is needed.

To one who has these difficulties in mind, the canonical derivation of Euler's equations seems largely illusory. It rests on axioms that need further justification, possibly through a kinetic–molecular theory of matter, or through the empirical success of their consequences. Accordingly, one should expect the historical genesis of Euler's equations to have been a difficult, roundabout process. Indeed, seventeen years elapsed between Daniel Bernoulli's first attempt at applying a general dynamic principle to fluid motion and Leonhard Euler's strikingly modern derivation of the equations named after him.

Another fact makes the history of early hydrodynamics even more intricate and interesting: the basic physico-mathematical tools of the modern derivation of Euler's equations were not originally available. In the early eighteenth century, there was no concept of a dimensional quantity, no practice of writing vector equations (even in the so-called Cartesian form), no concept of a velocity field, and no calculus of partial differential equations. The idea of founding a domain of physics on a system of general equations rather than on a system of general principles expressed in words did not exist.

Any historical investigation of the origins of hydrodynamics requires a *tabula rasa* of quite a few familiar notions of today's physicist. The purpose of the present chapter is to show how these notions gradually emerged together with modern hydrodynamics; it is not to determine who the main founder of this new science was. Excessive concern with priority questions leads to misinterpretations of the goals and concepts of the actors, and it harbors the myth of sudden, individual discovery. In contrast, the present chapter describes a long, multifaceted process in which fluid motion was gradually subjected to general dynamics, with a concomitant evolution of the principles of dynamics and with extensions of the classes of investigated flows.[2]

The first attempt at applying a general dynamical principle to fluid motion occurred in Daniel Bernoulli's *Hydrodynamica* of 1738. The principle was the conservation of live forces, expressed in terms of Huygens' pendulum paradigm. The main problem of fluid motion was efflux, or the parallel-slice flow through an opening on a vessel. The result was a geometrical expression of laws that implicitly contained the one-dimensional version of Euler's equations. This approach did not require the concept of internal pressure. Daniel Bernoulli nonetheless extended the concept of wall pressure to moving fluids, and derived 'Bernoulli's law' for this pressure. These inaugural achievements are described in Section 1.1.

Section 1.2 is devoted to the *Hydraulica* that Daniel's father Johann Bernoulli published in 1742. The standard problem was still parallel-slice efflux, approached through the

---

[2]In the history found in his *Méchanique analitique* [1788] pp. 436–7, Lagrange made d'Alembert the founder of hydrodynamics. He did not even mention Euler's name, although he no doubt appreciated his contributions (the second edition ([1811/15] vol. 2, p. 271) has the sentence: 'It is to Euler that we owe the first general formulas for the motion of fluids, founded on the laws of their equilibrium, and presented with the simple and luminous notation of partial differentials.' In disagreement, Truesdell ([1954] p. cxxv$_n$) writes: 'It seems that much of what d'Alembert is commonly credited with having done is taken from the simple and clear attributions of Lagrange, for I have searched for it in vain in d'Alembert's own works.' Grimberg [1998] has identified gaps and flaws in Truesdell's reading of d'Alembert.

pendulum analogy. The dynamical principle was now Newton's second law, together with a rule for replacing the gravities and the accelerations of the various parts of the system with equivalent gravities and accelerations acting on one part only. Again, this method did not require the concept of internal pressure. Johann Bernoulli nonetheless defined the internal pressure as some sort of contact force between successive slices of fluid, and gave its value in a generalization of Bernoulli's law to non-permanent flow. A key point of his success was his awareness of two contributions to the acceleration of a fluid slice: the velocity variation per time unit at a given height (our $\partial v/\partial t$); and the velocity variation per time unit due to the change of section (our $v\partial v/\partial z$, z being a coordinate in the direction of parallel motion). His style was more algebraic than his son's, with recourse to dimensional quantities including the acceleration g of gravity. But his reliance on partial differentials was only implicit.

Section 1.3 is devoted to the contributions of Jean le Rond d'Alembert. In 1743/44, the French philosopher and geometer rederived the results of the Bernoullis by means of a new principle of dynamics, according to which a moving system must be in equilibrium with respect to fictitious forces obtained by subtracting from the real (external) forces acting on the parts of the system the product of their mass and their acceleration. In the particular case of parallel flow, or for the oscillations of a compound pendulum, this method is equivalent to those of the Bernoullis. D'Alembert, however, was innovative in explicating the partial differentials in the expression of the fluid acceleration. Most importantly, his method enabled him to consider two-dimensional flows (with infinitely many degrees of freedom), whereas the Bernoullis were confined to flows with only one degree of freedom.[3]

D'Alembert achieved this tremendous generalization in his memoir on winds of 1747 and in his memoir on fluid resistance of 1749. There he obtained particular cases of Euler's equations, for the two-dimensional or axially-symmetric flow of an incompressible fluid. More precisely, his equations were those we would now obtain by eliminating the pressure from Euler's equations (vorticity equation). The reason for this peculiarity is that d'Alembert's principle essentially short-cuts the introduction of internal contact forces such as pressures. D'Alembert nevertheless had a concept of internal pressure, which he used in his expression of Bernoulli's law. In his memoir on winds he even indicated an alternative route to the equations of fluid motion, by balancing the pressure gradient, the gravity, and the inertial force at any point of the fluid.

Soon after studying d'Alembert's memoirs, Leonhard Euler showed how to derive completely general equations of fluid motion through a similar method, by applying Newton's second law to each fluid element and taking into account the pressure from the surrounding fluid. This achievement is described in Section 1.4. The clarity and modernity of Euler's approach has lent itself to the myth of a sudden emergence of Euler's hydrodynamics in Euler's magic hands. In reality, Euler struggled for many years to develop a satisfactory theory of fluid motion. He only reached his aim after integrating decisive contributions by Johann Bernoulli and by d'Alembert. His famous memoir of 1755 did not reflect sudden, isolated inspiration.[4]

[3]New insights into d'Alembert's fluid dynamics and mathematical methods are found in Grimberg [1998] (thesis directed by Michel Paty).

[4]Truesdell [1954] has a detailed, competent analysis of Euler's memoirs on fluid mechanics.

Nor should Euler's memoir be regarded as the last word on the foundations of perfect-fluid mechanics. As explained in Section 1.5, the last section in this chapter, Lagrange offered an alternative foundation to Euler's equations, based on the general principles of his analytical mechanics. He also specified the boundary conditions, without which Euler's equations would largely remain an empty formal scheme; he obtained a fundamental theorem about the existence of a velocity potential; and he gave a general method of approximation for solving the equations of narrow flows. Due to these advances, he could prove the approximate validity of the old hypothesis of parallel-slice motion and solve the problem of small waves on shallow water. Together with Euler's fundamental memoirs, these brilliant results were the starting-point for most of later hydrodynamics.

1.1   Daniel Bernoulli's *Hydrodynamica*

In 1738 the Swiss physician and geometer Daniel Bernoulli published his *Hydrodynamica, sive de viribus et motibus fluidorum commentarii* (hydrodynamics, a dissertation on the forces and motions of fluids). He coined the word *hydrodynamica* to announce a new, unified approach of hydrostatics and hydraulics. Although he did not create the modern science of fluid motion, his treatise marks a crucial transition: with novel and uniform methods, it solved problems that belonged to a long-established tradition.[5]

Since Greek and Roman hydraulics, an important problem of fluid motion was the flow of water from a vessel through an opening or a short pipe. The Renaissance and the seventeenth century saw the first experimental studies of this problem, as well as the first attempts to subject it to the laws of mechanics. Other topics of practical interest were the working of hydraulic machines and waterwheels, and ship resistance. Topics of philosophical interest were the elasticity of gases and Cartesian vortices. The *Hydrodynamica* covered all these subjects, except fluid resistance, which Bernoulli probably judged to be beyond the grasp of contemporary mathematics. His newest results concerned efflux. He also introduced the concept of work (*vis absoluta*) done by hydraulic machines, and he inaugurated the kinetic theory of elastic fluids.[6]

1.1.1   *The principle of live forces*

The basic principle on which Daniel Bernoulli based his hydrodynamics was what he called 'the equality of potential ascent and actual descent'. He thus alluded to Christiaan Huygens' study of the center of oscillation of a compound pendulum in the celebrated *Horologium Oscillatorium* of 1673. In modern terms, we would say that Huygens obtained the length of the simple pendulum that is equivalent to a given compound pendulum by equating the kinetic energy of the system of oscillating masses at a given instant to its sign-reversed potential energy. In Bernoulli's terms, the *potential ascent* means 'the vertical altitude which the center of gravity of the system would reach if the several particles, converting their velocities upward, are considered to rise as far as possible.' The *actual descent* denotes 'the vertical altitude through which the center of gravity has descended after the several particles have been brought to rest.' The potential ascent corresponds to

---

[5]Cf. Dugas [1950] pp. 274–8, Truesdell [1954] pp. XXIII–XXXI, Mikhailov [2002].

[6]On early hydraulics, cf. Rouse and Ince [1957] Chaps 2–9, Garbrecht [1987]. On D. Bernoulli's *Hydrodynamica*, cf. Mikhailov [1999], [2002]; Calero [1996] pp. 422–59.

our kinetic energy divided by the total weight, and the actual descent to the sign-reversed potential energy divided by the total weight.[7]

For a modern reader, Bernoulli's text is much harder to penetrate than this simple identification would suggest. The main difficulty comes from the lack of a theory of the combination of dimensional quantities, and the now archaic appeal to Euclidean proportions and equivalent lengths. The modern concept of dimensional quantities emerged at the turn of the nineteenth century, and found its first systematic formulation in Fourier's theory of heat.[8] A full history of early hydrodynamics would necessarily take into account this important transformation in the writing style of physico-mathematical equations. Modernized notation is nevertheless used in what follows, because the main points to be made resist this perversion of the original text.

In order to appreciate the daringness of Daniel Bernoulli's approach, one must remember that until the nineteenth century energy considerations were very rarely used in mechanics and elsewhere. Gottfried Wilhelm Leibniz's principle of the conservation of *vis viva*, which had both Huygenian and Cartesian roots, had little impact because the concept of live force (roughly our kinetic energy) was usually interpreted as a metaphysical threat to the Newtonian concept of accelerating force. The most significant exception to this general attitude was Daniel Bernoulli's father Johann, who used Leibniz's principle to ease the solution of various mechanical problems. Father and son also agreed with Leibniz that every apparent loss of live force in the universe was a dissimulation of live force in small-scale motions. They even believed that potential forms of live force should be reducible to invisible motions, as exemplified in Daniel's kinetic explanation of gas pressure.[9]

As the compound pendulum was the implicit paradigm of the Bernoullis' use of the conservation of live forces, some of Huygens' treatment must be recalled. Consider a pendulum made of two point masses A and B, rigidly connected to a massless rod that can oscillate around the suspension point O (see Fig. 1.1). In modern notation, the equality of the potential ascent to the actual descent reads:

$$\frac{m_A(v_A^2/2g) + m_B(v_B^2/2g)}{m_A + m_B} = z_G, \tag{1.3}$$

where m denotes a mass, v a velocity, g the acceleration of gravity, and z_G the descent of the gravity center of the two masses measured from the highest elevation of the pendulum during its oscillation. This equation leads to a first-order differential equation for the angle θ that the suspending rod makes with the vertical. The comparison of this equation with that of a simple pendulum then yields the expression $(a^2 m_A + b^2 m_B)/(a m_A + b m_B)$ for the length of the equivalent simple pendulum (with $a = OA$ and $b = OB$).[10]

1.1.2 *Efflux*

As Daniel Bernoulli could not fail to observe, there is a close analogy between this problem and the hydraulic problem of efflux, as long as the fluid motion occurs by parallel slices. Under the latter hypothesis, the velocity of the fluid particles that belong to the same

---

[7]D. Bernoulli [1738] pp. 11, 30.   [8]Cf. Ravetz [1961].
[9]Cf. Costabel [1983], Séris [1987].   [10]Cf. Vilain [2000] pp. 32–6.

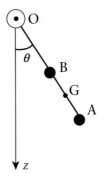

Fig. 1.1.   Compound pendulum.

section of the fluid is normal to and uniform through the section. Moreover, if the fluid is incompressible and continuous (no cavitation), then the velocity in one section of the vessel completely determines the velocity in all other sections. The problem is thus reduced to the fall of a connected system of weights with one degree of freedom only, just as it is for the case of a compound pendulum.

This analogy inspired Daniel Bernoulli's treatment of efflux. Consider, for instance, a vertical vessel with a section S depending on the downward vertical coordinate z (see Fig. 1.2). A mass of water falls through this vessel by parallel, horizontal slices. The continuity of the incompressible water implies that the product Sv is a constant through the fluid mass. The equality of the potential ascent and the actual descent implies that at every instant[11]

$$\int_{z_0}^{z_1} \left(\frac{v^2}{2g}\right) S \, dz = \int_{z_0}^{z_1} zS \, dz, \qquad (1.4)$$

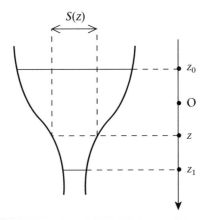

Fig. 1.2.   Parallel-slice flow of water in a vertical vessel.

[11]D. Bernoulli gave a differential, geometric version of this relation ([1738] pp. 31–5).

where z_0 and z_1 denote the (changing) coordinates of the two extreme sections of the fluid mass, the origin of the z-axis coincides with the position of the gravity center of this mass at the beginning of the fall, and the units are chosen so that the density of the fluid is *one*. As v is inversely proportional to the known function S of z, this equation yields a relation between z_0 and $v_0 = \dot{z}_0$, which can be integrated to give the motion of the highest fluid slice, and so forth. Bernoulli's investigation of efflux amounted to a repeated application of this procedure to vessels of various shapes.

The simplest sub-case of this problem is that of a broad container with a small opening of section s on its bottom (see Fig. 1.3). As the height h of the water varies very slowly, the escaping velocity quickly reaches a steady value u. As the fluid velocity within the vessel is negligible, the increase of the potential ascent in the time dt is simply given by the potential ascent $(u^2/2g)sudt$ of the fluid slice that escapes through the opening at velocity u. This quantity must be equal to the actual descent $hsudt$. Therefore, the velocity u of efflux is the velocity $\sqrt{2gh}$ of free fall from the height h, in conformity with the law formulated by Evangelista Torricelli in 1644.[12]

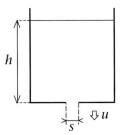

Fig. 1.3.   Idealized efflux through small opening (without *vena contracta*).

1.1.3   *Bernoulli's law*

Bernoulli's most innovative application of this method concerned the pressure exerted by a moving fluid on the walls of its container, a topic of importance for the physician and physiologist that he was. Previous writers on hydraulics and hydrostatics had only considered the hydrostatic pressure due to gravity. In the case of a uniform gravity g, the pressure per unit area on a wall portion was known to depend only on the depth h of this portion below the free water surface. According to the law enunciated by Simon Stevin in 1605, it is given by the weight gh of a water column (of unit density) that has a unit normal section and the height h. In the case of a moving fluid, Bernoulli defined and derived the 'hydraulico-static' wall pressure as follows.[13]

---

[12]D. Bernoulli [1738] p. 35. This reasoning assumes a parallel motion of the escaping fluid particle. Therefore, it only gives the velocity u beyond the contraction of the escaping fluid vein that occurs near the opening (Newton's *vena contracta*, cf. Lagrange [1788] pp. 430–1, Smith [1998]). On Torricelli's law and early derivations, cf. Blay [1985], [1992] pp. 331–352.

[13]D. Bernoulli [1738] pp. 258–60. Mention of physiological applications is found in Bernoulli to Shoepflin, 25 Aug. 1734, in D. Bernoulli [2002] p. 89: 'Hydraulico-statics will also be useful to understand animal economy with respect to the motion of fluids, their pressure on vessels, etc.'

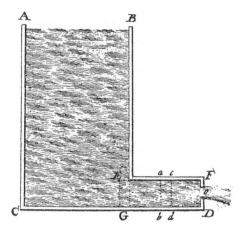

Fig. 1.4.   Daniel Bernoulli's figure accompanying
his derivation of the velocity-dependence of pressure ([1738] plate).

The section S of the vertical vessel ABCG of Fig. 1.4 is supposed to be much larger than
the section s of the appended tube EFDG, which is itself much larger than the section ε of
the hole o. Consequently, the velocity u of the water escaping through o is $\sqrt{2gh}$. Owing to
the conservation of the flux, the velocity v within the tube is $(\varepsilon/s)u$. Bernoulli goes on:[14]

> If in truth there were no barrier FD, the final velocity of the water in the same tube
> would be [s/ε times greater]. Therefore, the water in the tube tends to a greater
> motion, but its pressing [*nisus*] is hindered by the applied barrier FD. By this pressing
> and resistance [*nisus et renisus*] the water is compressed [*comprimitur*], which com-
> pression [*compressio*] is itself kept in by the walls of the tube, and thence these too
> sustain a similar pressure [*pressio*]. Thus it is plain that the pressure [*pressio*] on the
> walls is proportional to the acceleration … that would be taken on by the water if
> every obstacle to its motion should instantaneously vanish, so that it were ejected
> directly into the air.

Based on this intuition, Bernoulli imagined that the tube was suddenly broken at ab,
and made the wall pressure P proportional to the acceleration dv/dt of the water at this
instant. According to the principle of live forces, the actual descent of the water during the
time dt must be equal to the potential ascent it acquires while passing from the large
section S to the smaller section s, plus the increase of the potential ascent of the portion
EabG of the fluid. This gives (again, the fluid density is *one*)

$$hsv\ dt = \left(\frac{v^2}{2g}\right)sv\ dt + bs\ d\left(\frac{v^2}{2g}\right), \tag{1.5}$$

where $b = Ea$. The resulting value of the acceleration dv/dt is $(gh - \frac{1}{2}v^2)/b$. The wall
pressure P must be proportional to this quantity, and it must be identical to the static
pressure gh in the limiting case $v = 0$. It is therefore given by the equation

$$P = gh - \frac{1}{2}v^2, \tag{1.6}$$

[14]D. Bernoulli [1738] pp. 258–9. Translated in Truesdell [1954] p. XXVII. The *compressio* in this citation
perhaps prefigures the internal pressure later introduced by Johann Bernoulli.

which means that the pressure exerted by a moving fluid on the walls is lower than the static pressure, the difference being half the squared velocity (times the density). Bernoulli illustrated this effect in two ways (see Fig. 1.5): by connecting a narrow vertical tube to the horizontal tube EFDG, and by letting a vertical jet surge from a hole in this tube.

The modern reader may here recognize Bernoulli's law. In fact, Bernoulli did not quite write eqn (1.6), because he chose the ratio s/ε rather than the velocity v as the relevant variable. Also, he only reasoned in terms of *wall* pressure, whereas modern physicists apply Bernoulli's law to the *internal* pressure of a fluid.

There were other limitations to Bernoulli's hydrodynamics, of which he was largely aware. He knew that in some cases part of the live force of the water went to eddying motion, and he even tried to estimate this loss in the case of a suddenly enlarged conduit. He was also aware of the imperfect fluidity of water, although he decided to ignore it in his reasoning. Most importantly, he knew that the hypothesis of parallel slices only held for narrow vessels and for gradual variations of their section. But his method confined him to this case, since it is only for systems with one degree of freedom that the conservation of live forces suffices to determine the motion.[15]

1.2 Johann Bernoulli's *Hydraulica*

In 1742, Johann Bernoulli published his *Hydraulica*, with an antedate (1732) that made it predate his son's treatise. Although he had been the most ardent supporter of Leibniz's principle of live forces, he now regarded this principle as an indirect consequence of more fundamental laws of mechanics. His aim was to base hydraulics on an incontrovertible, Newtonian expression of these laws. To this end he adapted a method that he had invented in 1714 to solve the paradigmatic problem of the compound pendulum.

1.2.1 *Translation*

Consider again the pendulum of Fig. 1.1. According to Johann Bernoulli, the gravitational force $m_B g$ acting on B is equivalent to a force $(b/a)m_B g$ acting on A, because, according to

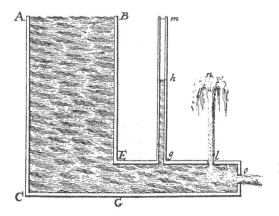

Fig. 1.5. Effects of the velocity dependence of pressure according to Daniel Bernoulli ([1738] plate).

[15]D. Bernoulli [1738] pp. 12 (eddies), 124 (enlarged conduit), 13 (imperfect fluid).

the law of levers, two forces that have the same moment have the same effect. Similarly, the 'accelerating force' $m_B b\ddot{\theta}$ of the mass B is equivalent to an accelerating force $(b/a)m_B b\ddot{\theta} = m_B(b/a)^2 a\ddot{\theta}$ at A. Consequently, the compound pendulum is equivalent to a simple pendulum with a mass $m_A + (b/a)^2 m_B$ located on A and subjected to the effective vertical force $m_A g + (b/a)m_B g$. It is also equivalent to a simple pendulum of length $(a^2 m_A + b^2 m_B)/(am_A + bm_B)$ oscillating in the gravity g, in conformance with Huygens' result. In summary, Johann Bernoulli reached the equation of motion by applying Newton's second law to a fictitious system obtained by replacing the forces and the momentum variations at any point of the system with equivalent forces and momentum variations at one point of the system. This replacement, based on the laws of equilibrium of the system, is what Bernoulli called 'translation' in the introduction to his *Hydraulica*.[16]

Now consider the canonical problem of water flowing by parallel slices through a vertical vessel of varying section (see Fig. 1.2). Johann Bernoulli 'translates' the weight $gS\mathrm{d}z$ of the slice $\mathrm{d}z$ of the water to the location z_1 of the frontal section of the fluid. This gives the effective weight $S_1 g\mathrm{d}z$, because, according to a well-known law of hydrostatics, a pressure applied at any point of the surface of a confined fluid is uniformly transmitted to any other part of the surface of the fluid. Similarly, Bernoulli translates the 'accelerating force' (momentum variation) $(\mathrm{d}v/\mathrm{d}t)S\mathrm{d}z$ of the slice $\mathrm{d}z$ to the frontal section of the fluid, with the result $(\mathrm{d}v/\mathrm{d}t)S_1\mathrm{d}z$. He then obtains the equation of motion by equating the total translated weight to the total translated accelerating force:

$$S_1 \int_{z_0}^{z_1} g \, \mathrm{d}z = S_1 \int_{z_0}^{z_1} \frac{\mathrm{d}v}{\mathrm{d}t}\mathrm{d}z. \tag{1.7}$$

1.2.2 Gorge

For Johann Bernoulli the crucial point was the determination of the acceleration $\mathrm{d}v/\mathrm{d}t$. Previous authors, he contended, had failed to derive correct equations of motion from the general laws of mechanics because they were only aware of one contribution to the acceleration of the fluid slices, namely, that which corresponds to the instantaneous change of velocity at a given height z, or $\partial v/\partial t$ in modern terms. They ignored the acceleration due to the broadening or to the narrowing of the section of the vessel, which Bernoulli called *gurges* (gorge). In modern terms, he identified the convective component $v\partial v/\partial z$ of the acceleration. Note that his use of partial derivatives was only implicit: due to the relation $v = (S_0/S)v_0$, he could split v into a time-dependent factor v_0 and a z-dependent factor S_0/S, and thus express the total acceleration as $(S_0/S)(\mathrm{d}v_0/\mathrm{d}t) - (v_0^2 S_0^2/S^3)(\mathrm{d}S/\mathrm{d}z)$.[17]

---

[16]J. Bernoulli [1714], [1742] p. 395. In modern terms, Johann Bernoulli's procedure amounts to equating the sum of moments of the applied forces to the sum of moments of the accelerating forces (which is the time derivative of the total angular momentum). Cf. Vilain [2000] pp. 448–50.

[17]J. Bernoulli [1742] pp. 432–7. He misleadingly called the two parts of the acceleration the 'hydraulic' and the 'hydrostatic' components. Truesdell's translation of *gurges* ([1954] p. XXXIII) as 'eddy' seems inadequate (although it does have this meaning in classical latin), because Bernoulli only meant the velocity difference between successive layers of the fluid. In his treatise on the equilibrium and motion of fluids ([1744] p. 157), d'Alembert interpreted J. Bernoulli's expression of the acceleration in terms of two partial differentials.

Thanks to the *gurges*, Johann Bernoulli successfully applied eqn (1.7) to various cases of efflux and retrieved his son's results.[18] He also offered a novel approach to the pressure of a moving fluid on the sides of its container. This pressure, he asserted, was simply the pressure, or *vis immaterialis*, that contiguous fluid parts exerted on one another, just as two solids in contact act on each other:[19]

> The force that acts on the side of the channel through which the liquid flows ... is nothing but the force that originates in the force of compression through which contiguous parts of the fluid act on one another.

Accordingly, Bernoulli divided the flowing mass of water into two parts separated by the section $z = \zeta$. Following the general idea of 'translation', the pressure that the upper part exerts on the lower part is

$$P(\zeta) = \int_{z_0}^{\zeta} \left(g - \frac{dv}{dt} \right) dz. \tag{1.8}$$

More explicitly, this is

$$P(\zeta) = \int_{z_0}^{\zeta} g \, dz - \int_{z_0}^{\zeta} v \frac{\partial v}{\partial z} dz - \int_{z_0}^{\zeta} \frac{\partial v}{\partial t} dz = g(\zeta - z_0) - \frac{1}{2} v^2(\zeta) + \frac{1}{2} v^2(z_0) - \frac{\partial}{\partial t} \int_{z_0}^{\zeta} v \, dz. \tag{1.9}$$

Johann Bernoulli thus obtained (in a widely different notation) a generalization of his son's law to unsteady parallel-slice flow.[20]

Johann Bernoulli interpreted the relevant pressure as an *internal* pressure analogous to the tension of a thread or the mutual action of contiguous solids in connected systems. Yet he did not rely on this new concept of pressure to establish the equation of motion (1.7). He only introduced this concept as a shortcut to the velocity dependence of wall pressure.[21]

1.3. D'Alembert's fluid dynamics

1.3.1 *The principle of dynamics*

In 1743, the French geometer and philosopher Jean le Rond d'Alembert published his influential *Traité de dynamique*, which subsumed the dynamics of connected systems under a few general principles.[22] The first illustration he gave of his approach was Huygens'

---

[18]D'Alembert later explained this agreement, see pp. 14–15.   [19]J. Bernoulli [1742] p. 442.

[20]J. Bernoulli [1742] p. 444. His notation for the internal pressure was π. In the first section of his *Hydraulica*, which he communicated to Euler in 1839, he only treated the steady flow in a suddenly enlarged tube. In his enthusiastic reply (5 May 1739, in Euler [1998] pp. 287–95), Euler treated the vertical, accelerated efflux from a vase of arbitrary shape with the same method of 'translation', not with the later method of balancing gravity with the internal pressure gradient, contrary to Truesdell's claim ([1954] p. XXXIII). Bernoulli subsequently wrote his second part, adding only obliqueness of the vessel to Euler's treatment.

[21]For a different view, cf. Truesdell [1954] p. XXVI, Calero [1996] pp. 460–74.

[22]D'Alembert began to read a memoir on the same theme at the Académie des Sciences on 24 Nov. 1742. A few months earlier, he had read two memoirs on the refraction of solids moving in fluids of variable density (a then classical approach to the refraction of light). His treatment did not involve any explicit fluid mechanics, for he

compound pendulum.[23] As we saw, Johann Bernoulli's solution to this problem leads to the equation of motion

$$m_A g \sin\theta + \left(\frac{b}{a}\right) m_B g \sin\theta = m_A a\ddot\theta + \left(\frac{b}{a}\right) m_B b\ddot\theta, \tag{1.10}$$

which may be rewritten as

$$a(m_A g \sin\theta - m_A a\ddot\theta) + b(m_B g \sin\theta - m_B b\ddot\theta) = 0. \tag{1.11}$$

This last equation is the condition of equilibrium of the pendulum under the action of the forces $m_A \mathbf{g} - m_A \boldsymbol\gamma_A$ and $m_B \mathbf{g} - m_B \boldsymbol\gamma_B$ acting, respectively, on A and B. In d'Alembert's terminology, the products $m_A \mathbf{g}$ and $m_B \mathbf{g}$ are the motions impressed (per unit time) on the bodies A and B under the sole effect of gravitation (without any constraint). The products $m_A \boldsymbol\gamma_A$ and $m_B \boldsymbol\gamma_B$ are the actual changes of their (quantity of) motion (per unit time). The differences $m_A \mathbf{g} - m_A \boldsymbol\gamma_A$ and $m_B \mathbf{g} - m_B \boldsymbol\gamma_B$ are the parts of the impressed motions that are destroyed by the rigid connection of the two masses through the freely-rotating rod. Accordingly, d'Alembert saw in eqn (1.11) a consequence of a general dynamic principle from which the motions destroyed by the connections should be in equilibrium.[24]

How d'Alembert arrived at this principle is not known. In 1703 Jacob Bernoulli, the elder brother of Johann, had derived the center of oscillation of the compound pendulum through the same method. D'Alembert did not refer to this source. Perhaps he found his inspiration while meditating on the compound pendulum. Or he could have deduced the principle from a new philosophy of motion, as is suggested by the presentation given in the *Traité de dynamique*.[25]

D'Alembert based dynamics on three laws, which he regarded as necessary consequences of the principle of sufficient reason. The first law is that of inertia, according to which a freely-moving body moves with a constant velocity in a constant direction. The second law stipulates the vector superposition of motions impressed on a given body. According to the third law, two (ideally rigid) bodies come to rest after a head-on collision if and only if their velocities are inversely proportional to their masses. From these three laws and some further recourse to the principle of sufficient reason, d'Alembert believed he could derive a complete system of dynamics without recourse to the older, obscure

assumed from the start a generalization of Newton's resistance formula for the impact of fluid on a solid segment. It did not involve any dynamics either, for he computed the deflection of the solid on the basis of Newton's second law. Cf. Académie Royale des Sciences, *Procès-verbaux* **60** (1741) pp. 369–404, 424–38; **61** (1742) pp. 126–33, 349–56 (text of the memoirs on refraction, also in Part III of d'Alembert [1744] with little change); *Procès-verbaux* **61** (1742) p. 424 (mention that D'Alembert has read a memoir on a new principle of dynamics). D'Alembert was not the only one to feel the need of a systematization of mechanics at that time; in the *Mémoires* of the same year, ([1742] pp. 1–52), Clairaut published his own 'general and direct principle' of dynamics (pp. 21–2), based on the introduction of internal forces (such as thread tension).

[23]D'Alembert [1743] pp. 69–70.

[24]Cf. Vilain [2000] pp. 456–9. D'Alembert [1743] reproduced and criticized Johann Bernoulli's derivation on p. 71.

[25]On Jacob Bernoulli as a source, cf. Lagrange [1788] pp. 176–7, 179–80, Dugas [1950] pp. 233–4, Vilain [2000] pp. 444–8. Jacob Hermann's treatment of the compound pendulum in his *Phoronomia* (1716) and Euler's early treatment of the same problem (1734) read like convoluted statements of Jacob Bernoulli's method.

concept of force as the cause of motion. He defined force as the motion impressed on a body, that is, the motion that a body would take if this force were acting alone without any impediment. The third law then implies that two contiguous bodies subjected to opposite forces are in equilibrium. Consequently, d'Alembert regarded statics as a particular case of dynamics in which the various motions impressed on the parts of the system mutually cancel each other.[26]

Based on this concept, d'Alembert derived the principle of virtual velocities, according to which a connected system subjected to various forces remains in equilibrium if the work of these forces vanishes for any infinitesimal motion of the system that is compatible with the connections.[27] As for the principle of dynamics, he regarded it as a self-evident consequence of his dynamic concept of equilibrium. In general, the effect of the connections in a connected system is to destroy part of the motion that is impressed on its components by means of external agencies. The rules of this destruction should be the same whether the destruction is total or partial. Hence, equilibrium should hold for that part of the impressed motions that is destroyed through the constraints. This is d'Alembert's principle of dynamics.

Stripped of d'Alembert's philosophy of motion, this principle stipulates that a connected system in motion should be, at any time, in equilibrium with respect to the fictitious forces $\mathbf{f} - m\gamma$, where $\mathbf{f}$ denotes the force applied on the mass point m of the system, and γ the acceleration of this mass point. As a simple example, consider two masses m_A and m_B hanging on the two sides of a massless pulley by an inextensible, massless thread (see Fig. 1.6). According to d'Alembert's principle, the forces $m_A g - m_A \gamma_A$ and $m_B g - m_B \gamma_B$ should be in equilibrium, and therefore should be equal. This condition, together with the kinematic condition $\gamma_A + \gamma_B = 0$, yields the equation of motion of the system. Compared to other treatments of the same problem, the essential advantage of d'Alembert's method

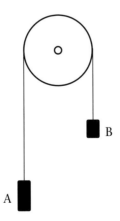

Fig. 1.6.   Simple connected system for illustrating d'Alembert's principle.

---

[26]D'Alembert [1743] pp. xiv–xv, 3. Cf. Hankins [1968], Fraser [1985].

[27]The principle of virtual velocities was first stated generally by Johann Bernoulli and thus named by Lagrange [1788] pp. 8–11. Cf. Dugas [1950] pp. 221–3, 320. The term work is of course anachronistic.

is that it does not require the introduction of the subtle (and obscure for d'Alembert) concept of the tension of the thread. It directly gives the equations of motion of the system if only the conditions of equilibrium are known.

1.3.2 *Efflux revisited*

At the end of his treatise on dynamics, d'Alembert considered the hydraulic problem of efflux through the vessel of Fig. 1.2. His first task was to determine the condition of equilibrium of the fluid when subjected to an altitude-dependent gravity $g(z)$. For this purpose he considered an intermediate slice of the fluid, and required the pressure from the fluid above this slice to be equal and opposite to the pressure from the fluid below this slice. According to a slight generalization of Stevin's hydrostatic law, these two pressures are given by the integral of the variable gravity $g(z)$ over the relevant range of elevation. Hence the equilibrium condition reads:[28]

$$S(\zeta) \int_{z_0}^{\zeta} g(z) \, dz = -S(\zeta) \int_{\zeta}^{z_1} g(z) \, dz, \tag{1.12}$$

or

$$\int_{z_0}^{z_1} g(z) \, dz = 0. \tag{1.13}$$

According to d'Alembert's principle, the motion of the fluid under a constant gravity g must be such that the fluid is in equilibrium under the fictitious gravity $g(z) = g - dv/dt$, where dv/dt is the acceleration of the fluid slice at the elevation z. Hence follows the equation of motion

$$\int_{z_0}^{z_1} \left(g - \frac{dv}{dt} \right) dz = 0, \tag{1.14}$$

which is the same as Johann Bernoulli's eqn (1.7).

D'Alembert further proved that this equation implied the conservation of live forces in Daniel Bernoulli's form. To this end, he inserted the product Sv, which does not depend on z, in the above equation. This gives

$$\int_{z_0}^{z_1} v \frac{dv}{dt} S \, dz = \int_{z_0}^{z_1} gvS \, dz. \tag{1.15}$$

As the two integrals can be regarded as sums over moving slices of fluid, this equation is equivalent to

---

[28]D'Alembert [1743] pp. 183–6.

$$\frac{\mathrm{d}}{\mathrm{d}t} \int_{z_0}^{z_1} \frac{1}{2} v^2 S \,\mathrm{d}z = \frac{\mathrm{d}}{\mathrm{d}t} \int_{z_0}^{z_1} gzS \,\mathrm{d}z, \qquad (1.16)$$

which is the differential version of Daniel Bernoulli's eqn (1.4).

D'Alembert intended his new solution of the efflux problem to illustrate the power of his principle of dynamics. He clearly relied on the long-known analogy with a connected system of solids. Yet he believed this analogy to be imperfect. Whereas in the case of solids the condition of equilibrium was derived from the principle of virtual velocities, in the case of fluids d'Alembert believed that only experiments could determine the condition of equilibrium. As he explained in his treatise of 1744 on the equilibrium and motion of fluids, the interplay between the various molecules of a fluid was too complex to allow for a derivation based on the only a priori known dynamics, that of individual molecules.[29]

In this second treatise, d'Alembert provided a similar treatment of efflux, including his earlier derivations of the equation of motion and the conservation of live forces, with a slight variant: he now derived the equilibrium condition (1.13) by setting the pressure acting on the bottom slice of the fluid to zero.[30] Presumably, he did not want to base the equations of equilibrium and motion on the concept of internal pressure, in conformance with his general avoidance of internal contact forces in his dynamics. His statement of the general conditions of equilibrium of a fluid, as found at the beginning of his treatise, only required the concept of wall pressure. Yet, in a later section of his treatise, d'Alembert introduced 'the pressure at a given height'

$$P(\zeta) = \int_{z_0}^{\zeta} \left(g - \frac{\mathrm{d}v}{\mathrm{d}t} \right) \mathrm{d}z, \qquad (1.17)$$

just as Johann Bernoulli had done, and for the same purpose of deriving the velocity dependence of wall pressure.[31]

In the rest of his treatise, d'Alembert solved problems similar to those in Daniel Bernoulli's *Hydrodynamica*, with nearly identical results. The only important difference concerned cases involving abrupt decrease of the fluid velocity. Whereas Daniel Bernoulli believed that dissipation into smaller-scale motion occurred in such cases, d'Alembert still applied the conservation of live forces. Daniel Bernoulli disagreed with these and a few other changes. In a contemporary letter to Euler he expressed his exasperation over d'Alembert's treatise:[32]

> I have seen with astonishment that apart from a few little things there is nothing to be seen in his hydrodynamics but an impertinent conceit. His criticisms are puerile indeed, and show not only that he is no remarkable man, but also that he never will be.

---

[29] D'Alembert [1744] pp. viii–ix.   [30] *Ibid.* pp. 19–20.   [31] *Ibid.* p. 139.

[32] D. Bernoulli to Euler, 7 July 1745, quoted in Truesdell [1954] p. XXXVIIn. Truesdell approves ([1954] p. XXXVII): 'D'Alembert's method makes no contribution and has had no permanent influence in fluid mechanics.'

1.3.3   *The cause of winds*

In this judgment, Daniel Bernoulli overlooked the fact that d'Alembert's hydrodynamics, being based on a general dynamics of connected systems, lent itself to generalizations beyond parallel-slice flow. In a prize-winning memoir of 1746 on the cause of winds, d'Alembert offered striking illustrations of the power of this approach. As thermal effects were beyond the grasp of contemporary mathematical physics, he focused on a cause that is now known to be negligible: the tidal force exerted by the Moon and the Sun. For simplicity, he confined his analysis to the case of a constant-density layer of air covering a spherical globe with uniform thickness. He further assumed that fluid particles originally on the same vertical line remained so in the course of time (owing to the thinness of the air layer) and that the vertical acceleration of these particles was a negligible fraction of gravity, and he neglected second-order quantities with respect to the fluid velocity and to the elevation of the free surface. His strategy was to apply his principle of dynamics to the motion induced by the tidal force $\mathbf{f}$ and the force of gravity $\mathbf{g}$ (for unit density), both of which depend on the location on the surface of the Earth.[33]

Calling $\boldsymbol{\gamma}$ the absolute acceleration of the fluid particles, the principle requires that the fluid layer should be in equilibrium under the force $\mathbf{f} + \mathbf{g} - \boldsymbol{\gamma}$. From earlier theories on the shape of the Earth (regarded as a rotating liquid spheroid), d'Alembert borrowed the equilibrium condition that the net force should be perpendicular to the free surface of the fluid. He also required that the volume of vertical cylinders of fluid should not be altered by their motion, in conformance with his constant-density model. As the modern reader would expect, from these two conditions d'Alembert derived some sort of momentum equation, and some sort of continuity equation. But he did it in a rather opaque manner. Some features, such as the lack of specific notation for partial differentials or the abundant recourse to geometrical reasoning, disconcert modern readers only.[34] Others were problematic to his contemporaries: he often omitted steps and introduced special assumptions without warning. Also, he directly treated the utterly difficult problem of fluid motion on a spherical surface without preparing the reader with simpler problems.

Suppose, with d'Alembert, that the tide-inducing luminary orbits above the equator (with respect to the Earth).[35] Using the modern terminology for spherical coordinates, denote by θ the colatitude of a given point of the terrestrial sphere with respect to an axis pointing toward the orbiting luminary (this is the geographical longitude), ϕ the longitude measured from the meridian above which the luminary is orbiting (this is *not* the geographical longitude), η the elevation of the free surface of the fluid layer over its equilibrium position, v_θ and v_ϕ the θ- and ϕ-components of the fluid velocity with respect to the Earth, h the depth of the fluid in its undisturbed state, and R the radius of the Earth (see Fig. 1.7).

---

[33]D'Alembert [1747]. D'Alembert treated the rotation of the Earth, the Sun's attraction, and the Moon's attraction as small perturbing causes whose effects on the shape of the fluid surface simply added (*ibid.* pp. xvii, 47). Consequently, he overlooked the Coriolis force in his analysis of the tidal effects (*ibid.* p. 65, he announces that he will be reasoning as if it were the luminary that rotates around the Earth).

[34]D'Alembert used a purely geometrical method to study the free oscillations of an ellipsoidal disturbance of the air layer.

[35]The Sun and the Moon actually do not, but the *variable* part of their action is proportional to that of such a luminary.

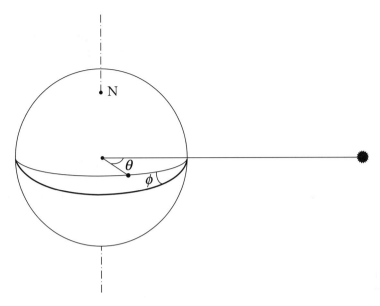

Fig. 1.7.   Spherical coordinates for d'Alembert's atmospheric tides. The fat line represents the visible part of the equator, over which the luminary is orbiting. N is the North pole.

D'Alembert first considered the simpler case $\phi \approx 0$, for which he expected the component v_ϕ to be negligible. To first order in η and v, the conservation of the volume of a vertical column of fluid yields

$$\frac{1}{h}\dot{\eta} + \frac{\partial v_\theta}{R\partial\theta} + \frac{v_\theta}{R\tan\theta} = 0, \tag{1.18}$$

which means that an increase in the height of the column is compensated for by a narrowing of its base (the dot denotes the time derivative for a fixed point on the Earth's surface). Since the tidal force $\mathbf{f}$ is much smaller than the gravity force, the vector sum $\mathbf{f}+\mathbf{g}-\boldsymbol{\gamma}$ makes an angle $(f_\theta - \gamma_\theta)/g$ with the vertical. To first order in η, the inclination of the fluid surface over the horizontal is $\partial\eta/R\partial\theta$. Therefore, the condition that $\mathbf{f}+\mathbf{g}-\boldsymbol{\gamma}$ should be perpendicular to the surface of the fluid is approximately equivalent to[36]

$$\gamma_\theta = f_\theta - g\frac{\partial\eta}{R\partial\theta}. \tag{1.19}$$

As d'Alembert noted, this equation of motion can also be obtained by equating the horizontal acceleration of a fluid slice to the sum of the tidal component f_θ and of the difference between the pressures on both sides of this slice. Indeed, the neglect of the

---

[36]D'Alembert [1747] pp. 88–9 (formulas A and B). The correspondence with d'Alembert's notation is given by $\theta \to u$, $v_\theta \to q$, $d\eta/d\theta \to -v$, $R/h\omega \to \varepsilon$, $\omega/Rg \to b^2/2a$, $R/gK \to 3S/4pd^3$ (with $f = -K\sin 2\theta$).

vertical acceleration implies that, at a given height, the internal pressure of the fluid varies as the product $g\eta$. Hence d'Alembert was aware of two routes to the equation of motion, namely, through his dynamic principle, and through an application of the momentum law to a fluid element subjected to the pressure of contiguous elements. In some sections he favored the first route, in others the second.[37]

In his expression of the time variations $\dot{\eta}$ and $\dot{v}_\theta$, d'Alembert considered only the forced motion of the fluid for which the velocity field and the free surface of the fluid rotate together with the tide-inducing luminary at the angular velocity $-\omega$. Then the values of η and v_θ at the colatitude θ and at the time $t + dt$ are equal to their values at the colatitude $\theta + \omega dt$ and at the time t. This gives

$$\dot{v}_\theta = \omega \frac{\partial v_\theta}{\partial \theta}, \ \dot{\eta} = \omega \frac{\partial \eta}{\partial \theta}. \tag{1.20}$$

D'Alembert equated the relative acceleration $\dot{v}_\theta$ with the acceleration γ_θ, for he neglected the second-order convective terms, and judged the absolute rotation of the Earth irrelevant (he was aware of the centripetal acceleration, but treated the resulting permanent deformation of the fluid surface separately; and he overlooked the Coriolis acceleration). With these substitutions, his equations (1.18) and (1.19) become ordinary differential equations with respect to the variable θ.

D'Alembert eliminated η from these two equations, and integrated the resulting differential equation for Newton's value $-K \sin 2\theta$ of the tide-inducing force f_θ. In particular, he showed that the phase of the tides (concordance or opposition) depended on whether the rotation period $2\pi/\omega$ of the luminary was smaller or larger than the quantity $2\pi R/\sqrt{gh}$, which he had earlier shown to be identical to the period of the free oscillations of the fluid layer.[38]

In another section of his memoir, d'Alembert extended his equations to the case when the angle ϕ is no longer negligible. Again, he had the velocity field and the free surface of the fluid rotate together with the luminary at the angular velocity $-\omega$. Calling $R_{\omega dt}$ the operator for the rotation of angle ωdt around the polar axis, and $\mathbf{v}(P, t)$ the velocity vector at point P and at time t, we have

$$\mathbf{v}(P, t + dt) = R_{\omega dt}\mathbf{v}(R_{\omega dt}P, t). \tag{1.21}$$

Expressing this relation in spherical coordinates, d'Alembert obtained

$$\dot{v}_\theta = \omega \left(\frac{\partial v_\theta}{\partial \theta} \cos \phi - \frac{\partial v_\theta}{\partial \phi} \frac{\sin \phi}{\tan \theta} - v_\phi \sin \phi \sin \theta \right), \tag{1.22}$$

---

[37]D'Alembert [1747] pp. 88–9. He represented the internal pressure by the weight of a vertical column of fluid. In his discussion of the condition of equilibrium (*ibid.* pp. 15–16), he introduced the balance of the horizontal component of the external force acting on a fluid element and the difference in the weight of the two adjacent columns as 'another very easy method' for determining the equilibrium. In the case of tidal motion with $\phi \approx 0$, he directly applied this condition of equilibrium to the 'destroyed motion' $\mathbf{f} + \mathbf{g} - \boldsymbol{\gamma}$. In the general case (*ibid.* pp. 112–13), he used the perpendicularity of $\mathbf{f} + \mathbf{g} - \boldsymbol{\gamma}$ to the free surface of the fluid.

[38]The elimination of η leads to the easily integrable equation
$(gh - R^2\omega^2) \, dv_\theta + gh \, d(\sin \theta)/ \sin \theta - R^2\omega K \sin \theta d(\sin \theta) = 0.$

$$\dot{v}_\phi = \omega\left(\frac{\partial v_\phi}{\partial\theta}\cos\phi - \frac{\partial v_\phi}{\partial\phi}\frac{\sin\phi}{\tan\theta} + v_\theta\sin\phi\sin\theta\right). \qquad (1.23)$$

For the same reasons as before, d'Alembert identified these derivatives with the accelerations γ_θ and γ_ϕ. He then applied his dynamic principle to obtain

$$\gamma_\theta = f_\theta - g\frac{\partial\eta}{R\,\partial\theta}, \qquad \gamma_\phi = -g\frac{\partial\eta}{R\sin\theta\,\partial\phi}. \qquad (1.24)$$

Lastly, he obtained the continuity condition

$$\dot{\eta} = \omega\left(\frac{\partial\eta}{\partial\theta}\cos\phi - \frac{\partial\eta}{\partial\phi}\frac{\sin\phi}{\tan\theta}\right) = -\left(\frac{\partial v_\theta}{\partial\theta} + \frac{v_\theta}{\tan\theta} + \frac{\partial v_\phi}{\sin\theta\partial\phi}\right), \qquad (1.25)$$

in which the modern reader may recognize the expression of a divergence in spherical coordinates.[39]

D'Alembert judged the resolution of this system to be beyond his capability. The purpose of this section of his memoir was to illustrate the power and generality of his method for deriving hydrodynamic equations. For the first time, he gave the complete equations of motion of an incompressible fluid in a genuinely bidimensional case. Thus emerged the velocity field and the corresponding partial derivatives with respect to two independent spatial coordinates. D'Alembert pioneered the application to the dynamics of continuous media of the earlier calculus of differential forms by Alexis Fontaine and Leonhard Euler. His notation of course differed from the modern one. Where we now write $\partial f/\partial x$ (following Gustav Kirchhoff), Fontaine wrote df/dx, and d'Alembert wrote A, with $df = A\,dx + B\,dy + \ldots$.

1.3.4   *The resistance of fluids*

In 1749, d'Alembert competed for another Berlin prize on the resistance of fluids, and failed: the Academy judged that none of the competitors had reached the point of comparing his theoretical results with experiments. D'Alembert did not deny the importance of this comparison for the improvement of ship design. However, he judged that the relevant equations could not be solved in the near future, and that his memoir deserved consideration for its methodological innovations. In 1752, he published an augmented translation of this memoir as a book.[40]

Compared with the earlier treatise on the equilibrium and motion of fluids, the first important difference was a new formulation of the laws of hydrostatics. In 1744, d'Alembert started with the uniform and isotropic transmissibility of pressure by any fluid (from

---

[39]D'Alembert [1747] pp. 111–14 (equations E, F, G, H, I). To complete the correspondence given in footnote 36, take $\phi \to A$, $v_\phi \to \eta$, $\gamma_\theta \to \pi$, $\gamma_\phi \to \varphi$, $g/R \to p$, $\partial\eta/\partial\theta \to -\rho$, $\partial\eta/\partial\phi \to -\sigma$, $\partial v_\theta/\partial\theta \to r$, $\partial v_\theta/\partial\phi \to \lambda$, $\partial v_\phi\,\partial\theta \to \gamma$, $\partial v_\phi/\partial\phi \to \beta$. D'Alembert has the ratio of two sines instead of the product in the last term in each of eqns (1.22) and (1.23). An easy, modern way to obtain these equations is to rewrite eqn. (1.21) as $\dot{\mathbf{v}} = [(\boldsymbol{\omega}\times\mathbf{r})\cdot\nabla]\mathbf{v} + \boldsymbol{\omega}\times\mathbf{v}$, with $\mathbf{v} = (0, v_\theta, v_\phi)$, $\mathbf{r} = (R, 0, 0)$, $\boldsymbol{\omega} = \omega(\sin\theta\sin\phi, \cos\theta\sin\phi, \cos\phi)$, and $\nabla = (\partial_r, \partial_\theta/R, \partial_\phi/R\sin\theta)$ in the local basis.

[40]D'Alembert [1752] p. xxxviii. For an insightful study of d'Alembert's work on fluid resistance, cf. Grimberg [1998]. See also Calero [1996] Chap. 8.

one part of its surface to another). He then derived the standard laws of this science, such as the horizontality of the free surface and the depth dependence of wall pressure, by qualitative or geometrical reasoning. In contrast, in his new memoir he relied on a mathematical principle borrowed from Alexis-Claude Clairaut's memoir of 1743 on the shape of the Earth. According to this principle, a fluid mass subjected to a force density $\mathbf{f}$ is in equilibrium if and only if the integral $\int \mathbf{f} \cdot d\mathbf{l}$ vanishes over any closed loop within the fluid and over any path whose ends belong to the free surface of the fluid.[41]

D'Alembert regarded this principle as a mathematical expression of his earlier principle of the uniform transmissibility of pressure. If the fluid is globally in equilibrium, he reasoned, then it must also be in equilibrium within any narrow canal of section ε belonging to the fluid mass. For a canal beginning and ending on the free surface of the fluid, the pressure exerted by the fluid on each of the extremities of the canal must vanish. According to the principle of uniform transmissibility of pressure, the force $\mathbf{f}$ acting on the fluid within the length $d\mathbf{l}$ of the canal exerts a pressure $\varepsilon \mathbf{f} \cdot d\mathbf{l}$ that is transmitted to both ends of the canal (with opposite signs). As the sum of these pressures must vanish, so does the integral $\int \mathbf{f} \cdot d\mathbf{l}$. This reasoning and a similar one for closed canals establish d'Alembert's new principle of equilibrium.[42]

Applying this principle to an infinitesimal loop, d'Alembert obtained (the Cartesian coordinate form of) the differential condition

$$\nabla \times \mathbf{f} = \mathbf{0}, \tag{1.26}$$

as Clairaut had already done. Combining it with his principle of dynamics, and confining himself to the steady motion ($\partial \mathbf{v}/\partial t = \mathbf{0}$, so that $\boldsymbol{\gamma} = (\mathbf{v} \cdot \nabla)\mathbf{v}$) of an incompressible fluid, he obtained the two-dimensional, Cartesian coordinate version of

$$\nabla \times [(\mathbf{v} \cdot \nabla)\mathbf{v}] = \mathbf{0}, \tag{1.27}$$

which means that the fluid must formally be in equilibrium with respect to the convective acceleration. D'Alembert then showed that this condition was met whenever $\nabla \times \mathbf{v} = \mathbf{0}$. Confusing a sufficient condition with a necessary one, he concluded that the latter property of the flow held generally.[43]

This property nonetheless holds in the special case of motion investigated by d'Alembert, that is, the stationary flow of an incompressible fluid around a solid body when the flow is uniform far away from the body (see Fig. 1.8). In this limited case, d'Alembert gave a correct proof, of which a modernized version follows.[44]

Consider two neighboring lines of flow beginning in the uniform region of the flow and ending in any other part of the flow, and connect the extremities through a small segment.

---

[41]D'Alembert [1752] pp. 14–17. On the figure of the Earth, cf. Todhunter [1873]. On Clairaut, cf. Passeron [1995]. On Newton's and MacLaurin's partial anticipations of Clairaut's principle, cf. Truesdell [1954] pp. XIV–XXII.

[42]As is obvious to the modern reader, this principle is equivalent to the existence of a single-valued function (P) of which $\mathbf{f}$ is the gradient and which has a constant value on the free surface of the fluid. The canal equilibrium results from the principle of solidification, the history of which is discussed in Casey [1992].

[43]D'Alembert [1752] art. 78. The modern hydrodynamicist may recognize in eqn (1.27) a particular case of the vorticity equation. The condition $\nabla \times \mathbf{v} = \mathbf{0}$ is that of irrotational flow.

[44]For a more literal rendering of d'Alembert's proof, cf. Grimberg [1998] pp. 43–8.

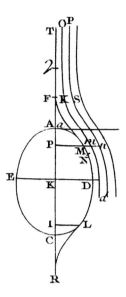

Fig. 1.8. Flow around a solid body according to d'Alembert ([1752] plate).

According to d'Alembert's principle together with the principle of equilibrium, the integral $\oint (\mathbf{v} \cdot \nabla)\mathbf{v} \cdot \mathbf{dr}$ vanishes over this loop. Using the identity

$$(\mathbf{v} \cdot \nabla)\mathbf{v} = \nabla \left(\frac{1}{2}v^2\right) - \mathbf{v} \times (\nabla \times \mathbf{v}), \tag{1.28}$$

this implies that the integral $\oint (\nabla \times \mathbf{v}) \cdot (\mathbf{v} \times \mathbf{dr})$ also vanishes. The only part of the loop that contributes to this integral is that corresponding to the small segment joining the endpoints of the two lines of flow. Since the orientation of this segment is arbitrary, $\nabla \times \mathbf{v}$ must vanish.

D'Alembert thus derived the condition

$$\nabla \times \mathbf{v} = \mathbf{0} \tag{1.29}$$

from his dynamic principle. He also obtained the continuity condition

$$\nabla \cdot \mathbf{v} = 0 \tag{1.30}$$

by requiring the constancy of the volume of a given element of fluid during its motion. More exactly, he obtained the special expressions of these two conditions in the cylindrically-symmetric case and in the two-dimensional case. In order to solve this system of two partial differential equations in the two-dimensional case, he noted that the two conditions meant that the forms $u\,dx + v\,dy$ and $v\,dx - u\,dy$ were exact differentials. This property holds, he ingeniously noted, if and only if $(u - iv)(dx + i\,dy)$ is an exact differential. This means that u and $-v$ are the real and imaginary parts of a (holomorphic) function of the complex variable $x + iy$. They must also be such that the velocity is uniform at infinity and tangent to the body along its surface. D'Alembert struggled to meet these boundary conditions through power-series developments, to little avail.[45]

The ultimate goal of this calculation was to determine the force exerted by the fluid on the solid, which is the same as the resistance offered by the fluid to the motion of a body with a velocity opposite to that of the asymptotic flow.[46] D'Alembert expressed this force as the integral of the fluid's pressure over the whole surface of the body. The pressure is itself given by the line integral of $-d\mathbf{v}/dt$ from infinity to the wall, in conformance with d'Alembert's earlier derivation of Bernoulli's law. This law still holds in the present case, because $-d\mathbf{v}/dt = -(\mathbf{v} \cdot \nabla)\mathbf{v} = -\nabla(\frac{1}{2}v^2)$. Hence the resistance could be determined, if only the flow around the body was known.[47]

In 1749, d'Alembert did not know enough about this flow to reach definite conclusions on the resistance. A few years later, he realized that, for a head–tail symmetric body, a solution of his differential equations was possible in which the fluid velocity was the same at the front and at the rear of the body (up to a sign change). Bernoulli's law gives zero resistance for this solution, since the head pressure exactly balances the tail pressure. As d'Alembert knew, his equations only admit one solution. Therefore, the flow is unique and symmetric, and the resistance must vanish. D'Alembert concluded:[48]

> Thus I do not see, I admit, how one can satisfactorily explain by theory the resistance of fluids. On the contrary, it seems to me that the theory, developed in all possible rigor, gives, at least in several cases, a strictly vanishing resistance; a singular paradox which I leave to future geometers for elucidation.

D'Alembert thus shed doubts on a theory to the construction of which he had much contributed. Through his dynamic principle and his equilibrium principle, he had obtained hydrodynamic equations for the steady flow of an incompressible fluid that we may retrospectively identify as the continuity equation, the condition of irrotational flow, and Bernoulli's law. Admittedly, he only wrote these equations for the axially-symmetric and two-dimensional cases that were relevant to the fluid-resistance problem. The modern reader may wonder why he did not try to write general equations of fluid motion in

[45]D'Alembert [1752] pp. 60–2. Here d'Alembert discovered the Cauchy–Riemann condition for u and $-v$ to be the real and imaginary components, respectively, of an analytic function in the complex plane, as well as a powerful method to solve Laplace's equation $\Delta u = 0$ in two dimensions. In [1761] p. 139, d'Alembert introduced the complex potential $\varphi + i\psi$ such that $(u - iv)(dx + i\,dy) = d(\varphi + i\psi)$. The real part φ of this potential is the velocity potential introduced by Euler in 1752; its imaginary part ψ is the so-called stream function, which is a constant on any line of current, as d'Alembert noted.

[46]D'Alembert gave a proof of this equivalence, which he did not regard as obvious.

[47]D'Alembert had already discussed fluid resistance in part III of his treatise of 1744. There he used a molecular model in which momentum was transferred by impact from the moving body to a layer of hard molecules. He believed, however, that this molecular process would be negligible if the fluid molecules were too close to each other, for instance, when fluid was forced through the narrow space between the body and a containing cylinder. In this case ([1744] pp. 205–6), he assumed a parallel-slice flow and computed the fluid pressure on the body through Bernoulli's law. For a head–tail symmetric body, this pressure does not contribute to the resistance if the flow has the same symmetry. After noting this difficulty, d'Alembert evoked the observed stagnancy of the fluid behind the body to retain only the Bernoulli pressure on the prow.

[48]D'Alembert [1768] p. 138. In his memoir of 1749, besides the Bernoulli pressure, d'Alembert evoked a velocity-proportional friction of the fluid on the body, and the *ténacité* of the fluid, according to which a certain (velocity-independent) force was required to separate the fluid molecules from each other at the prow of the body ([1752] pp. 106–8). For a modern, more general derivation of the paradox, see Appendix A.

Cartesian coordinate form. The answer is plain: he was following an older tradition of mathematical physics according to which general principles, rather than general equations, were applied to specific problems.

D'Alembert obtained his basic equations without recourse to the concept of pressure. Yet he had a concept of internal pressure, which he used to derive Bernoulli's law. Then we may wonder why he did not pursue the other approach sketched in his theory of winds, that is, the application of Newton's second law to a fluid element subjected to a pressure gradient. Plausibly, he favored a derivation that was based on his own principle of dynamics and thus avoided obscure internal forces.

D'Alembert knew well, however, that his equilibrium principle was simply the condition of uniform integrability for the force density $\mathbf{f}$. Had he cared to introduce the integral, say P, he would have found the equilibrium equation $\mathbf{f} = \nabla P$ that makes P the internal pressure. Applying his dynamic principle, he would have reached the equation of motion

$$\mathbf{f} - \rho \frac{d\mathbf{v}}{dt} = \nabla P, \tag{1.31}$$

which is simply Euler's equation. But he did not proceed along these lines, and rather wrote equations of motion that did not involve internal pressure.[49]

1.4   Euler's equations

1.4.1   *The Latin memoir*

Unlike d'Alembert, the Swiss geometer and Berlin Academician Leonhard Euler did not believe that a new dynamic principle was necessary for continuous or connected systems, and he had no objection to internal forces. In 1740, he congratulated Johann Bernoulli for having 'determined most accurately the pressure in every state of the water.' In 1750, he claimed that the true basis of continuum mechanics was Newton's second law applied to the infinitesimal elements of bodies. Among the forces acting on the elements, he included 'connection forces' acting on the boundary of the elements. In the case of fluids, these internal forces were to be identified with the pressure. The acceleration of the fluid elements therefore depended on the combined effect of the pressure gradient and external forces (gravity), as noted by d'Alembert in his memoir on winds. In hydraulic writings of 1750/51, Euler thus obtained the differential version

$$\frac{dv}{dt} = g - \frac{dP}{dz} \tag{1.32}$$

of Johann Bernoulli's equation (1.8) for parallel-slice efflux.[50]

[49]In this light, d'Alembert's later neglect of Euler's approach should not be regarded as a mere expression of rancor.

[50]Euler to J. Bernoulli, 18 Oct. 1740, in Euler [1998] pp. 386–9; Euler [1750] p. 90 (the main purpose of this paper was the derivation of the equations of motion of a solid). On the hydraulic writings, cf. Truesdell [1954] pp. XLI–XLV. These included Euler's evaluation of the pressure in the pipes that were being built to feed the fountains of Sanssouci (Euler [1752]), nicely discussed in Eckert [2002]. There Euler used the generalization (1.9) of Bernoulli's law to non-permanent flow, which he derived from eqn (1.32). As Eckert explains, the failure of the fountains project and an ambiguous letter from the King of Prussia to Voltaire have led to the myth of Euler's incapacity in concrete matters.

In a Latin memoir of 1752, probably stimulated by the two memoirs of d'Alembert he had reviewed for the Berlin Academy, Euler obtained the general equations of fluid motion for an incompressible fluid in terms of the internal pressure P and the Cartesian coordinates of the velocity $\mathbf{v}$. For this purpose, he simply applied Newton's second law to a cubic element of fluid subjected to the gravity $\mathbf{g}$ and to the pressure P acting on the cube's faces. By a now familiar reasoning, this procedure yields (for unit density)

$$\frac{\partial \mathbf{v}}{\partial t} + (\mathbf{v} \cdot \nabla)\mathbf{v} = \mathbf{g} - \nabla P. \tag{1.33}$$

Euler also obtained the continuity equation

$$\nabla \cdot \mathbf{v} = 0, \tag{1.34}$$

and eliminated P from the equation of motion to obtain

$$\left[\frac{\partial}{\partial t} + (\mathbf{v} \cdot \nabla) \right] (\nabla \times \mathbf{v}) - [(\nabla \times \mathbf{v}) \cdot \nabla]\mathbf{v} = \mathbf{0}. \tag{1.35}$$

Interestingly, Euler repeated d'Alembert's mistake of regarding $\nabla \times \mathbf{v} = \mathbf{0}$ as a necessary condition for the validity of the former relation, whereas it is only a sufficient condition. This error allowed him to introduce what later fluid theorists called the velocity potential, that is, the function $\varphi(\mathbf{r})$ such that $\mathbf{v} = \nabla\varphi$. Equation (1.33) may then be rewritten as

$$\frac{\partial}{\partial t}(\nabla\varphi) + \frac{1}{2}\nabla(v^2) = \mathbf{g} - \nabla P. \tag{1.36}$$

Spatial integration of this equation yields a generalization of Bernoulli's law:

$$P = \mathbf{g} \cdot \mathbf{r} - \frac{1}{2}v^2 - \frac{\partial \varphi}{\partial t} + C, \tag{1.37}$$

where C is a constant (time dependence can be absorbed into the velocity potential). Lastly, Euler applied this equation to the flow through a narrow tube of variable section to retrieve the results of the Bernoullis.[51]

Although Euler's Latin memoir contained the basic hydrodynamic equations for an incompressible fluid, the form of exposition was still in flux. Euler often used specific letters (coefficients of differential forms) for partial differentials rather than Fontaine's notation, and measured velocities and acceleration in gravity-dependent units. He proceeded gradually, from the simpler two-dimensional case to the fuller three-dimensional case. His derivation of the continuity equation was more intricate than we would now expect. In addition, he erred in believing in the general existence of a velocity potential. These characteristics make Euler's Latin memoir a transition between d'Alembert's fluid dynamics and the fully-modern foundation of this science found in the French memoirs of 1755.[52]

---

[51]Euler [1752] pp. 154–7.

[52]Cf. Truesdell [1954] pp. LXII–LXXV.

1.4.2   *The French memoirs*

The first of these memoirs is devoted to the equilibrium of fluids, both incompressible and compressible. Euler presumably realized that his new hydrodynamics contained a new hydrostatics based on the following principle: the action of the contiguous fluid on a given, internal element of fluid results from an isotropic, normal pressure P exerted on its surface. The equilibrium of an infinitesimal element subjected to this pressure and to the force density $\mathbf{f}$ of external origin then requires

$$\mathbf{f} - \nabla P = \mathbf{0}. \tag{1.38}$$

As Euler showed, all known results of hydrostatics follow from this simple mathematical law.[53]

In his next memoir, Euler obtained the general hydrodynamic equations for compressible fluids:

$$\nabla \cdot \rho\mathbf{v} + \frac{\partial \rho}{\partial t} = 0 \tag{1.39}$$

for the continuity condition, and 'Euler's equation'

$$\frac{\partial \mathbf{v}}{\partial t} + (\mathbf{v} \cdot \nabla)\mathbf{v} = \frac{1}{\rho}(\mathbf{f} - \nabla P), \tag{1.40}$$

to which a relation between pressure, density, and heat must be added for completeness. Euler now realized that $\nabla \times \mathbf{v}$ did not necessarily vanish, for example in the case of vortex flows. In a sequel to this memoir, he showed that Bernoulli's law nonetheless remained valid along the stream lines of any steady flow of an incompressible fluid. Indeed, owing to the identity

$$(\mathbf{v} \cdot \nabla)\mathbf{v} = \nabla\left(\frac{1}{2}v^2\right) - \mathbf{v} \times (\nabla \times \mathbf{v}), \tag{1.41}$$

the integration of the convective acceleration term along a line of flow eliminates $\nabla \times \mathbf{v}$ and contributes the $\frac{1}{2}v^2$ term of Bernoulli's law.[54]

Euler deplored the difficulty of solving his equations. He could not really handle any problem that was not accessible to earlier methods, although he devoted much space to the general conditions of integrability. His true achievement was a strikingly modern and crystal-clear expression of the foundations of hydrodynamics. Present derivations of the fundamental equations follow Euler's original procedures very closely. Unlike the earlier hydrodynamic writings of d'Alembert and of the Bernoullis, Euler's memoirs are immediately intelligible to the modern reader. They mark the emergence of a new style of mathematical physics in which fundamental equations take the place of fundamental principles.

Yet we should not underestimate Euler's debts to his predecessors. Euler himself paid tribute to the Bernoullis and to d'Alembert, despite his obscure role in d'Alembert's failure

---

[53]Euler [1755*a*] p. 5.

[54]Euler [1755*b*] pp. 63, 65; [1755*c*] 117. Cf. Truesdell [1954] pp. LXXV–C.

to win the Berlin prize on winds.[55] These authors anticipated essential features of Euler's approach. Johann Bernoulli had a concept of internal pressure, and some sort of convective derivative (the *gurges*). D'Alembert had particular cases of the partial differential equations of continuity and motion, as well as the general idea of deriving the equations of motion by balancing acceleration, external forces, and pressure gradient. Euler's role was to prune unnecessary and unclear elements in the abundant writings of his predecessors, and to combine the elements he judged most fundamental in the clearest and most general manner.

1.5 Lagrange's analysis

1.5.1 *Methods of resolution*

In his celebrated memoir of 1781 on fluid motion, Joseph Louis Lagrange judged that the foundations of this subject had been sufficiently established by d'Alembert and his followers. But he deplored the lack of efficient, rigorous methods for solving practical questions of fluid motion. Having already done much work on the integration of partial differential equations, he knew that the general integral of this kind of equation depended on an arbitrary function that could only be determined through the boundary conditions. A first condition for the determination of specific flows was a clear and complete statement of the boundary conditions.[56]

Already known were the condition that the velocity of the fluid on the walls of its container should be parallel to the walls, and the condition that the pressure on the free surface should be equal to the external pressure. Lagrange added the condition that a fluid particle initially on the free surface of the fluid should retain this property 'so that the fluid does not divide itself but always forms a continuous mass.' If $f(\mathbf{r}, t) = 0$ is the equation of the fluid surface, this condition implies

$$\frac{\partial f}{\partial t} + (\mathbf{v} \cdot \nabla)f = 0 \tag{1.42}$$

on the surface.[57]

In order to ease the resolution of Euler's equation, Lagrange systematically introduced the velocity potential φ, which reduces the number of unknown functions from three to one. It was therefore important to him to determine the condition under which this potential existed. The following, important theorem answered this question: *whenever the motion of an incompressible fluid is prompted by forces that derive from a potential (gravity or external pressure), a velocity potential exists.*

In order to prove this, Lagrange multiplied Euler's equation (1.40) by $\mathbf{dr}$ to obtain

$$\frac{\partial \mathbf{v}}{\partial t} \cdot \mathbf{dr} + (\nabla \times \mathbf{v}) \cdot (\mathbf{v} \times \mathbf{dr}) = \frac{1}{\rho}\mathbf{f} \cdot \mathbf{dr} - \frac{\mathrm{d}P}{\rho} - \mathrm{d}\left(\frac{v^2}{2}\right). \tag{1.43}$$

---

[55]Cf. Grimberg [1998] pp. 8–10.   [56]Cf. Truesdell [1955] pp. XC–CV.

[57]Lagrange [1781] p. 704. On later criticism of this condition, cf. Truesdell [1955] p. XCI.

If the pressure is a function of density only (which is, of course, the case for an incompressible fluid) and if $\mathbf{f}/\rho$ derives from a potential, then the right-hand side of this equation is an exact differential. After noting this, Lagrange applied his favorite method, power-series development, to the functions $\mathbf{v}(t)$ and $\nabla \times \mathbf{v}(t)$. As the left-hand side of eqn (1.43) must be an exact differential, the vanishing of the coefficients of $\nabla \times \mathbf{v}(t)$ up to order n implies that the $(n + 1)$th coefficient of $\mathbf{v} \cdot d\mathbf{r}$ is an exact differential or, equivalently, that the $(n + 1)$th coefficient of $\nabla \times \mathbf{v}(t)$ vanishes. As, by hypothesis, $\mathbf{v} \cdot d\mathbf{r}$ is an exact differential for $t = 0$, the first term of the development of $\nabla \times \mathbf{v}(t)$ must vanish. By induction, all other terms must then vanish. Therefore, $\nabla \times \mathbf{v}(t)$ vanishes at any time and there exists a velocity potential at any positive time.[58]

Although Lagrange seems to have believed that the conditions of his theorem were met for most flows in nature, he gave one example in which they were not, namely tidal motion (since the Coriolis forces do not derive from a potential). Lagrange also (incorrectly) argued that the velocity potential existed for small motions in which the second-order term $(\mathbf{v} \cdot \nabla)\mathbf{v}$ could be neglected. As either this condition or that of the previous theorem seemed to hold in many cases of motion, Lagrange believed he could restrict his analysis to potential flows without much loss of generality. He gave the propagation of sound in a compressible fluid as an example of the applicability of the second condition. He gave the motion of an incompressible fluid under the sole effect of gravity as an example of the applicability of the first condition.[59]

In the latter case, the equations for the velocity potential were still too complicated to allow integration in finite terms. Lagrange assumed one of the dimensions of the fluid to be very small, so that one of the coordinates of the fluid particles could be taken to be much smaller than the other coordinates. Then the velocity potential could be expressed as a power series with respect to this coordinate. Lagrange thus obtained the parallel-slice solution of the efflux problem in a first approximation, and also corrections depending on higher powers of the width of the vessel. Most originally, he showed that small surface disturbances on shallow water obeyed the equations of a vibrating string with a propagation velocity $\sqrt{gh}$, where h is the depth of the water.[60]

Lagrange's equations and boundary conditions for the velocity potential of an incompressible fluid were the invariable basis of much of nineteenth-century hydrodynamics, for instance, the theories of waves by Poisson, Cauchy, Stokes, Boussinesq, Korteweg, and de Vries. The resolution of these equations is intimately bound to the development of potential theory and Fourier analysis. To cite only two examples, Cauchy reinvented Fourier analysis in his memoir on waves, and Stokes obtained important theorems for the potential, which his friend Kelvin transposed to electric and magnetic contexts.[61]

---

[58]Lagrange [1781] pp. 714–17; Lagrange to d'Alembert, 15 Apr. 1781, in Lagrange [1867–1892] vol. 13, pp. 362–6. This proof only holds if the function $\mathbf{v}(t)$ is analytical. Cauchy [1827a] has the first rigorous, general proof.

[59]*Ibid.* pp. 713–18, 721–3, 728.   [60]*Ibid.* pp. 728–48. Cf. Chapter 2, pp. 35–37.

[61]Cf. Wise [1981], Darrigol [2000] pp. 128–9, Darrigol [2003].

1.5.2   *Continuum dynamics*

Lagrange returned to hydrodynamics when he wrote his *Méchanique analitique* of 1788. As is well known, he there obtained the general equations for the dynamics of connected systems by combining the principle of virtual velocities and d'Alembert's principle. From d'Alembert's viewpoint, fluids could not be treated on the same footing, because their internal composition was too complex to allow a deduction of the condition of equilibrium; this condition had to be obtained empirically. In his analytical mechanics, Lagrange prided himself on eliminating this asymmetry between solid and fluid dynamics. He first showed that the condition of equilibrium of an incompressible fluid derived from the principle of virtual velocities applied to an ideal continuum.

According to this principle, the moment (virtual work) of the force density **f** acting within a fluid mass and of the pressure $\overline{P}$ exerted on the free surface of the fluid must vanish for any displacement $\delta\mathbf{r}$ of the fluid particles that satisfies the condition of incompressibility $\nabla \cdot \delta\mathbf{r} = 0$. Through Lagrange's method of multipliers, this condition is equivalent to

$$\int (\mathbf{f} \cdot \delta\mathbf{r} + \lambda \nabla \cdot \delta\mathbf{r})d\tau - \int \delta\mathbf{r} \cdot \overline{P}d\mathbf{S} = 0, \tag{1.44}$$

where $\lambda(\mathbf{r})$ is the Lagrange multiplier, and the displacement $\delta\mathbf{r}$ is now arbitrary, except on solid walls where it must be parallel to the walls. Integrating by parts the λ term, this gives

$$\int (\mathbf{f} - \nabla\lambda) \cdot \delta\mathbf{r} \, d\tau + \int (\lambda - \overline{P})\delta\mathbf{r} \cdot d\mathbf{S} = 0. \tag{1.45}$$

Hence $\mathbf{f} - \nabla\lambda$ must vanish within the fluid, and the parameter λ must be equal to the external pressure on the free surface. This is equivalent to Euler's equilibrium condition, the parameter λ playing the role of the internal pressure.[62]

For a compressible fluid, there is no constraint on the displacement $\delta\mathbf{r}$ (save for parallelism on solid walls), but the moment of the internal forces of elasticity must be added to the moment of the external forces. As the 'elasticity' P tends to increase the volume $d\tau$ of the particles of fluid, Lagrange wrote this new moment as

$$\int P\delta(d\tau) = \int P(\nabla \cdot \delta\mathbf{r})d\tau \tag{1.46}$$

Consequently, the condition of equilibrium has the same form as in the case of incompressibility, and the 'elasticity' P plays the role of Euler's internal pressure. D'Alembert's principle, combined with this condition, yields Euler's equations of fluid motion. In all, Lagrange's purely analytical approach to the equilibrium and motion of fluids led to the same set of fundamental equations as Euler's more intuitive approach. With a mathematical subtlety that prevented large diffusion, he subsumed the conditions of equilibrium of a continuum under a general principle of statics.[63]

---

[62]Lagrange [1788] pp. 139–45, 438–41 (case of motion). A similar procedure was previously given in Lagrange [1761] pp. 435–59. Cf. Truesdell [1954] p. CXXIV.

[63]Lagrange [1788] pp. 155–7, 492–93. A similar procedure is found in Lagrange [1761] pp. 459–68 (although at that time Lagrange used a generalization of a variational principle by Euler instead of d'Alembert's principle).

1.5.3 *The Lagrangian picture*

Combining d'Alembert's principle and the principle of virtual velocities, Lagrange obtained Euler's equation in the form

$$\mathbf{f} - \rho \boldsymbol{\gamma} - \nabla P = \mathbf{0}. \tag{1.47}$$

In his memoir of 1781, Lagrange followed d'Alembert's and Euler's original method of characterizing the fluid motion through the velocity $\mathbf{v}$ as a function of the points of space $\mathbf{r}$ and the time t. This leads to the 'Eulerian' form (1.1) of Euler's equation. In the *Méchanique analitique*, however, Lagrange judged that 'a distinct idea of the nature of this equation' required another representation, in which the position $\mathbf{r}$ of the fluid particles is regarded as a function of time and of their position $\mathbf{R}$ at the origin of time.[64]

Multiplying eqn (1.47) by the differential $d\mathbf{r}$ yields

$$(\mathbf{f} - \rho \boldsymbol{\gamma}) \cdot d\mathbf{r} - dP = 0, \tag{1.48}$$

or, in terms of the coordinates R_1, R_2, R_3, and t,

$$\left(\mathbf{f} - \rho \frac{\partial^2 \mathbf{r}}{\partial t^2} \right) \cdot \frac{\partial \mathbf{r}}{\partial R_i} - \frac{\partial P}{\partial R_i} = 0. \tag{1.49}$$

This is the so-called 'Lagrangian' form of Euler's equation. For an incompressible fluid, the continuity condition further requires that the transformation $\mathbf{R} \rightarrow \mathbf{r}$ locally conserves volumes, that is,[65]

$$\det \left(\frac{\partial r_i}{\partial R_j} \right) = 1. \tag{1.50}$$

As Lagrange noted, this form of the equations of fluid motion is more complex than the Eulerian form. Despite its name, it was not invented by Lagrange.[66] Euler introduced it in his theory of the motion of rivers of 1751, in his theory of sound of 1759, and Laplace used it in his theory of tides of 1776. In these cases the neglect of second-order terms with respect to $\mathbf{r} - \mathbf{R}$ turns the equations into more manageable ones.

An interesting question is why the Eulerian picture historically preceded the Lagrangian one. There may be no simple answer, however. Daniel Bernoulli naturally focused on the fluid's velocity, since he based his analysis on the principle of live forces. His father, who did not rely on this principle, still focused on velocity, presumably because it was the main quantity of interest in the efflux problem. The same could be said for d'Alembert's treatise on fluids, with its classical emphasis on efflux. In his memoirs on winds and on fluid resistance, velocity was again the most relevant quantity, the more so because the flow was steady (or uniformly rotating in the wind case). Perhaps one should instead wonder why Euler introduced the Lagrangian picture in an acoustic context. The answer may be that

---

[64] Lagrange [1788] p. 442.   [65] *Ibid.* p. 283.

[66] *Ibid.* p. 280. Cf. Truesdell [1954] pp. CXIX–CXXIII. Lagrange's earliest discussion of this picture is in Lagrange [1761] pp. 448–52.

earlier solutions of problems of elastic motion, such as d'Alembert's for vibrating strings or Lagrange's for sound propagation, were formulated in terms of the displacements of the particles of the system that determined the elastic response.

The vanity of seeking the true founder of hydrodynamics should now be clear. Although Euler's name is legitimately attached to the equations of motion of inviscid fluids, his contribution should only be regarded as one step toward the end of a long formative process. Particular cases of his equations or closely-related statements already appeared in the works of the Bernoullis. D'Alembert invented a general method through which the equations of any problem of fluid motion could be formulated, and obtained the first partial differential equations of fluid mechanics. Euler brilliantly capitalized on these earlier achievements. Lagrange offered alternative foundations, and powerful methods for solving the equations.

An essential element of this evolution was the recurrent analogy between the efflux from a vase and the fall of a compound pendulum. Any dynamic principle that solved the latter problem also solved the former. Daniel Bernoulli appealed to the conservation of live forces, Johann Bernoulli to Newton's second law together with the idiosyncratic concept of *translatio*, and d'Alembert to his own dynamic principle of the equilibrium of destroyed motions. With this more general principle and his taste for partial differentials, d'Alembert leapt from parallel-slice flows to higher problems that involved two-dimensional anticipations of Euler's equations. His method implicitly contained a completely general derivation of these equations, as Lagrange later showed. Another important element was the concept of internal pressure. So to say, the door on the way to general fluid mechanics opened with two different keys, namely, d'Alembert's principle, or the concept of internal pressure. D'Alembert and Lagrange used the first key, and introduced internal pressure only as a derivative concept. Euler used the second key, and ignored d'Alembert's principle. As Euler guessed (and as d'Alembert suggested *en passant*), Newton's old second law applies to the volume elements of the fluid, if only the pressure of fluid on fluid is taken into account. Euler's equations derive from this deceptively simple consideration.

2

WATER WAVES

> Of all the beautiful forms of water waves that of Ship Waves is perhaps most
> beautiful, if you can compare the beauty of such beautiful things. The subject of
> ship waves is certainly one of the most interesting in mathematical science.[1]
> (William Thomson, August 1887)

As d'Alembert and Euler admitted, one could well know how to write the basic equations
of hydrodynamics without knowing how to apply them to concrete problems. In the case of
fluid resistance, this gap could only be filled in the twentieth century. Yet there is one kind
of problem that earlier fluid theorists could solve to their satisfaction, namely, the motion
of waves on the free surface of water. In 1781, Lagrange wrote the basic equations of water
waves, and solved them in the simplest case of small waves on shallow water. His
nineteenth-century followers determined the celerity of small, plane, monochromatic
waves on water of constant depth, the pattern of waves created by a local action on the
water surface, the shape of oscillatory or solitary waves of finite size, and the effect of
friction, wind, and a variable bottom on the size and shape of the waves.[2]

There is, however, a puzzling contrast between the conciseness and ease of the modern
treatment of these topics, and the long, difficult struggles of nineteenth-century physicists
with them. For example, a modern reader of Poisson's old memoir on waves finds a
bewildering accumulation of complex calculations where he would expect some rather
elementary analysis. The reason for this difference is not any weakness of early nineteenth-
century mathematicians, but our overestimation of the physico-mathematical tools that
were available in their times. It would seem, for instance, that all that Poisson needed to
solve his particular wave problem was Fourier analysis, which Joseph Fourier had introduced a few years earlier. In reality, Poisson only knew a raw, algebraic version of Fourier
analysis, whereas modern physicists have unconsciously assimilated a physically 'dressed'
Fourier analysis, replete with metaphors and intuitions borrowed from the concrete wave
phenomena of optics, acoustics, and hydrodynamics. In our mind, a Fourier component is
no longer a mere coefficient in an algebraic development, it is a periodic wave that may
interfere with other waves in a manner we can easily imagine.

The transition from a dry mathematical analysis to a genuinely physico-mathematical
analysis occurred gradually in the nineteenth century, through reversible analogies between
different domains of physics. It concerned not only Fourier analysis, but also the theory of

---

[1]Thomson [1887/] p. 410.

[2]Nineteenth-century wave theorists did not understand the random, statistical character of ocean waves, nor
the mechanisms responsible for their formation. Progress on these difficult questions only occurred in the 1950s,
cf. Kinsman [1965].

ordinary differential equations, potential theory, perturbative methods, Cauchy's method of residues, etc. The modern recourse to such mathematical techniques involves a great deal of implicit knowledge that only becomes apparent in comparisons with older usage.

The motivation for the introduction of more powerful tools of analysis was mainly experimental. Most water-wave phenomena were known well before they could be explained. In most cases, they were discovered in connection with navigation problems. Not surprisingly, the wave theorists after Poisson and Cauchy shared an interest in the rational development of navigation. Waves were relevant to several aspects of this science, namely: tide prediction, ship rolling, ship resistance, harbor safety, the wearing of canals, etc. British natural philosophers such as Airy, Stokes, Thomson, Rayleigh, and Lamb were evidently more concerned with these questions than their continental counterparts. They did most to bring the theory of water waves to the service of sea and canal travel, although there were a few French contributions in Saint-Venant's wake.[3]

Section 2.1 is devoted to the theories of waves developed between 1775 and 1825 by the four French mathematicians Laplace, Lagrange, Poisson, and Cauchy, mostly for the sake of mathematics, on the basis of the new hydrodynamics. Section 2.2 is devoted to Scott Russell's many instructive experiments on waves of various kinds, including his now famous and then infamous solitary wave, in the context of British Association sponsored research on ship design. Section 2.3 presents Airy's wave theory of tides and his critical analysis of Russell's results. Section 2.4 deals with the problem of finite waves of permanent shape, as studied by Stokes, Boussinesq, and Rayleigh. It also includes Boussinesq's treatment of the evolution of an arbitrary swell, through which he arrived (in 1877) at the equation which is now attributed to Korteweg and de Vries [1895]. Section 2.5, the last section in this chapter, concerns the application of optical or acoustic ideas of interference to the explanation of water-wave phenomena. Due to such innovations, Stokes, Reynolds, and Rayleigh forged the concept of group velocity, Rayleigh solved the problem of waves created by a drifting fishing line, and Kelvin computed the pattern of ship waves, thereby inventing the celebrated method of stationary phase.

2.1 French mathematicians

2.1.1 *Laplace's attempt*

In 1775/76, Pierre-Simon de Laplace published his celebrated theory of tides, based on the hydrodynamics of Jean le Rond d'Alembert. Laplace represented the oceans as a layer of perfect liquid of variable depth on a uniformly-rotating spheroid, subjected to the variable attraction of the Moon and the Sun. Applying d'Alembert's principle of dynamics to the fluid particles, and neglecting the vertical acceleration of the water as well as any quantity of second order with respect to the fluid velocity, he obtained the fundamental equations of tidal motion. As will appear in a moment, the former approximation requires the depth of the water to be small compared to the length over which the tidal elevation varies sensibly; the latter approximation requires the tidal elevation to be much smaller than the depth.[4]

---

[3]Saint-Venant [1888] provides the most competent and thorough history of the water-wave problem to date. See also Craik [2004] for French and British contributions before 1850, and Craik [2005] for Stokes's contributions.

[4]Cf. Cartwright [1999] chap. 6.

For a modern reader, it is obvious that Laplace's equations are those for the propagation of small waves in shallow water, with an additional term corresponding to the Coriolis force and an external force density corresponding to the lunar and solar perturbations. Laplace could not state so much, since at that time the theory of water waves remained to be developed. He did realize, however, that his derivation of the tidal equations opened the road to the simpler problem of the propagation of small disturbances in a large pond of uniform depth. Laplace knew Isaac Newton's analogy between water waves and the oscillations of a fluid in a U-shaped tube, which gave a propagation velocity proportional to the square root of the length of the wave, but he judged this argument to be 'very uncertain'. His own theory rested on well-established mechanical principles.[5]

Laplace focused on free propagation, which only occurs if the cause of the wave is localized in space and time. The obvious example is a stone thrown into a pond. In order to ease calculation, Laplace considered a narrow canal instead of a pond, and the emersion of a solid body instead of its impact:[6]

> The simplest manner to conceive the formation of waves is to imagine an arbitrary curve, dipped into the fluid to a very small depth and held in this state until all the fluid is in equilibrium; when this curve is thereafter withdrawn from the canal, it is clear that the fluid will tend to retrieve its equilibrium state by forming successive waves.

Laplace then used the so-called Lagrangian picture, in which the fluid motion is described by giving the position $(X + \xi, \ Y + \eta)$ of a particle of the fluid at time t as a function of its position (X, Y) at the origin of time (the moment when the curve is withdrawn). To first order in ξ and η, the incompressibility of water implies the continuity equation

$$\frac{\partial \xi}{\partial X} + \frac{\partial \eta}{\partial Y} = 0. \tag{2.1}$$

According to d'Alembert's principle of dynamics, the work of the sum of inertial, gravitational, and pressure forces during a virtual displacement $d(X + \xi, \ Y + \eta)$ of the position of a fluid particle at any given time must vanish. Taking the ordinate axis to be vertical and directed upwards, this gives

$$\frac{\partial^2 \xi}{\partial t^2} d(X + \xi) + \frac{\partial^2 \eta}{\partial t^2} d(Y + \eta) + g \, d(Y + \eta) + \frac{dP}{\rho} = 0, \tag{2.2}$$

where g is the acceleration of gravity, ρ the density of water, and P the pressure. To first order, this equation makes $(\partial^2 \xi / \partial t^2) \, dX + (\partial^2 \eta / \partial t^2) \, dY$ an exact differential, so that

$$\frac{\partial^2}{\partial t^2} \left(\frac{\partial \xi}{\partial Y} - \frac{\partial \eta}{\partial X} \right) = 0. \tag{2.3}$$

As the expression in parenthesis and its first time derivative vanish identically for $t = 0$, it must vanish at any time. Together with the continuity equation, this gives

$$\frac{\partial^2 \eta}{\partial X^2} + \frac{\partial^2 \eta}{\partial Y^2} = 0. \tag{2.4}$$

Laplace first took η to be a function of Y and t only, multiplied by $\cos kX$. Then the differential equation (2.4) and the boundary condition $\eta = 0$ at the bottom $Y = 0$ of the canal further restrict η to the form

$$\eta = a(t) \sinh kY \cos kX. \tag{2.5}$$

The function $a(t)$ is determined through the condition that, for a virtual displacement along the water surface, the pressure does not vary. Using eqn (2.2), assuming the form $Y = h + \varepsilon \cos kX$ for the water surface at $t = 0$, and retaining only terms of first order in ξ, η, and ε, this implies that

$$\frac{\partial^2 \xi}{\partial t^2} + g \frac{\partial \eta}{\partial X} = \varepsilon g k \sin kX \tag{2.6}$$

for $Y = h$. The derivation of this equation with respect to X and the continuity equation (2.1) yield

$$-\frac{\partial^2}{\partial t^2} \frac{\partial \eta}{\partial Y} + g \frac{\partial^2 \eta}{\partial X^2} = \varepsilon g k^2 \cos kX \tag{2.7}$$

for $Y = h$. Substituting the form (2.5) for η then leads to the equation

$$\frac{d^2 a}{dt^2} k \cosh kh + agk^2 \sinh kh = -\varepsilon g k^2. \tag{2.8}$$

The only solution of this equation that agrees with the vanishing of a and da/dt for $t = 0$ is

$$a = \frac{\varepsilon}{\sinh kh} (\cos \omega t - 1), \tag{2.9}$$

with

$$\omega^2 = gk \tanh kh. \tag{2.10}$$

The corresponding elevation of the water surface above its original height h is, at the same order of approximation,

$$\sigma(X,t) = \varepsilon \cos kX + \eta(X,h;t) = \varepsilon \cos kX \cos \omega t. \tag{2.11}$$

Laplace thus obtained what we would now call a standing wave, as a consequence of his seeking a factored solution. The modern reader may wonder why he did not also find a solution of the form $\sin kX \sin \omega t$ and superpose it with the former solution to get the progressive form $\cos(kX - \omega t)$. The reason is that the initial condition of zero velocity imposes the cosine form of the time dependence. Hence Laplace did not reach the progressive sine solution for the free propagation of small disturbances on water of finite depth, although he came very close to it from a formal point of view.

The rest of Laplace's analysis was unfortunately flawed. To proceed from a sine-shaped disturbance to a disturbance caused by local emersion, Laplace could not rely on Fourier

synthesis, which was unknown at that time. Instead, he truncated the sine function
by taking $\sigma(X,0) = \varepsilon(\cos kX - \cos k\alpha)$ for $|X| \leq \alpha$, and $\sigma(X,0) = 0$ for $|X| \geq \alpha$. In what
he called 'a delicate application of the calculus of partial differentials', he then rewrote
the product $\cos kX \cos \omega t$ in the expression (2.11) for $\sigma(X,t)$ as $\frac{1}{2}[\cos(kX - \omega t)+$
$\cos(kX + \omega t)]$, and replaced the latter cosines by their truncated values. This gives a
propagation of the depression toward the two extremities of the X-axis, and without
deformation. The propagation velocity ω/k only depends on the depth of water and on
the spatial period of the truncated cosine (roughly determined by the curvature of the
originally immersed solid). As the calculus of partial differentials was still in its infancy,
Laplace did not realize that the truncated wave no longer satisfied his differential
equations.[7]

2.1.2 *Lagrangian foundations*

In his memoir of 1781, Lagrange addressed the problem of water waves in a most elegant
manner, with no mention of Laplace's earlier analysis. As already mentioned in the
previous chapter, his purpose was to apply the methods of analytical mechanics to
hydrodynamics, and thus solve a large class of useful problems, including the traditional
efflux from a vase and the less-explored water-wave problem.[8]

In these two problems, the fluid is (nearly) incompressible and the gravity is a constant
g. Lagrange based his analysis on eqn (1.37):

$$\frac{\partial \varphi}{\partial t} = \mathbf{g} \cdot \mathbf{r} - \frac{P}{\rho} - \frac{(\nabla \varphi)^2}{2} + C \qquad (2.12)$$

for the velocity potential φ, which he knew to exist whenever the motion was started from
rest by the sole effect of gravity and external pressures. He then assumed that the fluid
mass never left the space between two mutually-close parallel planes, so that a power
development of the potential with respect to the perpendicular coordinate could be used.
This condition is met in Bernoulli's problem of efflux from a narrow vase, as well as in the
propagation of surface disturbances in shallow water. In the latter case, Lagrange's
method is simply illustrated by assuming two dimensions only, a flat horizontal bottom,
and velocity and surface disturbances so small that terms involving their second powers
can be neglected.[9]

At the lowest non-trivial order, the expansion of the potential has the form

$$\varphi(x,y,t) = \varphi_0(x,t) + y\varphi_1(x,t) + y^2\varphi_2(x,t), \qquad (2.13)$$

where x is the horizontal coordinate and y is the vertical one. The incompressibility of
water gives

$$\frac{\partial^2 \varphi}{\partial x^2} + \frac{\partial^2 \varphi}{\partial y^2} = 0, \qquad (2.14)$$

[7]*Ibid.* p. 307.

[8]Lagrange [1781]. Lagrange did not mention Euler's memoirs, although they were probably a major source of
inspiration. Cf. Grattan-Guinness [1990] pp. 664–5.

[9]*Ibid.* pp. 728–48.

so that $\varphi_0'' + 2\varphi_2 = 0$ (primes denote derivation with respect to x). The vanishing of the vertical velocity at the bottom $y = 0$ implies that $\varphi_1 = 0$. To summarize, the potential must have the form

$$\varphi = \varphi_0 - \frac{1}{2}y^2\varphi_0''. \tag{2.15}$$

The equation of the surface is obtained by making P a constant and neglecting second-order terms in eqn (2.12):

$$\frac{\partial\varphi}{\partial t} + g(y - h) = 0. \tag{2.16}$$

The condition that a particle of the surface should remain on the surface yields

$$\frac{\partial^2\varphi}{\partial t^2} + \dot{x}\frac{\partial^2\varphi}{\partial t\partial x} + \dot{y}\frac{\partial^2\varphi}{\partial t\partial y} + g\dot{y} = 0, \tag{2.17}$$

where $(\dot{x}, \dot{y})$ is the velocity of the fluid particle. To first order, this gives

$$\frac{\partial^2\varphi}{\partial t^2} + g\frac{\partial\varphi}{\partial y} = 0 \quad \text{for} \quad y = h. \tag{2.18}$$

Combining this condition with eqn (2.15), we obtain

$$\frac{\partial^2\varphi_0}{\partial t^2} - gh\frac{\partial^2\varphi_0}{\partial x^2} = 0. \tag{2.19}$$

The general integral of this equation, which d'Alembert had given in his theory of vibrating strings, is

$$\varphi_0(x,t) = f(x - ct) + g(x + ct), \tag{2.20}$$

where f and g are two arbitrary (differentiable) functions, and

$$c = \sqrt{gh}. \tag{2.21}$$

According to eqn (2.16), the elevation of the water surface has the same form. The two components represent the distortionless propagation of any (small) perturbation with the velocities $+c$ and $-c$.

Lagrange concluded his analysis with a speculative extension to waves on deep water. He argued that the 'tenacity and the mutual adherence' of the particles of water confined the agitation to a superficial layer of water, the thickness of which would depend on the propagation velocity through formula (2.21).[10]

Like Laplace, Lagrange selected physics problems according to the possibilities of mathematical analysis. Both mathematicians came to the water-wave problem after realizing that mathematical procedures they had designed in other contexts, namely tides and

---

[10]Lagrange [1781]. Lagrange did not mention Euler's memoirs, although they were probably a major source of inspiration. Cf. Grattan-Guinness [1990] p. 748.

efflux, applied to this problem. They both found that their mathematics only gave limited solutions of the wave problem: standing sine waves for Laplace, and small-depth solutions for Lagrange. They both tried to overcome these limitations by speculative moves that later proved illegitimate.

2.1.3 *Poisson's thorny, but thorough analysis*

In the following thirty years, mathematical analysis progressed so much that the flaws of Laplace's and Lagrange's theories of waves became obvious. On 27 December 1813, an Academic committee including Legendre, Poinsot, Laplace, Biot, and Poisson made 'the waves at the surface of an indefinitely deep liquid' the subject of the Academy prize for the year 1816. Laplace wrote the announcement:[11]

> A ponderable fluid mass, primitively at rest, and indefinitely deep, is set into motion under the effect of a given cause. It is asked to determine, after a given time, the form of the external surface of the fluid and the velocity of every of the molecules situated on this surface.

This was his old problem of 1776, in a slightly more general form.

Laplace's brilliant disciple Siméon Denis Poisson, who belonged to the prize committee, wrote the first memoir on this subject that reached the Academy. He was one of the first Polytechnicians, with an unusual capacity for labyrinthine mathematical analysis and a deep interest in fundamental physics.[12]

In his memoir, Poisson first recalled the earlier contributions by Newton, Laplace, and Lagrange. He judged Newton's siphon analogy to be 'insufficiently founded'. Laplace's solution of 1776, he politely noted, only applied to an initial sine-shaped form of the water surface, and could not be truncated to yield a solution of the local-perturbation problem. Lagrange's solution of 1782 was correct for small depth, but its extension to large depth was illegitimate. In order to prove the latter point, Poisson appealed to 'the principle of the homogeneity of quantities', probably borrowed from Fourier's theory of heat. This early dimensional argument went as follows.[13]

Poisson, like Laplace, assumed that the waves were produced by the sudden withdrawal of a partially-immersed body. In infinitely-deep water, the only 'lines' of the problem are the breadth l of the original depression of the water surface, and the product gt^2, where g is the acceleration of gravity and t is the time of observation. The distance traveled by a wave summit at time t must therefore be a homogenous function of l and gt^2. If this distance is independent of l, then it must be proportional to gt^2 and the wave is accelerated like a free-falling body. If the wave has constant velocity, this distance must be proportional to $t\sqrt{gl}$. Therefore, Lagrange's assumption of waves traveling at a constant velocity, independent of their mode of production, is impossible. Whether the waves produced by emersion

---

[11]Cf. the *Procès-verbaux* of the Académie des Sciences **5** (1812–1815) pp. 262, 292, 546, 556, 595, and the statement in Cauchy [1827a] p. 1.

[12]Poisson's memoir was read on 2 October 1815, and a sequel on 18 December 1815. It was published in 1818 in a volume dated 1816. A summary of the main conclusions appeared in the *Annales de chimie et de physique* (Poisson [1817b]). Cf. Grattan-Guinness [1990] pp. 666–74, Dahan, [1989a]. For a modern treatment, cf. Lamb [1932] pp. 384–98. On Poisson's physics in general, cf. Arnold [1983].

[13]Poisson [1816] pp. 71–5.

travel with constant velocity, with constant acceleration, or else with variable acceleration can only be decided by calculation.

Having thus dismissed Lagrange's approach to deep-water waves, Poisson adopted Lagrange's equations for the velocity potential φ. In the two-dimensional case, and for a small perturbation of the fluid surface, these equations are eqn (2.14), namely

$$\frac{\partial^2 \varphi}{\partial x^2} + \frac{\partial^2 \varphi}{\partial y^2} = 0,$$

within the fluid mass, $\partial \varphi / \partial y = 0$ at the bottom $y = 0$, and eqn (2.18):

$$\frac{\partial^2 \varphi}{\partial t^2} + g \frac{\partial \varphi}{\partial y} = 0 \quad \text{for} \quad y = h.$$

Poisson, now imitating Laplace's procedure, sought factored solutions of the form $\cosh ky \, \cos k(x - a) \sin \omega t$ or $\cosh ky \cos k(x - a) \cos \omega t$. The boundary condition (2.18) requires that eqn (2.10) holds, namely[14]

$$\omega^2 = gk \tanh kh.$$

Poisson then obtained the most general solution by superposition of the factored solutions. Using Fourier's identity (without naming Fourier)

$$f(x) = \frac{1}{\pi} \int \int f(a) \cos k(x - a) \, da \, dk \tag{2.22}$$

and eqn (2.16), namely

$$\frac{\partial \varphi}{\partial t}(x, \, h;t) + g(y - h) = 0,$$

for the fluid surface, he easily found that the superposition[15]

$$\varphi = -\frac{g}{\pi} \int_{-\infty}^{+\infty} f(a) \, da \int_{0}^{+\infty} dk \frac{\cosh ky}{\cosh kh} \cos k(x - a) \frac{\sin \omega_k t}{\omega_k} \tag{2.23}$$

met the initial conditions of zero velocity ($\varphi = 0$) and surface shape $y = h + f(x)$. The corresponding elevation $\sigma(x,t)$ of the water surface above the level h is

$$\sigma = \frac{1}{\pi} \int_{-\infty}^{+\infty} f(a) \, da \int_{0}^{+\infty} dk \cos k(x - a) \cos \omega_k t. \tag{2.24}$$

Poisson then studied the behavior of these two double integrals in the case of large depth, for which $\omega_k = \sqrt{gk}$. He did this in a purely mathematical manner, by cleverly combining changes of variables, integration by parts, and power series developments. To

[14]Poisson [1816] p. 82.   [15]Poisson [1816] p. 92.

give a first idea of these 'rather thorny transformations' consider the first integral in the expression (2.23) of the potential. It is a linear combination of terms of the form

$$\xi = \int_0^{+\infty} e^{-g\gamma k} \frac{\sin \omega_k t}{\omega_k} g \, dk = \int_0^{+\infty} 2e^{-\gamma\omega^2} \sin \omega t \, d\omega, \tag{2.25}$$

where γ is a linear combination of x, y, and a with complex-number coefficients. Derivation with respect to time yields

$$\dot\xi = \int_0^{+\infty} 2\omega e^{-\gamma\omega^2} \cos \omega t \, d\omega. \tag{2.26}$$

Integration by parts then yields

$$\dot\xi = \left[-\gamma^{-1} e^{-\gamma\omega^2} \cos \omega t\right]_0^{+\infty} - \gamma^{-1} t \int_0^{+\infty} e^{-\gamma\omega^2} \sin \omega t \, d\omega, \tag{2.27}$$

or

$$\gamma\dot\xi + \frac{1}{2} t\xi = 1. \tag{2.28}$$

The integral of this equation is

$$\xi = \gamma^{-1} e^{-t^2/4\gamma} \int e^{t^2/4\gamma} \, dt. \tag{2.29}$$

Poisson thus reached a familiar form, whose behavior for small and large times t he obtained through development in positive and negative powers, respectively, of t. He then computed the corresponding expression for the potential φ and the derived velocities, paying special attention to the case when the profile $f(a)$ of the disturbance is very narrow.[16]

Poisson's most detailed discussion of the wave pattern was based on the formula (2.24) for the surface disturbance. For a very narrow disturbance, the double integral in this formula may be replaced by the simpler expression

$$\sigma = \frac{A}{\pi} \int_0^{+\infty} dk \cos kx \cos t\sqrt{gk} = \frac{2A}{\pi g} \int_0^{+\infty} \omega \, d\omega \cos\left(\frac{\omega^2 x}{g}\right) \cos \omega t, \tag{2.30}$$

where A is the area of a vertical section of the original disturbance. Poisson astutely rewrote the last integral as $\sigma = (A/\pi g)(I_+ + I_-)$, where

---

[16] *Ibid.* pp. 93–107; Poisson [1817a] p. 85 (thorny).

$$I_\pm = \int_0^{+\infty} \omega \, d\omega \cos\left(\frac{\omega^2 x}{g} \pm \omega t\right) = \int_0^{+\infty} \omega \, d\omega \cos\left[\frac{x}{g}\left(\omega \pm \frac{gt}{2x}\right)^2 - \frac{gt^2}{4x}\right]. \tag{2.31}$$

For obvious symmetry reasons, it is sufficient to consider the case $x > 0$. Putting

$$w = \sqrt{\frac{x}{g}}\left(\omega \pm \frac{gt}{2x}\right) \tag{2.32}$$

and

$$\alpha^2 = \frac{gt^2}{4x}, \tag{2.33}$$

we obtain

$$I_\pm = \frac{g}{x} \int_{\pm\alpha}^{+\infty} dw(w \mp \alpha)\cos(w^2 - \alpha^2) = \mp \frac{g\alpha}{x} \int_{\pm\alpha}^{+\infty} \cos(w^2 - \alpha^2) \, dw \tag{2.34}$$

and

$$\sigma = \frac{2A\alpha}{\pi x} \int_0^\alpha \cos(w^2 - \alpha^2) \, dw = \frac{2A\alpha}{\pi x}\left[\cos\alpha^2 \int_0^\alpha \cos w^2 \, dw + \sin\alpha^2 \int_0^\alpha \sin w^2 dw\right]. \tag{2.35}$$

The modern reader may recognize the Fresnel integrals that appear in the theory of diffraction. Poisson, who had no such knowledge, developed these integrals in powers of α and gave numerical estimates of the position of the first extrema of σ. As he noted, these extrema occur for well-defined values of $\alpha = \sqrt{gt^2/4x}$. Therefore, the crests of the waves move with the acceleration of gravity.[17]

For large values of α, the two integrals in the last expression for σ differ little from their limit $\frac{1}{2}\sqrt{\pi/2}$. Hence the surface profile is approximately given by

$$\sigma = \frac{A\alpha}{x\sqrt{\pi}} \cos\left(\alpha^2 - \frac{\pi}{4}\right). \tag{2.36}$$

The behavior of this function is mostly given by the fast oscillations of the cosine, with an amplitude increasing linearly in time and decreasing with distance as $x^{-3/2}$. Maxima approximately correspond to $\alpha^2 = \pi/4 + 2n\pi$, where n is an integer. The distance λ between two consecutive crests at a given time, which Poisson calls wavelength, is given by $\lambda \partial\alpha^2/\partial x = 2\pi$, or $\lambda = 8\pi x^2/gt^2$. The period of the oscillations at a given place is such

---

[17]Poisson [1816] pp. 108–14. Without any comment, Poisson ignored the indefinite contribution $\int_{\pm\alpha}^{+\infty} wdw \cos(w^2 - \alpha^2) = \frac{1}{2}\sin(+\infty)$ to the integral (2.34). This indetermination results from the use of a singular distribution $f(a) = A\delta(a)$ for the initial surface deformation. Convolution with a regular profile eliminates the indefinite, infinitely-oscillating terms.

that $\tau \partial \alpha^2 / \partial t = 2\pi$, or $\tau = 4\pi x/gt$. As Poisson noted, the period is a function of the wavelength only, namely $\tau = \sqrt{2\pi\lambda/g}$.[18]

Poisson also considered the more general case in which α is still large but the width of the original disturbance is no longer negligible. He assumed a truncated parabolic profile

$$f(a) = \frac{h(l^2 - a^2)}{l^2} \quad \text{for } |a| \leq l,\tag{2.37}$$

and performed the integration over a explicitly in the formula (2.24) for σ. This led him, after painstaking consideration of the variation rates of the various factors in the remaining integral, to the formula

$$\sigma = \tilde{f}\left(\frac{gt^2}{4x^2}\right)\frac{\alpha}{x\sqrt{\pi}}\cos\left(\alpha^2 - \frac{\pi}{4}\right),\tag{2.38}$$

where

$$\tilde{f}(k) = \int_{-\infty}^{+\infty} f(a)e^{-ika}\,\mathrm{d}a = \frac{4hl^4}{k^3}(\sin kl - kl\cos kl).\tag{2.39}$$

This new factor involves the sine and the cosine of $\alpha^2 l/x$, which oscillate much slower than the $\cos(\alpha^2 - \pi/4)$ factor, as long as the distance x is much larger than the width l of the original perturbation.[19]

As Poisson noted, the crests of the modulating envelope travel at a constant velocity, since the maxima of $\tilde{f}$ occur for definite values of the dimensionless ratio glt^2/x^2. Poisson described the resulting wave pattern as *ondes dentelées* (dentate waves, see Fig. 2.1). This expression indicates that he regarded the envelope as physically more important than its accelerated corrugation. A tooth (local peak), Poisson reasoned, corresponds to a fixed value of α and therefore decreases like $1/x$ as it moves away from the origin; however, an anti-node corresponds to a fixed value of $gt^2l/4x^2$ and therefore decreases more slowly, as $1/\sqrt{x}$. This is why Poisson believed the anti-nodes to be more visible than the teeth.[20]

In the last sections of his memoir, Poisson obtained similar results in the more realistic, three-dimensional case. To a modern reader, much of his lengthy essay seems uselessly complicated and overly abstract. It must be recalled, however, that Poisson was discovering, or at least perfecting, much of the calculus he needed for his problem. Most importantly, he could not benefit from the physico-mathematical language later developed in the context of wave optics and acoustics. At that time, Fourier analysis and synthesis still

---

[18] *Ibid.* pp. 113–14, 119–20. This means that, in the vicinity of a distant point, the progressive sine wave solution with $\omega_k = \sqrt{gk}$ approximately represents the traveling disturbance. Poisson, who did not have the modern propensity to favor sine wave solutions, did not make this remark.

[19] *Ibid.* pp. 115–18.

[20] *Ibid.* pp. 119–26. At a given distance x, only the first oscillations of the water surface are unaffected by the finite width of the generating perturbation. After a time of order $x/\sqrt{gl}$, the modulation of these oscillations begins. Their amplitude, which originally grew linearly in time, now oscillates between limits that ultimately decrease as t^{-3}.

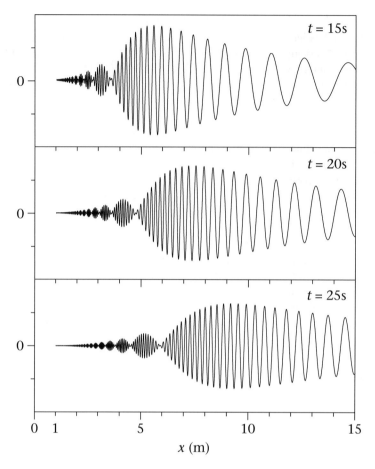

Fig. 2.1.   Computer drawing of Poisson's "dentate waves" caused by a local, parabolic disturbance of the water surface (with exaggerated vertical scale). The faster oscillations travel with a constant acceleration, their slower modulation travel at constant velocity.

were—despite Fourier's intentions—mostly formal operations. They did not carry with them the series of images and metaphors that later physicists learned together with them. Notions such as monochromatic wave and constructive/destructive interference were lacking. As we will see in a moment, these notions not only eased the expression of Poisson's results, but they also suggested more expedient demonstrations. One author of this simplification, Horace Lamb, professed a 'deep admiration' for Poisson's memoir on waves.[21]

---

[21]Lamb [1904] p. 372.

2.1.4 Cauchy's prize-winning memoir

Being himself an Academician and a member of the prize committee, Poisson could not compete for the Academy's prize on waves. A young but already important mathematician, Augustin Cauchy, won the prize.[22] The original text of his memoir was published eleven years later in the *Mémoires des savants étrangers*, with a few appendices taking into account Poisson's contribution. The overlap between Poisson's and Cauchy's memoirs is considerable, even though they worked independently. They both used Lagrange's velocity potential and the relevant differential equations; they both considered a local perturbation of the fluid surface; and they both solved the equations through Fourier analysis. This last point is the most remarkable because Cauchy, unlike Poisson, was not aware of Fourier's theory of heat when he submitted his memoir. He simply reinvented the reciprocal relation between a function and its Fourier transform.[23]

From a mathematical point of view, Cauchy was more systematic and more rigorous than Poisson. In particular, he carefully attended to the existence conditions for various kinds of solutions of his differential equations. A major novelty of his memoir was a rigorous proof of Lagrange's theorem regarding the existence of the velocity potential.[24]

For this purpose, Cauchy used the Lagrangian form of the equations of motion. Denoting by x_i the coordinates at time t of the fluid particle that has the coordinates X_i at time zero, F_i the components of the force density acting within the fluid, P the pressure, and ρ the density, these equations read (in anachronistic tensor notation):

$$\rho \ddot{x}_i \, dx_i = F_i \, dx_i - dP. \tag{2.40}$$

If the fluid is incompressible and if the force density **F** derives from a potential, $\ddot{x}_i \, dx_i$ must be an exact differential. With respect to the coordinates X_i, this implies

$$\frac{\partial}{\partial X_i}\left(\dot{v}_k \frac{\partial x_k}{\partial X_j}\right) - \frac{\partial}{\partial X_j}\left(\dot{v}_k \frac{\partial x_k}{\partial X_i}\right) = 0. \tag{2.41}$$

Permutations of the partial derivatives then lead to

$$\frac{\partial}{\partial t}\left(\frac{\partial v_k}{\partial X_i}\frac{\partial x_k}{\partial X_j} - \frac{\partial v_k}{\partial X_j}\frac{\partial x_k}{\partial X_i}\right) = 0, \tag{2.42}$$

or, by integrating from time zero to time t,

[22]Cf. Belhoste [1991] pp. 87–91, Grattan-Guinness [1990] pp. 674–81, Dahan [1989a]. In July 1815, a month before Poisson submitted his first memoir on waves, Cauchy read a note containing the main results of his theory, namely, the constant acceleration of the waves, the decrease of the height of a wave during its propagation, and the increase of the distance between two successive waves; cf. Académie des Sciences, *Procès-verbaux* **5** (1812–1815) p. 530, Cauchy [1827a] p. 188. Bruno Belhoste notes ([1991] pp. 297–8) that Cauchy also investigated the production of waves at the interface between a compressible and an incompressible fluid. This unpublished manuscript is inserted in the *Cahier sur la théorie des ondes* belonging to Madame de Pomyers.

[23]Cauchy [1827a]. On Cauchy's ignorance of Fourier, cf. Cauchy [1818] and Cauchy [1827a] p. 291.

[24]Cauchy [1827a] pp. 35–43. Cauchy's rigor was not flawless: although he was aware that eqn (2.18) only held for $y = h$, he used reasoning that implicitly assumed its validity for any y and thus derived the equation $\partial^4 \varphi / \partial t^4 + g^2 \partial^2 \varphi / \partial x^2 = 0$ (*ibid.* pp. 52–3). Fortunately, this assumption happens to be correct in the case of infinite depth, the only one treated in Cauchy's prize memoir. Cauchy corrected this slip in an appendix to the final publication (*ibid.* pp. 173–4). Cf. Craik [2004] p. 6.

$$\frac{\partial v_k}{\partial X_i}\frac{\partial x_k}{\partial X_j} - \frac{\partial v_k}{\partial X_j}\frac{\partial x_k}{\partial X_i} = \frac{\partial v_j^0}{\partial X_i} - \frac{\partial v_i^0}{\partial X_j}. \tag{2.43}$$

Using the identity $\partial v_k/\partial X_i = (\partial v_k/\partial x_l)(\partial x_l/\partial X_i)$, the incompressibility condition $\det(\partial x_i/\partial X_j) = 0$, and some algebra, Cauchy finally obtained the simple relation[25]

$$\omega_j(t) = \omega_i(0)\frac{\partial x_i}{\partial X_j}, \tag{2.44}$$

where $\omega_1 = \partial v_2/\partial x_3 - \partial v_3/\partial x_2$, and so forth. Consequently, if a velocity potential exists at time zero, the condition for its existence is maintained at any later time. This is Lagrange's theorem.

Another mathematical difference between Poisson's and Cauchy's memoirs was the latter's systematic recourse to dimensionless variables. For example, Cauchy rewrote eqn (2.30) in terms of the variables $\mu = gt^2 k$ and $\kappa = gt^2/2x$ to obtain

$$\sigma = \frac{A}{\pi gt^2}\int_0^{+\infty} d\mu\cos\frac{\mu}{2\kappa}\cos\mu^{1/2}. \tag{2.45}$$

Under this form, it is immediately clear that the wave crests correspond to definite values of $gt^2/2x$, so that their motion is uniformly accelerated. In general, Cauchy sought universality beyond the specific physics problems he was studying. He tried to extract formulas and structures that had intrinsic mathematical value and could eventually serve in other physical situations.[26]

Regarding the physical discussion of waves, the scope of Cauchy's differed from Poisson's. Like Laplace, Poisson confined his analysis to disturbances created by the sudden emersion of a solid body. He briefly indicated how the case of an impulsive pressure applied on a portion of the fluid surface could be included in his general formulas, but he did not pursue the analysis of this case any further. In contrast, Cauchy showed how the initial fluid velocity depended on the impulsive pressure, and thus reached a physical interpretation of the velocity potential as the internal impulsive pressure resulting from the external impulsion (for unit density). He also proved that the motion of the fluid at any instant could be regarded as being created from rest by impulsive pressures applied on its surface, a result important to later British hydrodynamicists.[27]

In other respects, Cauchy's physical discussion was less complete than Poisson's. Cauchy only described waves independent of the shape of the original disturbance,[28] whereas Poisson regarded the effect of this shape as the most perspicuous aspect of

---

[25]This is the Lagrangian expression of the fact, established by Helmholtz in 1858, that the convective derivative of the vorticity vanishes in an incompressible, Eulerian fluid. A much easier proof of the theorem (Lamb [1932] p. 17) is obtained by noting that $\dot{\mathbf{v}}\cdot d\mathbf{r} = \partial(\mathbf{v}\cdot d\mathbf{r})/\partial t - d(v^2/2)$ in the Lagrangian picture, for which $\mathbf{r}$ denotes the evolving position of a given fluid particle. As $\dot{\mathbf{v}}\cdot d\mathbf{r}$ is an exact differential at any time, if $\mathbf{v}\cdot d\mathbf{r}$ is an exact differential at time zero then it must be so at any later time.

[26]Cauchy [1827a] p. 88.   [27]Poisson [1816] p. 92; Cauchy [1827a] pp. 14–15.

[28]Cauchy [1827a] pp. 92–4 gave the validity condition for this.

Fig. 2.2.   Disturbed water surface with a convex profile, as imagined by Cauchy.

wave motion. After reading Poisson, Cauchy investigated this question more thoroughly than Poisson had done. He showed that, for any symmetric profile of the immersed body, the modulating envelope of the fast oscillations was the Fourier transform of the profile (eqns (2.38) and (2.39)). He confirmed Poisson's result for the parabolic profile and expressed it with the slightly more apt metaphor of *ondes sillonnées*. He also showed that, for convex profiles (as in Fig. 2.2), the modulating factor did not oscillate. In response to this nice theorem, Poisson argued that the only case of physical interest was the small-depth parabolic profile, because other profiles would not be consistent with the continuity of the fluid during the first instants of the motion.[29]

The comparison between Cauchy's and Poisson's memoirs suggests that Poisson was more concerned with physical meaning, and Cauchy with mathematical meaning. Poisson's physics nonetheless remained idealized physics. As we will see shortly, his and Laplace's emersion method for producing waves does not work in practice. Poisson did not perform any experiment. He contented himself with calling, in the introduction to his memoir, for an experimental confirmation of his theory.[30]

2.1.5 *Apparent confirmations*

In 1820, the Turin-based hydraulician George Bidone claimed to have confirmed Poisson's most striking prediction, namely, the uniformly-accelerated motion of the first waves created by a local perturbation of the water surface, as well as the numerical values of the accelerations of the two first waves ($0.3253g$ and $0.1183g$). Bidone operated with a 24-inch wide and 24-inch deep canal. He did not say how he measured the velocity of the waves, but he dwelt on the difficulty he encountered in applying the Laplace–Poisson emersion method for the production of waves. The immersed body did not instantly leave the water surface upon withdrawal as the two mathematicians had imagined. On the contrary, the water adhered to the body and followed it to a certain height until it violently fell down (see Fig. 2.3). Bidone believed he could circumvent this difficulty by attending to the two first waves only, which in his opinion were created before the fall of the raised water column. Apparently, he did not realize that Poisson's calculations did not apply to this impulsive excitation either. It is not clear how he reached such 'a marvelous agreement between theory and experiment.'[31]

In 1825, the Leipzig professor Ernst Heinrich Weber and his brother Wilhelm published a very thorough *Wellenlehre*, which summarized all previous theories of waves and

[29]Cauchy [1827*a*] note XVI, pp. 196, 220; Poisson [1829*b*]. Fourier [1818] recommended the investigation of a non-parabolic profile.

[30]Poisson [1816] p. 78.

[31]Bidone [1820] p. 25. Poisson [1829*b*] p. 571 noted Bidone's confirmation of the accelerated waves. Strangely, he did not comment on the failure of the emersion method, even though his new memoir was about the permissible profiles of the initial water surface.

Fig. 2.3.   The emersion of a parabolic solid according to Bidone ([1820] plate).

provided many astute, quantitative experiments on this matter. Their motivation was the
recent development of wave physics in acoustic and optical contexts, owing to the works of
Ernst Chladni, Félix Savart, Thomas Young, and Augustin Fresnel. They wanted to
provide the subject with a solid empirical basis, using water waves as an archetype of
wave motion. In their most extended series of experiments, they used two long, narrow
water tanks (see Fig. 2.4). They disturbed the water at one end of the tank, and obtained
'self-drawn' wave profiles by suddenly withdrawing a vertically- and longitudinally-
immersed board. They also measured the time a wave took to travel along the tank, and
visualized the internal fluid motion through suspended dust particles.[32]

   The Weber brothers became aware of Poisson's 'very important' theory of waves after
they had performed their experiments, but before the final editing of their treatise. As they
believed their observations to confirm some aspects of this theory, they included a
commentary of Poisson's paper in French. They approved his general description of the
wave pattern, with faint accelerated waves at the front, followed by constant-velocity
waves with a 'dentate' surface. They also confirmed the proportionality between the
period of oscillation and the square root of the wavelength.[33]

   These conclusions would not have resisted a more accurate reading of Poisson and more
adequate experiments. As Scott Russell later commented, the Webers' tank was too
narrow, too shallow, and too short to approximate the ideal conditions of frictionless
deep-water wave motion far from the source. In order to create their waves, the Webers
dipped a glass tube vertically into water, drew up the water by suction, and let it fall back.
This method differs widely from the static surface deformation imagined by Poisson. Most
fatally, the two brothers mistook Poisson's *ondes dentelées* to mean large waves with a
ruffled surface, whereas Poisson's formulas show that he meant what we would now call
modulated waves. What they actually observed was probably capillarity ripples superposed with gravity waves. The lack of figures and concise summaries in Poisson's memoir
favored the confusion. As Thomson put it in 1871:

> A great part of what they [Poisson and Cauchy] have to say would be much shortened
> even by the addition of graphic representations, and it would be much easier for any
> one (the authors I believe included) to understand the whole character of the
> phenomena investigated, with illustration like this of the chief function on which
> the expression of these depends.

---

[32]Weber and Weber [1825] pp. V, 105–17 (self-drawn profiles), 166–99 (velocity), 117–55 (visualization).
[33]*Ibid.* pp. 377–434.

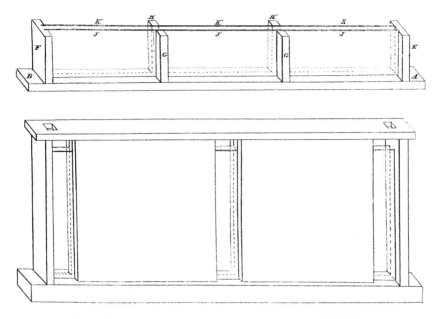

Fig. 2.4. The experimental tanks of the Weber brothers ([1825] plate).

What had become a common practice in the 1870s would, however, have seemed costly bad taste to a French mathematician of the early nineteenth century.[34]

2.2 Scott Russell, the naval engineer

2.2.1 *A horse's discovery*

In 1833, the Cambridge astronomer James Challis reviewed the present state of hydrodynamics for the British Association. Although he praised Poisson's and Cauchy's theories of waves and rejoiced over their verification by Bidone and the Webers, he concluded on a pessimistic note, lamenting over the stagnation of the more pressing problem of fluid resistance. The hydrodynamics of d'Alembert and Euler completely failed on this matter, since it yielded a vanishing resistance. Newton's old theory of resistance, based on individual impacts of the fluid molecules at the prow of the immersed body, at least explained the usually observed proportionality of the resistance with the square of the velocity. Yet even this simple law suffered exceptions. In particular, Challis referred to a 'singular fact' observed in canal navigation: for a speed of four or five miles per hour the hauled boat rose out of the water and the resistance was suddenly diminished.[35]

---

[34]Russell [1845] p. 25n; Weber and Weber [1825] p. 106; Thomson to Stokes, 20 Nov. 1871, *ST*.

[35]Challis [1833] p. 155. More will be said on Newton's theory in Chapter 7, pp. 264–265.

John Scott Russell, a young Glasgow engineer who specialized in steam power and naval architecture, knew well of this striking anomaly.[36] He later described it in vivid terms:

> As far as I am able to learn, the isolated fact was discovered accidentally on the Glasgow and Ardrossan Canal of small dimensions. A spirited horse in the boat of William Houston, Esq., one of the proprietors of the works, took fright and ran off, dragging the boat with it, and it was then observed, to Mr. Houston's astonishment, that the foaming stern surge which used to devastate the banks had ceased, and the vessel was carried on through water comparatively smooth, with a resistance very greatly diminished. Mr. Houston had the tact to perceive the mercantile value of this fact to the Canal Company with which he was connected, and devoted himself to introducing on that canal vessels moving with this high velocity.

There was indeed, in the 1830s, a system of fly-boats carrying passengers on two Scottish canals. A pair of horses drew each boat at a speed of about 10 miles per hour.[37]

Stimulated by Challis's interest in this paradox of fluid resistance, Scott Russell submitted his own simple solution at the Edinburgh meeting of the British Association in 1834. The motion of a boat through water, he reasoned, raised the pressure of the water at the bottom of the ship above its static value. This caused a partial emersion of the boat, and the observed decrease in resistance. Denoting by S and S' the transverse sections of immersion for velocities zero and v, respectively, Russell wrote the strange non-dimensional equation $S'v = S(v - v^2/2g)$, and inserted the resulting value of S' in the Newtonian resistance formula $R = S'v^2\rho/2$. Of this departure of the resistance law from a quadratic form, he said that he had found ample evidence in towing experiments.[38]

A fuller version of this argument displays Russell's crude misunderstanding of the laws of mechanics. There he derived the bottom pressure from the well-known front pressure of the Newtonian theory of resistance, artistically combined with the isotropy of pressure. In the rest of his reasoning, he seems to have confused the Archimedean displacement with the dynamic displacement $\rho S v$.[39]

2.2.2   *The great, solitary wave*

Russell was not a man to worry over such infractions of the laws of mechanics. He did, however, recognize that his consideration only gave a gradual correction to the Newtonian resistance, not the desired Houston jump. In order to understand this stronger anomaly, he attended to the fluid motion induced by the boat. One day, 'the happiest of [his] life', something unexpected happened:[40]

---

[36]For a biography, cf. Emmerson [1977]. On Russell and waves, cf. Bullough [1988]. On ship hydrodynamics in the nineteenth century, cf. the excellent Wright [1983].

[37]Russell [1839] p. 79. Cf. Thomson [1887f] pp. 418–20, with the lament: 'Is it possible not to regret the old fly-boats between Glasgow and the Ardrossan and between Glasgow and Edinburgh, and their beautiful hydro-dynamics, when, hurried along on the railway, we catch a glimpse of the Forth and Clyde Canal still used for slow goods traffic; or of some swampy hollows, all that remains of the Ardrossan Canal on which the horse and Mr. Houston and Scott Russell made their discovery?'

[38]Russell [1834]. Of course, Russell intended his formula to be used with fixed foot and pound units.

[39]Russell [1839] p. 57.

[40]Russell [1865], vol. 1, p. 217, [1839] p. 61.

In directing my attention to the phenomena of the motion communicated to a fluid
by the floating body, I early observed one very singular and beautiful phenomenon,
which is so important, that I shall describe minutely the aspect under which it first
presented itself. I happened to be engaged in observing the motion of a vessel at a
high velocity, when it was suddenly stopped, and a violent and tumultuous agitation
among the little undulations which the vessel had formed around it, attracted my
notice. The water in various masses was observed gathering in a heap of a well-
defined form around the centre of the length of the vessel. This accumulated mass,
raising at last to a pointed crest, began to rush forward with considerable velocity
towards the prow of the boat, and then passed away before it altogether, and
retaining its form, appeared to roll forward alone along the surface of the quiescent
fluid, a *large, solitary, progressive wave*. I immediately left the vessel, and attempted
to follow this wave on foot, but finding its motion too rapid, I got instantly on
horseback and overtook it in a few minutes, when I found it pursuing its solitary path
with a uniform velocity along the surface of the fluid. After having followed it for
more than a mile, I found it subside gradually, until at length it was lost among the
windings of the channel. This phenomenon I observed again and again as often as the
vessel, after having been put in rapid motion, was suddenly stopped; and the accompanying circumstances of the phenomenon were so uniform, and some consequences
of its existence so obvious and important, that I was induced to make *The Wave* the
subject of numerous experiments.

Russell soon suspected a connection between the existence of solitary waves and
Houston's resistance paradox. A few trials confirmed that 'the velocity of the motion of
the solitary wave had a peculiar relation to a certain well-defined point of transition in the
resistance of the fluid.' Russell performed the necessary experiments 'during the leisure of
two summers', 1834 and 1835, with the support of canal, naval, and academic authorities,
and with the help of 'two scientific friends' and 'a dozen hired assistance'. Four different
vessels were towed in canals of various depths at a velocity ranging between 3 and 15 miles
per hour. Horses provided the towing force, directly in 1834, and through a suspended-
weight regulator in 1835 (see Fig. 2.5). A dynamometer measured the resistance. Russell
found it to increase regularly until a certain critical velocity depending on the depth was
reached, then to suddenly diminish, and finally to increase again (see Fig. 2.6). The critical
velocity turned out to be identical to the velocity of the solitary wave for the given depth h.
With a gun-shooting friend and a chronometer, Russell measured the time that this wave
took to travel between two distant points. This gave him Lagrange's velocity formula $\sqrt{gh}$,
or more precisely $\sqrt{g(h + \sigma)}$, where σ is the height of the wave crest above the undisturbed
water surface.[41]

Russell also described how the shape of the water surface around the moving vessel
evolved with the velocity (see Fig. 2.7). For velocities inferior to the critical value, the
water level is raised around the prow, thus forming 'the great primary wave of displacement'. The resulting inclination of the vessel, Russell reasoned, increases its effective
transverse section of immersion and the corresponding resistance. When the velocity of
the vessel reaches the critical value, this wave has the velocity of a solitary wave. The push
from the vessel is no longer necessary for its progression. If the velocity is further

---

[41]Russell [1835*a*], [1837*b*], [1839] pp. 61 (quote), 47 (friends), 49–50.

Fig. 2.5. Russell's towing mechanics of 1835 (plate of Russell [1839], redrawn in Thomson [1887f]).

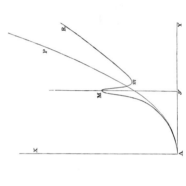

Fig. 2.6. Resistance as a function of towing velocity according to Russell [1839] p. 49.

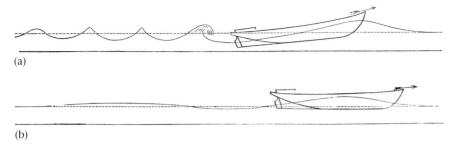

(a)

(b)

Fig. 2.7.   Positions of a canal boat towed at a velocity inferior to the critical velocity (a), superior to the critical
velocity (b) (Russell [1839] p. 70).

increased, the vessel catches up with its own wave, so as to be 'poised on its summit'. The
effective transverse section is much smaller, and so is the resistance.[42]

For subcritical velocities, Russell also noted the 'posterior wave of displacement',
namely, the depression of the water surface at the stern that necessarily accompanies its
rise at the prow. As water rushes into this depression from both sides, Russell reasoned, it
induces a series of oscillations of the water behind the vessel (see Fig. 2.7(a)). The violence
of these oscillations increases until the critical velocity is reached. They subside beyond
this velocity, because the posterior wave no longer exists.[43]

2.2.3   *Wave-lined vessels*

Whatever the value of this intuitive reasoning, it convinced Russell that the accumulation
of water at the prow of a vessel was a major obstacle to its progression.[44] In canals of small
depth this obstacle could be overcome by exceeding the critical velocity. For maritime
navigation, this cause of resistance necessarily grew with increased velocity. It could
compromise the high-speed, steam-powered navigation in which the city of Glasgow had
the highest stakes. Russell soon suggested a remedy, namely, to shape the prow of the
vessel according to hollow lines, so that it could enter the water without ruffling its surface.
Specifically, he recommended lines made of two parabolic arcs, for this shape would
induce a uniformly-accelerated motion of the water along the lines. As he later put it,
'There is a way of setting about the removal of the water from the place the ship wants to
enter, which is pleasant and profitable to both.' Russell noted that hollow lines had long
been used by pirates, to whom speed was essential. They occur spontaneously in a most
primitive mode of ship construction: binding the extremities of two planks, and separating
their middle part through a transverse beam. Russell only claimed to be first in showing
their theoretical superiority.[45]

---

[42]Russell [1835a], [1839] p. 40.   [43]Russell [1839] pp. 65–7.

[44]According to the modern understanding of ship resistance, the wave component derives from the waves that
propagate away from the ship, not from a direct action on the prow.

[45]Russell [1835b], [1837a], [1839] p. 51, [1865] vol. 1, p. 161 (quote and pirates). Cf. Wright [1983], pp. 71–80.

THE WAVE.

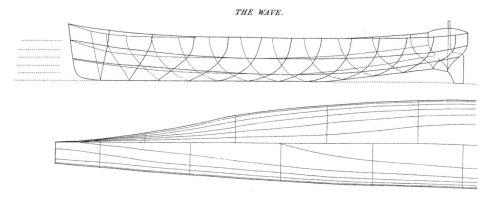

Fig. 2.8.   Russell's first hollow-line model: "The Wave" ([1839] plate).

In 1835, Russell built *The Wave*, a model with a 75-foot keel and a 6-foot beam, to test this new principle of ship construction (see Fig. 2.8). The following year he continued with a series of more important wave-lined vessels: the *Storm*, the *Skiff*, ... and the *Scott Russell*. In the early 1840s, while steering a British Association committee 'on the form of vessels', he performed some twenty thousand observations with models and full-scale vessels, ranging from 30 inches to 1300 tons. The wave profile came out best, though with some modification. Russell found the hollowness of parabolic lines to be excessive, and ultimately adopted sine-shaped lines for their analogy with the harmonic waves of Lagrange's theory. For the rest of his career, he pressed for the systematic use of the 'wave profile' and repeatedly denounced British conservatism in matters of ship design. In the mid-1850s he applied his wisdom to the *Great Eastern*, a monster metal vessel built for the Eastern Navigation Company.[46]

2.2.4   *Taming water waves*

Russell's investigation of the best form of ships went along with further studies of water waves. At the Bristol meeting of 1836, the British Association appointed a 'Committee on Waves' directed by Russell and John Robison. Russell gave a first report of this research at the Liverpool meeting of 1837. A section of this report was devoted to an attempt at explaining tides in terms of solitary waves, which will be discussed shortly. In most of his report, Russell described experiments he made in canals and in a 20-foot long and 1-foot broad experimental reservoir. Through a clever optical method, he established the $\sqrt{g(h + \sigma)}$ velocity formula for solitary waves. He found that these waves had a quasi-cycloidal form which was independent of the way they were produced. He described the induced motion of the fluid particles: 'By the transit of the wave the particles of the fluid

---

[46]Russell [1835*b*], [1841], [1842*a*], [1843*b*], [1865] vol. 1, pp. 210–11 (sine lines), [1852], [1865] vol. 1, p. XXX (denouncing), [1854], [1857] (*Great Eastern*). Cf. Emmerson [1977], Wright [1983] p. 80, who claims that Russell applied far less hollow lines to the *Great Eastern* than required by his theory, despite a lot of propaganda. In the later conceptions of ship resistance developed by Rankine and Froude, wave formation still played a role, though with mechanisms different from Russell's and in competition with two other forms of resistance, namely skin friction and eddy formation. Cf. Wright [1883] Chaps 5–7, and Chapter 7.

are raised from their places, transferred forwards in the direction of the motion of the wave, and permanently deposited at rest in a new place at a considerable distance from their original position', in opposition to 'second-order', oscillatory waves in which the particles oscillate around a fixed point. He found that two solitary waves 'cross[ed] each other without change of any kind'. He observed that sea waves, originally of second order, evolved into solitary waves after breaking on a gently sloped shore. He determined that the highest possible wave had a relative height σ equal to the depth h. Lastly, he performed a few measurements on sea waves. Owing to unfavorable weather conditions, these gave little more than the independence of the waves on the depth of the sea.[47]

In a later report, Russell confirmed the singular properties of solitary waves, extended his investigation to other sorts of waves, and compared his results with previous mathematical theories. As we will see shortly, the Astronomer Royal, George Biddell Airy, had already denied the existence of solitary waves and downgraded Russell's observations to a mere confirmation of Lagrange's shallow-water waves. Russell, who saw Airy's text just before sending his report to the printer, was naturally disappointed:[48]

> This paper I have long expected with much anxiety, in the hope that it would furnish a final solution of this difficult problem [the discrepancy between wave theory and wave phenomena], a hope justified by the reputation and position of the author, as well as by the clear views and elegant processes which characterize some of his former papers ... It is deeply to be deplored that the methods of investigation employed with so much knowledge, and applied with so much tact and dexterity, should not have led to a better result.

Russell insisted that his waves, unlike Lagrange's, had a definite shape for a given height, with a length about six times their height. New experiments performed 'after the best methods employed in inductive philosophy' confirmed this point. The disturbance produced by the injection of additional water at one end of his tank soon evolved, while propagating along the channel, into the perfectly stable form of the solitary wave (see Fig. 2.9). When the injection was irregular, a compound wave was produced which evolved into separate solitary waves (see Fig. 2.12). Using Weber's self-drawing method, Russell showed that the shape of the solitary height was perfectly determined for a given height and tended to a cusped shape when the maximal height was reached (see Fig. 2.11). Lastly, Russell confirmed his velocity formula for this wave, $\sqrt{g(h + \sigma)}$, instead of Lagrange's or Airy's formulas.[49]

2.2.5   The four orders

No one, Russell argued, had predicted or observed his great solitary wave before him: Lagrange's waves were too small compared to the depth of water; the mode of production of Poisson's and Cauchy's waves precluded solitary waves; and the Weber brothers believed that a positive wave never went without a correlative negative wave. In order to avoid confusion of his great wave with others' waves, Russell introduced the following four orders of waves (see Fig. 2.10).[50]

---

[47]Russell [1837c] pp. 423 (reservoir, quote), 424 (cycloid), 425 (crossing waves), 426 (sea waves).

[48]Russell [1845] pp. 27, 30; Airy [1845].   [49]Russell [1845] pp. 27 (quote), 33–4, 45–6.

[50]*Ibid.* pp. 23–5 (priority), 9 (orders).

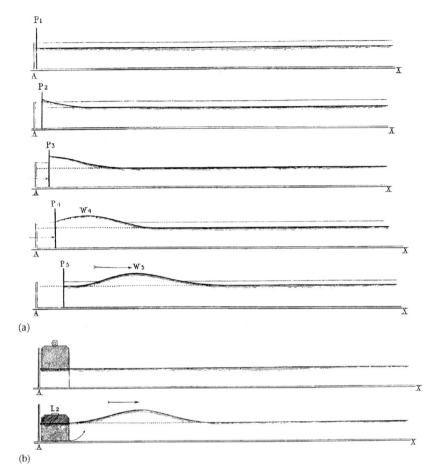

(a)

(b)

Fig. 2.9. Two ways of producing a solitary wave: through the displacement of a wall (a); through the immersion of a solid (b) (Russell [1845] plate).

System of Water Waves.

ORDERS.	FIRST.	SECOND.	THIRD.	FOURTH.
Designation.	Wave of translation	Oscillating waves.	Capillary waves.	Corpuscular wave.
Characters ...	Solitary	Gregarious.........	Gregarious.......	Solitary.
Species ... {	Positive	Stationary.........	Free.	
	Negative..................	Progressive	Forced.	
Varieties {	Free	Free.		
	Forced	Forced.		
Instances {	The wave of resistance.	Stream ripple	Dentate waves...	Water-sound wave.
	The tide wave	Wind waves.......	Zephyral waves.	
	The aërial sound wave.	Ocean swell.......		

Fig. 2.10.   Russell's wave orders ([1845] p. 9).

(i) *Waves of translation.* They involve mass transfer. Positive waves of this kind can be solitary. Negative ones are always accompanied by an undulating series of secondary waves (see Fig. 2.13).

(ii) *Oscillatory waves.* These do not involve mass transfer. They appear as groups of successively positive and negative waves. They are the most commonly seen waves, created by wind for instance. They can be progressive or standing.

(iii) *Capillary waves.* These only involve a minute-depth agitation of the water. They depend on the surface tension of the water.

(iv) *Corpuscular waves.* These are rapid successions of solitary waves. Sound waves are the prime example.

Although Russell focused on the first order, he also performed careful experiments on the second and third kind. For instance, he showed that the Kelland–Airy formula $c^2 = (g/k) \tanh kh$ accurately represented the velocity c of progressive oscillatory waves, even when their amplitude was not small.[51] He illustrated the evolution of such waves

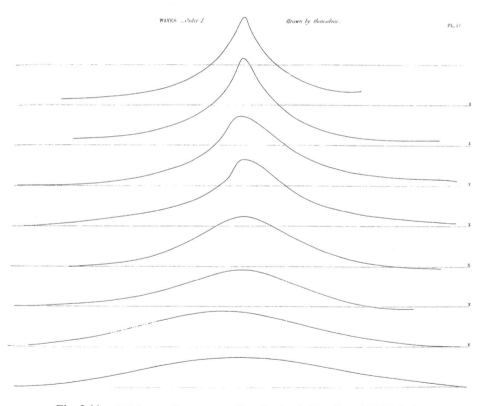

Fig. 2.11.   Self-drawn solitary-wave profiles of various heights (Russell [1845] plate).

---

[51]*Ibid.* p. 67.

when approaching a shore (see Fig. 2.14). He drew the shape of steady waves produced by an obstacle in the bed of the stream (see Fig. 2.15). He obtained a beautiful pattern of capillary waves by plunging a rod vertically in a stream of water (see Fig. 2.17).[52]

Strangest to Russell's readers must have been the fourth order of waves, supposed to represent sound waves. For any physicist at that time, sound corresponded to the propagation of small-amplitude vibrations through an elastic medium. No special kind of wave was needed. As appears from a posthumously published manuscript, Russell rejected this explanation for he believed it could not explain the ability of sound to propagate far from its source. From the fact that the sound of a tuning fork or the vibrations of a string could be heard at a non-negligible distance only if the fork or string was attached to a hollow case with an aperture, he inferred that sound was not the harmonic vibration of the fork and surrounding air but the repeated emission of solitary waves through the aperture of the case. As solitary waves are surface waves, Russell needed to imagine an open surface for the medium of propagation. For sound in water, the free water surface did the job. For sound in air, he imagined an ocean of air of large but finite depth around the Earth. Most daringly, he proposed that light was a wave of fourth order in an even larger ocean of ether.[53]

These suggestions only confirm Russell's ignorance of elementary principles of mechanics. The Royal Society never published the series of manuscripts it received from him on this theme. Yet the elite of British natural philosophers often praised Russell's early works on waves and ship forms, for they admired the quality of his experiments and the frequent validity of his intuitions.

2.3. Tides and waves

2.3.1 *Russell's illumination*

Between Russell's careful experiments on water waves and his hair-raising speculation on corpuscular waves, there was a middle ground which seems to have perplexed his learned supporters, namely, the notion that tides were essentially solitary waves of very large extent. As Russell recounts, he submitted this idea to William Whewell in 1835 together with a plan for observations. Whewell had then been working for several years on tidal observations and prediction, and was with John Lubbock, the leading British expert on this topic. He approved Russell's project, which thus became part of the duties of the 'Committee on Waves'.[54]

At the Liverpool meeting of 1837, Russell reported the tidal observations the committee had made on the rivers Dee (Cheshire) and Clyde (Scotland). He also promoted his own theory of tides. The general idea was to divide the problem into two parts: the general elevation of water in the Pacific and Atlantic Ocean as ruled by celestial mechanics, and the propagation of this elevation in smaller basins, channels, and rivers

---

[52]Russell (*ibid.* p. 78) was aware of similar observations by Poncelet ([1831] p. 78).

[53]Russell [1885].

[54]Russell [1837c] p. 420. In 1838 (*BAR* p. 20), Whewell praised Robison and Russell for 'highly valuable materials, likely to assist us in the further prosecution of the subject [the theory of tides].' On Lubbock, Whewell, and tides, cf. Deacon [1971] Chap. 12.

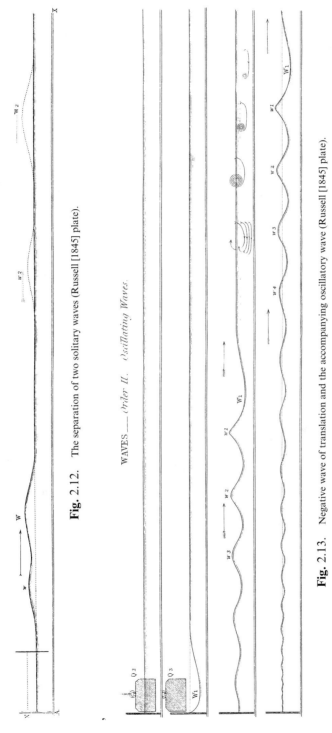

Fig. 2.12. The separation of two solitary waves (Russell [1845] plate).

WAVES — Order II. Oscillating Waves.

Fig. 2.13. Negative wave of translation and the accompanying oscillatory wave (Russell [1845] plate).

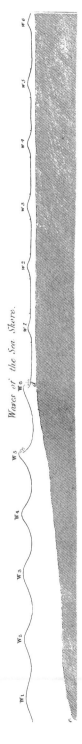

Fig. 2.14. Waves approaching a shore and evolving into solitary waves (Russell [1845] plate).

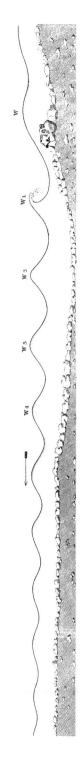

Fig. 2.15. Standing wave created by an obstacle in running water (Russell [1845] plate).

Fig. 2.16. The evolution of a compound solitary wave according to Russell ([1845] plate).

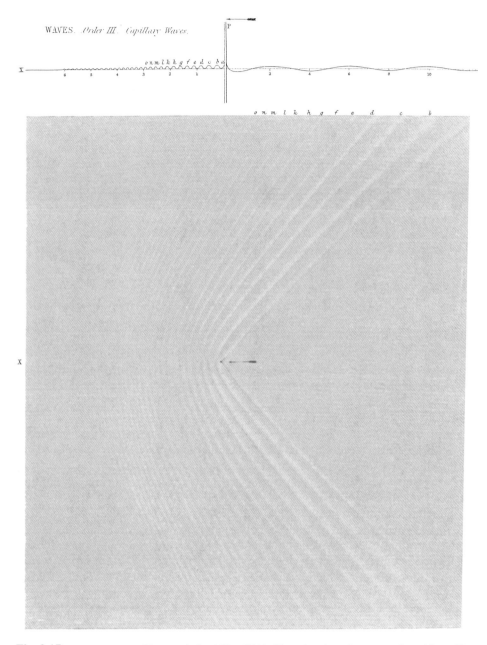

Fig. 2.17.   Waves generated by a vertical rod ($\varnothing = 1/16$ inch) moving along the water surface with a uniform velocity. The smaller waves in front of the rod are capillarity waves (Russell [1845] plate).

as ruled by terrestrial hydrodynamics. Russell described the latter mechanism as follows:[55]

> The *Tide Wave* appears to be ... identical with the great primary wave of translation; its velocity diminishes and increases with the depth of the fluid, and appears to approximate closely to the velocity due to half the depth of the fluid ... —The tide appears to be a compound wave, one elementary wave bringing the first part of the flood tide, another the high water, and so on: these move with different velocities according to the depth. On approaching shallow shores the anterior tide waves move more slowly in the shallow water, while the posterior waves, moving more rapidly, diminish the distance between two successive waves. The tide wave becomes thus dislocated, its anterior surface rising more rapidly, and its posterior surface descending more slowly than in deep water.—A tidal bore is formed when the water is so shallow at low water that the first waves of flood tide move with a velocity so much less than that due to the succeeding part of the tidal wave, as to be overtaken by the subsequent waves, or wherever the tide rises so rapidly, and the water on the shore or in the river is so shallow that the height of the first wave of the tide is greater than the depth of the fluid at that place.

Russell thus explained a few basic facts: that for river tides, the time of ebb is larger than the time of flood, with the difference increasing with the distance from the mouth of the river; that tides can be very different in nearby locations, that they depend on the bottom of the sea or the form of channels and rivers, and that strong river tides are often accompanied by a breaking surge or tidal bore (*mascaret* in French). In his later water-tank experiments, Russell verified that 'compound solitary waves' evolved during their propagation so that the front became steeper than the rear (see Fig. 2.16, p. 58).[56]

2.3.2   *From Newton to Whewell*

Although the idea that tides were a wave phenomenon was not as new as Russell suggested, it departed from the then current approaches to tide theory and prediction. The historical background of these approaches must first be recalled.[57]

In his *Principia* Newton gave the correct expression for the force that is responsible for tides, namely, the combined action of the Moon's and the Sun's attractions. His derivation of the resulting deformation of the surface of the oceans was only tentative and retrospectively erroneous. He seems to have adopted an equilibrium theory, with retardation due to friction. According to the pure equilibrium theory that Colin MacLaurin, Leonhard Euler, and Daniel Bernoulli developed in their competition for the 1740 prize of the French Academy, under every instantaneous configuration of the Moon and the Sun, the water surface takes the form it would have if the corresponding forces were acting permanently. Retaining only the lunar action in a first approximation, the net force exerted by the Moon on oceanic water is the Newtonian gravitational force, which is proportional to the inverse

---

[55]Russell [1837*c*] p. 426. Although Russell's identification of the tidal wave with a compound solitary wave makes little sense from a modern point of view, his theory appears to be similar to Partiot's more correct theory, discussed later on p. 82.

[56]Russell [1837*c*], [1838], [1845].   [57]The following account is based on Cartwright [1999].

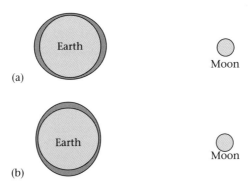

(a)

(b)

Fig. 2.18.   Tides on an ocean of uniform depth as given the equilibrium theory (a); as inferred from observed
    tides (b).

squared distance of the water from the Moon, minus the inertial force due to the acceleration of the Earth toward the Moon, which is proportional to the inverse squared distance
of the center of the Earth from the Moon. Therefore, this net force is a maximum at the
points closest to and furthest from the Moon. For a uniform ocean covering the whole
Earth, the resulting equilibrium surface (obtained by making the total potential of the
terrestrial and lunar forces a constant) has the form indicated in Fig. 2.18(a). Unfortunately, observed tides more closely correspond to the form indicated in Fig. 2.18(b).

For this reason, in 1776 Laplace proposed a dynamic theory of tides. Assuming that the
horizontal velocity of the water was the same on a vertical line, and neglecting second-
order quantities, he obtained the equations of motion (in modern notation)

$$\frac{\partial u}{\partial t} - 2\Omega v \cos \theta = -\frac{1}{R}\frac{\partial}{\partial \theta}(g\zeta - U - \delta U),$$
$$\frac{\partial v}{\partial t} + 2\Omega u \cos \theta = -\frac{1}{R \sin \theta}\frac{\partial}{\partial \phi}(g\zeta - U - \delta U) \tag{2.46}$$

where u and v are the velocity components along the meridians and the parallels, respectively, θ and ϕ are the colatitude and the longitude, respectively, Ω is the angular velocity of
the Earth, R is the radius of the Earth, ζ is the elevation of the water surface, U is the
combined gravitational potential from the Moon and the Sun, and δU is the gravitational
self-potential of the water. The first terms on the left-hand side of these equations
correspond to the acceleration of the water particles, and the second to the Coriolis
force (not yet named so, of course). The right-hand side corresponds to the sum of pressure
forces (depending on the elevation of the surface) and gravitational forces. These equations are to be solved in combination with the continuity equation

$$\frac{\partial}{\partial \theta}(uh \sin \theta) + \frac{\partial}{\partial \phi}(vh) + R \sin \theta \frac{\partial \zeta}{\partial t} = 0, \tag{2.47}$$

where h is the original depth of the water.[58]

[58]Laplace [1775/76].

Laplace decomposed the potential U from the Moon and Sun into three terms that had monthly, diurnal, and semi-diurnal variations, and then solved his equations through perturbative methods in the analytically simple case for which the depth h varies as the sine-squared of the latitude. As he himself realized, this assumption could not pass for a realistic representation of the oceans. In the last section of his memoir, he switched to a semi-empirical method in which the elevation of the water in one harbor was represented as a sum of sine functions with the frequencies of the perturbing forces. In modern terms, we would say that he understood that the forced oscillations of the water surface necessarily had the same spectrum as the perturbing forces, owing to the linearity of the basic equations.[59]

Laplace's memoir looked and still looks forbiddingly complex, not only because of the idiosyncratic notation and the elliptic style, but also because most of the developments were purely algebraic. Physical discussion was confined to the first assumptions and to the final results, whereas a modern tide-theorist would anticipate and comment on the intermediate algebraic steps by appealing to general notions of forced oscillations and wave propagation. That Laplace's equations in fact describe a wave motion modified by the Coriolis force is easily seen by combining them to get, for $\Omega = 0$ and constant h,

$$\frac{\partial^2 \zeta}{\partial t^2} - gh\Delta\zeta = -h\Delta(U + \delta U), \qquad (2.48)$$

where Δ is the two-dimensional Laplacian. Although Laplace must have recognized d'Alembert's equation of vibrating strings, he did not exploit this analogy in his theory of tides. Instead, he appended to this theory the water-wave calculations with which our story began.

Laplace's theory was alien to contemporary British physics, which remained dependent on older Newtonian methods and tended to ignore the newer French mathematical physics. In 1813, the founder of the wave theory of light, Thomas Young, judged that the theory of tides was too practically important to be treated with Laplace's abstruse methods. Instead of the learned calculus of partial differentials, he offered a simple analogy between the ocean and a pendulum:

> The oscillation of the sea and of lakes, constituting the tides, are subject to laws exactly similar to those of pendulums capable of performing vibrations in the same time, and suspended from points which are subjected to compound regular vibrations, of which the constituent periods are completed in half a lunar and half a solar day.

In modern words, he assimilated tides with the forced oscillations of harmonic oscillators subjected to the superposition of two periodic forces.[60]

In order to justify this analogy (perhaps suggested by Laplace's equations), Young first showed that, in a canal of constant depth h, long waves of small amplitude were propagated with the Lagrangian velocity $c = \sqrt{gh}$. If the canal was terminated by a wall at one end, standing waves occurred. If the canal had the finite length L, the period of the oscillations could only be an integral multiple of a fundamental period L / c, as in closed

---

[59]Kelvin's later tide-predicting machine was based on the same principle.

[60]Young [1823] p. 307. Euler [1740] anticipated the pendulum analogy and gave the first treatment of the harmonically driven harmonic oscillator.

organ pipes. Further assuming that the sea was equivalent to a canal along its greatest length, Young replaced it with a set of pendulums that had the same periods. He was thus left with the elementary problem of determining the response of a damped harmonic oscillator to a sinusoidal excitation.[61]

As is now known to any physics undergraduate, the general solution to this problem is the sum of a free oscillation that exponentially decreases in time owing to the damping force, and a forced oscillation whose amplitude varies as $(\omega_0^2 - \omega^2)^{-1}$ if the eigenfrequency ω_0 is not too close to the excitation frequency ω. When the former frequency exceeds the latter, the forced oscillations are in phase with the exciting ones. In the opposite case, the two oscillations are in opposition. This result has an immediate, fruitful application: for the known order of magnitude of the depth and size of the oceans, their fundamental period of oscillation is much larger than half a day, so that the phase of tidal oscillations is opposed to the phase of the inducing luminary (as in Fig. 2.18(b)). Through equally elementary reasoning, Young explained several other well-known properties of the tides.

Russell was apparently unaware of Young's insights when he proposed his wave conception of tides, but he knew about Whewell's successful program of tide observation and prediction. As befits the author of *The history of inductive sciences*, Whewell's approach was inductive:

> I believe the instances are comparatively few in the history of philosophy, in which the general laws of the phenomena have been pointed out by the theory before they had been gathered by observation. The law of the tides, thus empirically obtained, may be used either as tests of the extant theories, or as suggestions for the improvement of those portions of mathematical hydraulics on which the true theory must depend.

Like a Ptolemean astronomer, Whewell tried to fit the results of measurements into simple harmonic formulas. Such was the basis for his reduction of tides in a given port.[62]

In order to connect tides observed in different locations, Whewell followed Young's suggestion to draw 'cotidal maps' that represented lines of high water at successive hours on a day of full Moon (see Fig. 2.19). According to Young, 'these lines would indicate ... the directions of the great waves, to which that of the progress of the tides in succession must be perpendicular.' Although Whewell did not refer to Young and doubted the possibility of theoretically deriving these lines, he allowed himself to identify the cotidal line at a given time with 'the summit or ridge of the *tide-wave* at that time.' He described the global forced wave that followed the motion of the Moon and the Sun, as well as the freely-propagating waves in smaller open seas, basins, channels, and rivers. These waves progressed with the depth-dependent velocity that Lagrange had derived and the Weber brothers had verified.[63]

2.3.3 *Airy's wave theory of tides*

The Astronomer Royal, George Biddell Airy, was also unaware of Young's wave theory when he wrote the article 'Tides and waves' for an 1845 volume of the *Encyclopaedia*

---

[61]Young [1813], [1823].   [62]Whewell [1834] p. 19.

[63]Whewell [1833] pp. 148 (cotidal maps), 149 (tidewave), 212 (Lagrangian velocity); Young [1823] p. 293.

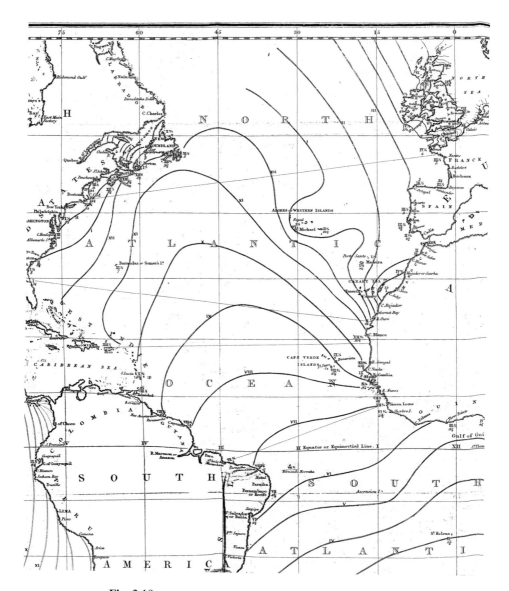

Fig. 2.19. A portion of Whewell's first cotidal map ([1833] plate).

metropolitana. However, he was familiar with Whewell's and Russell's tide studies. He did not cite Russell as a stimulus for his own theory, presumably because he had a poor opinion of Russell's theories in general. After noting the 'great value' of Russell's experiments, he warned the reader 'against attaching any importance to the theoretical expressions which are mingled with them in the original account.'[64]

[64]Airy [1845] p. 350.

As an eminent representative of the new generation of British natural philosophers who had thoroughly assimilated the methods of French mathematical physics, Airy was not only able to condemn Russell's loose theorizing but also to precisely assess the merits of Laplace's formidable calculations. While he found obscurities and even mistakes in this theory, his overall judgment was admiring:[65]

> We must allow [Laplace's theory] to be one of the most splendid works of the greatest mathematician of the past age. To appreciate this, the reader must consider, first, the boldness of the writer who, having a clear understanding of the gross imperfection of the methods of his predecessors, had also the courage deliberately to take up the problem on grounds fundamentally correct ...; secondly, the general difficulty of treating the motions of fluids; thirdly, the peculiar difficulty of treating the motions when the fluid covers an area which is not plane but convex; and, fourthly, the sagacity of perceiving that it was necessary to consider the Earth as a revolving body, and the skill of correctly introducing this consideration. This last point alone, in our opinion, gives the greater claim for reputation than the boasted explanation of the long inequality of Jupiter and Saturn.

Airy's main reason for abandoning Laplace's theory was not its mathematical difficulty nor any fundamental incorrectness in its assumptions, but the practical impossibility of solving the tidal equations for the actual form of the bottom of the sea. His own approach was based on the properties of canal waves. These directly concerned the behavior of river tides. They also shed light on oceanic tides, as far as an ocean could be replaced by a series of adjacent canals. Accordingly, Airy began with a thorough analysis of wave propagation in a canal. Lagrange's theory was too restrictive since it only applied to small, long waves. Cauchy's and Poisson's theories were even less relevant, since they supposed a mode of production of the waves that was never encountered in tide theory.[66]

Airy's analysis was based on the Lagrangian picture of fluid motion, as was Laplace's theory of 1776. Denote by X and Y the coordinates of the fluid particles when the fluid is at rest, and $X + \xi$ and $Y + \eta$ their coordinates when the fluid is in motion. As before, the X-axis lies along the bottom of the canal, and the Y-axis is vertical. Like Laplace, though with more elementary methods, Airy proved that the harmonic expressions

$$\xi = \varepsilon \cosh kY \cos kX \cos \omega t, \quad \eta = \varepsilon \sinh kY \sin kX \cos \omega t, \tag{2.49}$$

with $\omega^2 = gk \tanh kh$, satisfied the continuity equation, the equations of motion, and the boundary conditions as long as the motion was small. Unlike Laplace, he combined this solution with the other solution

$$\xi = -\varepsilon \cosh kY \sin kX \sin \omega t, \quad \eta = \varepsilon \sinh kY \cos kX \sin \omega t \tag{2.50}$$

to get the solution

$$\xi = \varepsilon \cosh kY \cos (kX - \omega t), \quad \eta = \varepsilon \sinh kY \sin (kX - \omega t), \tag{2.51}$$

---

[65] Airy [1845] p. 279.   [66] Ibid. pp. 280–1.

which propagates with the velocity $c = \omega/k$ such that

$$c^2 = \frac{g}{k}\tanh kh. \tag{2.52}$$

In this state of motion, the fluid particles perform elliptical oscillations that tend to circular ones for infinite depth. As Airy noted, this result agrees with the earlier observations of suspended solid particles made by the Webers and by Russell.[67]

2.3.4 *From river tides to ocean tides*

In the case of tides, the wavelength is much larger than the depth. Then the previous equations imply that the horizontal motion is sensibly the same from the surface to the bottom, and the vertical motion is comparatively very small.[68] Airy assumed this property to hold even in the case of river tides, for which the elevation of the water was no longer negligible compared to the depth. This enabled him to reach more exact, nonlinear equations of motion. He reasoned as follows.

The volume of the vertical slice of fluid lying between the planes X and $X + \delta X$ is $h\delta X$ in the undisturbed condition, and $(h + \sigma)[X + \delta X + \xi(X + \delta X) - X - \xi(X)]$ in the disturbed condition (σ denotes the elevation of the surface above its original height h). Therefore, the continuity of the fluid implies

$$\left(1 + \frac{\partial\xi}{\partial X}\right)\left(1 + \frac{\sigma}{h}\right) = 0. \tag{2.53}$$

The pressure on each side of the slice varies hydrostatically, since the vertical acceleration is neglected. Therefore, its longitudinal gradient only depends on the slope of the surface:

$$\frac{\partial P}{\partial X} = \rho g\frac{\partial\sigma}{\partial X}. \tag{2.54}$$

Newton's second law applied to the fluid slice then gives

$$\rho h\,\delta X\frac{\partial^2\xi}{\partial t^2} = -\frac{\partial P}{\partial X}\delta X(h + \sigma) = -\rho g\,\delta X(h + \sigma)\frac{\partial\sigma}{\partial X}. \tag{2.55}$$

Eliminating σ through the continuity equation, Airy finally obtained

$$\frac{\partial^2\xi}{\partial t^2} = gh\frac{\partial^2\xi}{\partial X^2}\left(1 + \frac{\partial\xi}{\partial X}\right)^{-3}. \tag{2.56}$$

Airy solved this equation perturbatively. The motion being $\xi_0 = \varepsilon\cos(\omega t - kX)$ in the lowest approximation, he obtained the next approximation ξ_1 by integrating the equation

$$\frac{\partial^2\xi_1}{\partial t^2} - gh\frac{\partial^2\xi_1}{\partial X^2} = -3gh\frac{\partial^2\xi_0}{\partial X^2}\frac{\partial\xi_0}{\partial X}, \tag{2.57}$$

---

[67] Airy [1845] pp. 290 (solution), 344 (Weber), 347 (Russell).   [68] *Ibid.* p. 294.

with the condition that for $X = 0$ the oscillation should still be $\varepsilon \cos \omega t$. This gives, for the corresponding elevation of the surface,

$$\sigma_1 = -a \sin(\omega t - kx) + \frac{3}{4} \frac{a^2}{h} kx \sin 2(\omega t - kx), \qquad (2.58)$$

with $a = kh\varepsilon$ and $x = X + \xi_1$.[69]

This solution represents the evolution of a tidal wave as it propagates from the mouth $x = 0$ along a flat, prismatic river without intrinsic current.[70] As is seen from Fig. 2.20, the front of the waves becomes steeper than the rear. This explains why the rise of the water takes more time than its descent at a station far from the mouth. Airy further derived the velocity of the wave crests (for which $d\sigma_1/dx = 0$) at the same approximation:

$$c = \sqrt{gh}\left(1 + \frac{3}{2}\frac{a}{h}\right). \qquad (2.59)$$

He found this formula to be compatible with the velocity measurement of high waves by the Webers and Russell, despite Russell's claim that the velocity of a solitary wave of height σ obeyed the formula $c = \sqrt{g(h + \sigma)}$.[71]

In the case of oceanic tides, the height of the waves is negligible compared to the depth, so that the continuity equation (2.53) and the equation of motion (2.55) can be linearized. However, the direct action of the Moon and the Sun is no longer negligible. The vertical component of this action amounts to a negligible modification of gravity. However, the equation of motion now includes the horizontal component F of this action:

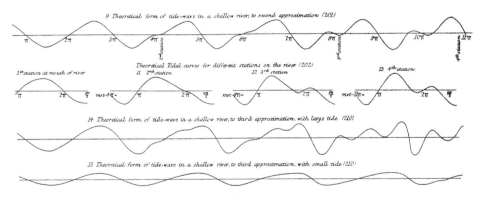

Fig. 2.20.   The evolution of a sine wave along a canal to second and third order in its amplitude (Airy [1845] plate).

[69] Airy [1845] pp. 297, 300.

[70] Airy believed the solution to be still valid far from the mouth. In reality, the consistency of the approximation requires that $x << h/ka$.

[71] Airy [1845] pp. 300–1. As Stokes, Saint-Venant, and Boussinesq later made clear, Airy's formula applies to the crest of long, non-permanent waves, whereas Russell's formula applies to permanent waves whose length is comparable to the depth of water. Simple derivations of Airy's formula are found in Lamb [1932] pp. 261–2, 278–80.

$$\frac{\partial^2 \xi}{\partial t^2} - gh\frac{\partial^2 \xi}{\partial x^2} = F. \tag{2.60}$$

Here F is the superposition of harmonic components with a latitude-dependent phase and amplitude. Airy determined the resulting forced oscillations for circular canals running along a parallel and along a meridian, and for canals closed at both ends. In each case, there are free oscillations at frequencies that are integral multiples of a fundamental frequency. The amplitude of the forced oscillations depends on how close these eigenfrequencies are to the frequencies of the tidal force F. Airy also introduced friction proportional to the velocity, and discussed the consecutive damping of free oscillations.[72]

In conclusion to this analysis, Airy admitted that his assumption of tidal canals of uniform depth and breadth was no more realistic than Laplace's assumption of an Earth-covering ocean with a special law of depth. The main advantage he saw in his method was that it permitted a more detailed consideration of the interplay of the lunar, solar, and frictional forces, since all the equations could be solved in finite terms through elementary analysis. In brief, his theory failed as much as Laplace in quantitative tide prediction, but it offered more qualitative insights.[73]

2.3.5   *The inverse method, for and against Russell*

Airy did not confine his study of waves to aspects relevant to tide theory. He also explained commonly-known properties of water waves, and some of Russell's more surprising results. As he was generally unable to integrate his hydrodynamic equations for the actual forces that produced the wave motion, he ingeniously inverted this procedure: he sought to compute, for a hypothetical form of fluid motion, the forces that would maintain this motion. This is much easier to do, since differentiations are involved instead of integrations. From the knowledge of these forces, he then inferred what the actual motion would be in their absence, or what additional action on the water could produce the hypothetical motion.

As a first example, consider the breaking of waves on a sloping shore. Airy computed the forces necessary to maintain a constant shape of the waves when they approach the shore. The result is forces that pull the tip of each wave in the direction opposite to that of their progression. Since in reality these forces do not act, the tips of the waves must bend forward, as should happen at the beginning of the breaking process. Another example is the swelling of waves under wind. Airy injected a swelling motion in the equations of motion. The resulting forces turn out to be pressures applied to the rear of the waves, as would naturally be expected for waves before the wind.[74]

A third example is the 'great primary wave', or forced wave that accompanies a canal boat in its motion. In this case, the horizontal disturbance ξ and the surface disturbance σ are functions of $x - vt$ only, where v is the velocity of the boat. In the small-long-wave approximation, the equation of motion (2.60) gives $F = (v^2 - gh)\xi''$, while the continuity equation (2.53) gives $\sigma = -h\xi'$. Therefore, the force that is necessary to maintain this motion has the same sign as the slope of the surface when the velocity of the boat is inferior

[72]Airy [1845] pp. 310–39. For a concise account of Airy's theory of oceanic tides, cf. Lamb [1932] pp. 267–73.
[73]*Ibid.* p. 363.   [74]*Ibid.* p. 314.

to that of free waves; it has the opposite sign in the reverse case, and it vanishes when the two velocities are equal. This conclusion agrees with the relative position of a canal boat and its forced wave, and with the drop in resistance in the critical case.[75]

Airy thus explained Russell's observations, but implicitly rejected his intuitive theory of solitary-wave riding. Through the same kind of argument, he dismissed Russell's solitary wave. For waves of finite height, the equation of motion is the nonlinear equation (2.56). Without additional force and for a disturbance propagating without any change of shape, it can only hold if the slope ξ' of the disturbance is a constant. As this slope must vanish at infinity, there is no such disturbance. Airy concluded that the solitary wave was mathematically impossible. What Russell had observed was a wave small enough for Lagrange's theory to apply approximately:[76]

> We are not disposed to recognize this wave [Russell's] as deserving the epithets 'great' or 'primary' ... and we conceive that, ever since it was known that the theory of shallow waves of great length was contained in the equation $\partial^2\xi/\partial t^2 = gh\partial^2\xi/\partial x^2$ ... the theory of the solitary wave has been perfectly well known.

As we have already seen, this authoritative judgment failed to disturb Russell's belief in the novelty of his solitary waves.

2.4 Finite waves

2.4.1 *Stokes's BA report*

In 1846 a new leader of British hydrodynamics, the Cambridge professor George Gabriel Stokes, reviewed the state of this field for the British Association. Since the previous report by Challis, there had been much British work on waves, in a good part stimulated by Russell's experiments. Stokes played down the importance of Poisson's and Cauchy's memoir: 'The mathematical treatment of such cases [waves produced by emersion] is extremely difficult; and, after all, motions of this kind are not those which it is most interesting to investigate.' In the wake of Russell's and Airy's works on waves, tides, and navigation, what had become most important was the study of 'simpler cases of wave motion, and those which are more nearly connected with the phenomena which it is most desirable to explain.'[77]

Among the simpler cases of motion, Stokes retained waves with a length much longer than the depth. As Lagrange, George Green, Philip Kelland, and Airy had shown, these waves propagated without deformation in a canal of constant section as long as their height was much smaller than the depth. Their velocity obeyed a simple formula. Green and Airy had computed their deformation for a slowly-varying canal depth or breadth.

---

[75] *Ibid.* pp. 349–50.

[76] *Ibid.* p. 346. As Stokes later noted, Airy overlooked the fact that his equation of motion applied to waves longer than those observed by Russell.

[77] Stokes [1846a] p. 161. This opinion echoed an earlier remark by Kelland ([1840] p. 497): 'I doubt much ... whether such men as Laplace and Lagrange would have been induced, with the expectation of joining experiment on her lower and more trodden fields, to reconsider and remodel their investigations; nor have I any reason to hope, that such men as Poisson and Cauchy will quit the delectable atmosphere in which they are involved, of abstruse analysis, for the more humble, but not less important task of endeavouring to treat the simpler problems in a manner not made general arbitrarily to lead to the most elegant formulae, but general to that extent, and in that mode, in which the problem in nature is so.'

Airy had shown how finite height affected their propagation. Stokes also dwelled on the fruitful application that Airy had given of this sort of wave to the theory of tides.[78]

Another case of special interest was given by 'waves which are propagated with a constant velocity and without change of form, in a fluid of uniform depth, the motion being in two dimensions and periodical.' By implicit analogy with monochromatic plane waves in optics, he regarded these waves as 'the *type* of oscillatory waves in general'. Green had given the expression $\sqrt{g/k}$ for the velocity of such waves in the case of infinite depth, and Kelland had anticipated Airy's results in the case of finite depth.[79]

Stokes then turned to the controversial issue of solitary waves. Stokes admitted that Russell's experiments made the *sui generis* character of solitary waves probable, but he denied that friction was the only cause of the decay of such waves. To sustain this opinion he did not use Airy's objection, which only excluded solitary waves of arbitrarily long length. Rather, he referred to recent calculations by Samuel Earnshaw. The Reverend mathematician had integrated the equations of motion for a wave of permanent shape that met a condition experimentally verified by Russell, namely, that fluid particles originally in the same vertical plane remained so during the passage of the wave. In Earnshaw's opinion, this result confirmed the existence of solitary waves. Stokes drew the opposite conclusion from the same calculation, for he noted that Earnshaw waves could not be connected to the surrounding fluid at rest without an absurd discontinuity of the velocity. As Stokes did not question the experimental truth of parallel-plane motion, he concluded that there was a necessary non-frictional decay of solitary waves.[80]

Not only did Stokes deny the properties of solitary waves that Russell judged most essential, but he also condemned—without naming Russell—applications of solitary waves to tides and to sound:[81]

> With respect to the importance of this peculiar wave ... it must be remarked that the term *solitary wave*, as so defined [as a phenomenon *sui generis*] must not be extended to the tide wave, which is nothing more ... than a very long wave, of which the form may be arbitrary. It is hardly necessary to remark that the mechanical theories of the solitary wave and the aërial sound wave are altogether different.

2.4.2   *Stokes on finite oscillatory waves*

In 1846, Stokes believed permanent, solitary waves of finite height to be impossible. But the existence of permanent, oscillatory waves of finite height remained plausible. Also, Russell had found that the (phase) velocity of oscillatory waves obeyed the Kelland–Airy formula (2.52) (for infinitely-small waves) even when the waves were no longer small with respect to the depth. Stimulated by this result and its apparent contradiction with Airy's velocity formula (2.59) for finite waves, Stokes sought a perturbative solution of Euler's

---

[78]Stokes [1846a] pp. 161–4 (long waves), 171–5 (tides); Lagrange [1781]; Green [1838]; Kelland [1840]; Airy [1845]. On early British wave theory, cf. Craik [2004] 8–24.

[79]Stokes [1846a] p. 164; Green [1839]; Kelland [1840]. Kelland believed the motion to have a form independent of the height of the waves, for he used erroneous boundary conditions.

[80]Stokes [1846a]: pp. 168–70; Earnshaw [1849] (read in Dec. 1845). Cf. Craik [2004] pp. 17–18.

[81]Stokes [1846a] p. 170.

equations that made the fluid velocity the gradient of a potential and a function of $x - ct$ and y only.[82]

As in Lagrange's theory of waves, the potential must satisfy eqn (2.14), namely

$$\frac{\partial^2 \varphi}{\partial x^2} + \frac{\partial^2 \varphi}{\partial y^2} = 0.$$

The equation of the surface is:

$$\frac{\partial \varphi}{\partial t} + \frac{(\nabla \varphi)^2}{2} + g(y - h) = 0. \tag{2.61}$$

The boundary condition at the bottom of the channel is $\partial \varphi / \partial y = 0$ when $y = 0$. The condition that a particle on the surface should remain on the surface is

$$\left(\frac{\partial}{\partial t} + \frac{\partial \varphi}{\partial x} \frac{\partial}{\partial x} + \frac{\partial \varphi}{\partial y} \frac{\partial}{\partial y} \right) \left(\frac{\partial \varphi}{\partial t} + \frac{(\nabla \varphi)^2}{2} + g(y - h) \right) = 0 \tag{2.62}$$

at any point of the surface. The general integral of eqn (2.14) that meets the first boundary condition is

$$\varphi = Cx + \sum_k \cosh ky (A_k \cos kx + B_k \sin kx). \tag{2.63}$$

The first term may be dropped as it represents a constant velocity. To first order in φ, the second boundary condition (2.62) and the condition that the velocity is a function of $x - ct$ and y only imply eqn (2.52), namely

$$c^2 = \frac{g}{k} \tanh kh,$$

for every term of the sum over k. Since there is only one value of k that meets this condition, the sum is reduced to a sine wave.[83]

As a corollary, the propagation of a solitary wave without change of form is impossible at first order. In modern terms, we would say that the dispersion (dependency of celerity on wavelength) of infinitely-small monochromatic water waves implies the spreading of wave packets. Stokes concluded:[84]

> Thus the degradation in the height of such waves, which Mr. Russell observed, is not to be attributed wholly, (nor I believe chiefly,) to the imperfect fluidity of the fluid ... but is an essential characteristic of a solitary wave. It is true that this conclusion depends on an investigation which applies strictly to indefinitely small motions only: but if it were true in general that a solitary wave could be propagated uniformly, without degradation, it would be true in the limiting case of indefinitely small motions; and to disprove a general proposition it is sufficient to disprove a particular case.

[82]Stokes [1847a]. Cf. Craik [2005].   [83]Ibid. [1847a]. pp. 199–204.

[84]Ibid. 204. This objection is invalid, because it assumes that the length of the waves is kept constant in the zero-amplitude limit, whereas for a solitary wave the length grows indefinitely when the amplitude tends to zero.

After this new blow to Russell's interpretation of the solitary wave, Stokes proceeded to give a theoretical justification of Russell's experimental results on oscillatory waves.[85] To second order in the amplitude a of the wave, the celerity of the waves still obeys the Kelland–Airy formula (2.52), in conformity with Russell's measurements. This result does not contradict Airy's formula (2.59), Stokes explained, because the latter assumes waves much longer than the depth, whereas the smallness of Stokes's perturbations is easily seen to contradict this condition.[86] The equation of the surface is

$$y = h + a\cos kx - \left[\frac{\cosh kh(2\cosh^2 kh + 1)}{4\sinh^3 kh}\right] ka^2 \cos 2kx. \tag{2.64}$$

For infinite depth and to third order, it is

$$y = h + a\cos kx - \frac{1}{2}ka^2\cos 2kx + \frac{3}{8}k^2 a^3 \cos 3kx, \tag{2.65}$$

in fair agreement with the trochoids that Russell had inferred from observations of high sea waves (see Fig. 2.21). To the same order, the deep-water celerity becomes

$$c = \sqrt{\frac{g}{k}}\left(1 + \frac{1}{2}k^2 a^2\right) \tag{2.66}$$

Lastly, Stokes found that, for high waves, the propagation of the waves was accompanied by a net flux of water. He even recommended taking into account this flux in the dead reckoning of the position of ships.[87]

2.4.3 *Gerstner's waves and ship rolling*

Stokes returned to water waves in the 1870s, when he had to write a memorandum on the measurement of waves for the Meteorological Council.[88] A good knowledge of the height

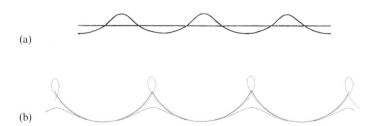

(a)

(b)

Fig. 2.21.   Wave of finite height according to Stokes's theory ([1847a] p. 212) (a); according to Russell's cycloidal interpretation of ocean waves (b).too pale?

---

[85]Stokes [1847a] pp. 205–8. To second order, Stokes also gave finite-depth results.

[86]*Ibid.* p. 209. Moreover, Airy dealt with a different problem, namely, the deformation of a wave that has a sine shape near the origin.

[87]Stokes [1847a] pp. 198–9, 208–9.   [88]Cf. Froude to Stokes, 17 Jan. 1873, in Stokes [1907].

and length of sea waves, he argued, was necessary for a proper control of ship rolling. This preoccupation and discussions with William Thomson—who was involved in similar questions—probably led him to improve his theory of high waves and to reflect on the highest possible wave. In 1880, he used the publication of the first volume of his collected papers as an opportunity to update his views on this topic.[89]

In the first place, Stokes expressed his opinion on an old theory of finite, oscillatory waves on infinitely-deep water that had become popular among naval engineers. This theory, published in 1802 by the Prague mathematics professor, engineer, and knight, Franz Joseph von Gerstner, assumed a circular motion of the fluid particles with a radius diminishing with the distance from the surface:[90]

$$x = X + k^{-1}e^{kY}\cos k(X - ct), \; y = Y - k^{-1}e^{kY}\sin k(X - ct), \tag{2.67}$$

where x and y are the coordinates at time t of the particle that has the mean coordinates X and Y, $2\pi/k$ is the wavelength, and c is the celerity of the wave. This motion is easily seen to satisfy the continuity condition and the equations of motion. The pressure for a given fluid particle is independent of time if and only if $c = \sqrt{g/k}$. It is then a function of Y only, so that the wave surface can be any of the lines for which Y is a negative constant. Fig. 2.22 illustrates the resulting waves for different values of this constant. The highest waves, for which the constant vanishes, have an infinitely-sharp edge. Their surface is a cycloid generated by a circle of radius k^{-1} rolling on the underside of the line $y = k^{-1}$. The other waves are trochoids with an eccentricity decreasing with their amplitude. Gerstner believed his waves to be the only ones compatible with the general principles of mechanics. In fact, as the Leipzig mathematician Ferdinand Moebius noted some twenty years later, Gerstner's derivation relies on the specific assumption that the pressure around any particle of the fluid remains the same in the course of time (whereas general principles require this to be true only for the particles at the free surface of the fluid).[91]

The Webers' *Wellenlehre* included a detailed analysis of Gerstner's waves. They found reasonable agreement with the observed motion of suspended particles, although the radius of the circular motions did not quite vary as Gerstner predicted. Their overall judgment was laudatory: 'Even if these conditions [for Gerstner's calculation to apply] are not completely met in reality, Gerstner's investigation remains not only interesting but also useful.' Russell, who became acquainted with Gerstner's waves through the Webers' book, found even better agreement with observation than the Webers had. His judgment was enthusiastic: 'Gerstner's theory is characterized by simplicity of hypothesis, precision of application, its conformity with the phaenomena, and the elegance of its results.'[92]

---

[89]Both Stokes and Thomson implicitly assumed a simple relation between observed ocean waves and the theory of finite waves of permanent shape. The modern, statistical theory of ocean waves contradicts this view: cf. Kinsman [1965].

[90]Gerstner [1802], [1804]. This motion has the same form as the large-depth limit of Airy's equations (2.51) for infinitesimal oscillatory waves. The only difference is that for Airy the surface of the water could only correspond to a large negative value of Y, whereas for Gerstner any negative value would do.

[91]Cf. Stokes [1880a] Lamb [1932] pp. 421–3; Weber and Weber [1825] p. 368 (for Moebius's remark). In Gerstner's original reasoning [1802], steady waves are investigated first, and a uniform translation is superposed onto these waves to yield progressive waves.

[92]Weber and Weber [1825] pp. 338–72, 368 (quote); Russell [1845] p. 368n.

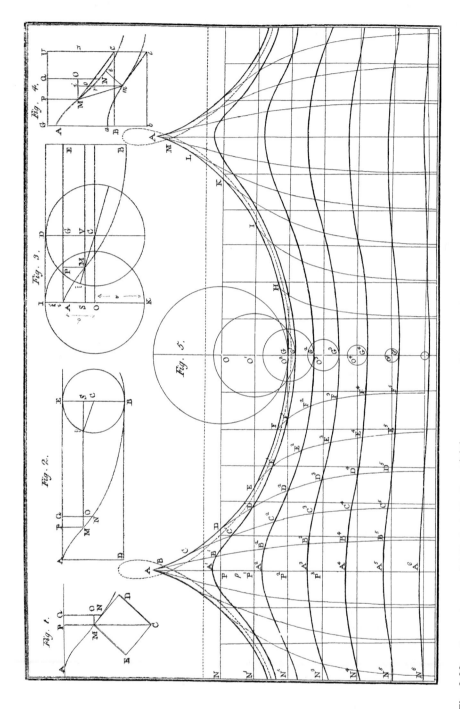

Fig. 2.22. Gerstner's waves ([1802] plate). The lines $A^iB^iC^i$ ... represent possible wave profiles; the circles represent the orbits of fluid particles; the remaining lines represent the successive forms of a line of particles that is vertical when passed by a crest or a through.

In the 1860s, British and French interest in ship rolling led to three rediscoveries of Gerstner's waves, by the Edinburgh engineering professor William Rankine, by the naval engineer William Froude, and by the Director of the Ecole du Génie Maritime Ferdinand Reech. When, in the early 1870s, the French leader in applied mechanics Adhémar Barré de Saint-Venant and his disciple Joseph Boussinesq became aware of Gerstner's theory, they fully endorsed it. As they noted, Gerstner's waves imply a rotational motion of the water and therefore cannot be regarded as being generated by pressures acting on a perfect liquid originally at rest. In their eyes, this fact did not preclude the application to sea waves, for the latter usually have a long history in which the imperfect fluidity of water plausibly plays a role. Stokes judged differently: in his view, only irrotational waves could be produced by natural causes. Consequently, these waves were worth analytical efforts, despite the much greater simplicity of Gerstner's waves.[93]

2.4.4 *From wedge-shaped waves to solitary waves*

Next to his dismissal of Gerstner's waves, Stokes inserted a supremely elegant proof that, if the crest of an irrotational wave has a sharp edge, then this edge necessarily makes an angle of 120°. As a fluid particle travels along the surface, its velocity (in a reference system in which the wave is stationary) must vanish at the angular points. At a short distance r from such a point, the velocity must vary as $\sqrt{r}$ according to Bernoulli's law. The irrotational character of the wave implies the existence of a velocity potential. As this potential is harmonic, it is the real part of a function of the complex variable $x + iy$ that can be developed in whole powers of this variable. Taking the origin of coordinates at the angular point, the potential behaves as the real part of a power of $x + iy$. In polar coordinates, this gives the form $\varphi \propto r^n \cos n\theta$. On the vertex, the normal velocity $\partial\varphi/\partial\theta$ must vanish, and the tangential velocity $\partial\varphi/\partial r$ must be proportional to $\sqrt{r}$. The latter condition implies $n = 3/2$. The former then requires that the angle of the vertex should be 120°.[94]

By 1880, Stokes believed that the highest possible wave (for a given wavelength) had this 120° cusped shape. Yet his correspondence with Thomson shows that a few months earlier he still hesitated. It also shows that he sought opportunities to verify this prediction:

> I have in mind when I have occasion to go to London to take a run down to Brighton if a rough sea should be telegraphed, that I may study the forms of waves about to break. I have a sort of imperfect memory that swells breaking on a sandy beach became at one phase very approximately wedge-shapes.

During the next summer, Thomson invited him 'to see and *feel* the waves' on his yacht. In the fall, Stokes wrote to his friend:

> You ask if I have done anything more about the greatest possible wave. I cannot say that I have, at least anything to mention mathematically. For it is not a very mathematical process taking off my shoes and stockings, tucking up my trousers as

[93]Rankine [1862]; Froude [1862]; Reech [1869]; Saint-Venant [1871*b*] Boussinesq [1877]: pp. 345–6; Stokes [1880*a*]. In principle, wind could exert a shear stress on the water surface and thus induce vorticity of the water. In reality, however, the normal pressures are more important and observed waves are very nearly irrotational, as Stokes expected; cf. Kinsman [1965].

[94]Stokes [1880*b*]. Twentieth-century experiments have confirmed the 120° cusps, cf. Kinsman [1965].

high as I could, and wading out into the sea to get in line with the crest of some small
waves that were breaking on a sandy beach.

These adventurous observations seemed to confirm the 120° edge for the highest possible
waves.[95]

From a theoretical point of view, what convinced Stokes of the existence of wedge-
shaped waves was a new perturbation method that enabled him, in the fall of 1879, to push
the calculation of finite oscillatory (and irrotational) waves to third order for finite depth
and to fifth order for infinite depth. The trick was to simplify the expression of the
boundary conditions by using the potential φ and Lagrange's stream function ψ (the
harmonic conjugate of φ) as independent variables instead of the coordinates x and y.
The calculations indicated that, for large amplitudes, the tip of the waves came closer to
the 120°-cusp shape when the order of perturbation increased. For the exact oscillatory
solutions, Stokes expected the cusp shape and divergent series to occur for a definite value
of the amplitude/wavelength ratio in the case of infinite depth, and for a definite value of
the amplitude/depth ratio in the case of finite depth. In the latter case he realized that the
waves 'tend[ed] to assume the character of a series of disconnected solitary waves.'[96]

In October 1879, the latter finding prompted him to write to Thomson: 'Contrary to an
opinion expressed in my [BA] report [of 1846], I am now disposed to think there is such a
thing as a solitary wave that can be theoretically propagated without degradation.'
Thomson disagreed: 'The more I think of it the more I am disposed to conclude that
there is no such thing as a steady free periodic series of waves in water of any depth. I can't
believe in the solitary wave.' This divergence of opinion came from Thomson's suspicion
that Stokes's series for finite waves never converged and only indicated *approximately*
steady waves. In the following years, there was indeed much controversy about the
convergence of these series. The story only ended in 1925, with Tullio Levi-Civita's
rigorous proof of the existence of finite waves of permanent shape.[97]

2.4.5   *Boussinesq on solitary waves*

Unknown to Stokes and Thomson, the mathematical existence of solitary waves had
already been argued twice—in 1871 by a remote French theorist, and in 1876 by a rising
star of British natural philosophy. The French investigator, Joseph Boussinesq, had been
working on open-channel theory for some time. In the steps of his mentor Saint-Venant,
he tried to subject every aspect of the motion of water in rivers and canals to mathematical
analysis.[98] He was aware of Russell's observations, and also of the more precise measurements of solitary waves performed by the French hydraulician Henry Bazin. He had
already written a long memoir on water waves of small height on water of constant
depth. In addition to results that could be found in earlier memoirs by Green, Kelland,

---

[95]Stokes to Thomson, 20 Sept. 1879, 11 Oct. 1879, 15 Sept. 1880, *ST*; Thomson to Stokes, 14 July 1880, *ST*.

[96]Stokes [1880c] pp. 320, 325. Stokes probably borrowed this method from Helmholtz [1868d], discussed later
on pp. 164–5.

[97]Stokes to Thomson, 6 Oct. 1879, *ST*; Thomson to Stokes, 10 Oct. 1879, *ST*; Levi-Civita [1925]. Cf. Lamb
[1932] p. 420.

[98]See Chapter 6, pp. 233–8.

and Airy (of which he was unaware), he offered a few preliminary considerations on waves of finite height that may have led him to reflect on Russell's wave.[99]

In his first derivation of the solitary wave, published in 1871 in the *Comptes rendus*, Boussinesq sought an approximate solution of Euler's equations that propagated at the constant speed c without deformation in a rectangular channel. His success in this difficult task depended on his special flair in estimating the relative importance of the various terms of his developments. His basic strategy was to develop the velocity components u and v in powers of the vertical distance y from the bottom of the channel, and to determine the coefficients of this development through the boundary conditions. Lagrange had already tried this route and written the resulting series of differential equations, but had found their integration to exceed the possibilities of contemporary analysis unless nonlinear terms were dropped. A century later, Boussinesq managed to include these terms.[100]

To second order in y, Lagrange's expression (2.15) of the velocity potential implies the form

$$u = \alpha - \frac{1}{2}\alpha'' y^2, \quad v = -\alpha' y \tag{2.68}$$

of the velocity components, where α is a function of x only and the primes denote derivation with respect to x. Denote by σ the elevation of the surface above its original height h. The conservation of flux in a reference system bound to the wave implies

$$\int_0^{h+\sigma} u \, dy = c\sigma. \tag{2.69}$$

The resulting constraint on the unknown function α is

$$\alpha(h + \sigma) - \frac{1}{6}\alpha''(h + \sigma)^3 = c\sigma. \tag{2.70}$$

Boussinesq solved this equation perturbatively. At the lowest order of approximation, the cubic term is dropped on the left-hand side, and σ is neglected with respect to h, so that $\alpha = c/h$. At the next order of approximation, the latter value of α is substituted into the cubic term, and σ is neglected with respect to h in this term only. This gives

$$\frac{\alpha}{c} = \frac{\sigma}{h + \sigma} + \frac{1}{6}\sigma'' h \tag{2.71}$$

and

$$\frac{u}{c} = \frac{\sigma}{h + \sigma} + \frac{1}{6}\frac{\sigma''}{h}(h^2 - 3y^2), \quad \frac{v}{c} = -\frac{\sigma'}{h}y. \tag{2.72}$$

Boussinesq then obtained the equation of the surface by substituting these expressions into the boundary condition[101]

---

[99]Bazin [1865]; Boussinesq [1872a].   [100]Boussinesq [1871a]; Lagrange [1781].

[101]The other boundary condition, that a particle of the surface should remain on the surface, is a consequence of eqn. (2.69).

$$u^2 + v^2 - 2\frac{\partial \varphi}{\partial t} + 2g(y - h) = 0 \quad \text{for } y = h + \sigma. \tag{2.73}$$

As the potential φ is a function of $x - ct$ only, $\partial \varphi / \partial t$ is the same as $-cu$.

In order to clarify subsequent approximations, it is convenient to introduce the dimensionless variables $\varepsilon = \sigma/h$, $\varepsilon' = \sigma'$, and $\varepsilon'' = h\sigma''$. Boussinesq assumed the wave to be small and gently sloped, and therefore treated ε, ε'/ε, and $\varepsilon''/\varepsilon$ as small quantities. He thus obtained the equation of the surface

$$c^2 = gh\left(1 + \frac{3}{2}\frac{\sigma}{h} + \frac{1}{3}\frac{h^2\sigma''}{\sigma}\right), \tag{2.74}$$

where terms in ε^2, $\varepsilon'^2/\varepsilon$, and ε'' and all smaller terms are neglected.[102] This equation may be rewritten as

$$\varepsilon'' = 3K\varepsilon - \frac{9}{2}\varepsilon^2, \quad \text{with } K = \frac{c^2}{gh} - 1. \tag{2.75}$$

A first integration yields

$$\varepsilon'^2 = 3\varepsilon^2(K - \varepsilon). \tag{2.76}$$

The maximum $\varepsilon' = 0$ of the corresponding curve is reached when $\varepsilon = K$. Consequently, the velocity of the wave is related to the height σ_M of its summit through

$$c = \sqrt{g(h + \sigma_M)}, \tag{2.77}$$

which is Russell's formula. Boussinesq then integrated a second time to reach

$$\frac{\sigma}{h} = \frac{2K}{1 + \cosh\left[\sqrt{3K}(x - ct)/h\right]}. \tag{2.78}$$

His plot of this curve is presented in Fig. 2.23.

A couple of months later, Boussinesq submitted to the French Academy a more general theory that gave the deformation of a small, gently-sloped, but otherwise arbitrary wave during its progression in a channel of constant depth.[103] His calculation was still based on Lagrange's development of the velocity potential in powers of y. To fourth order, this development has the form[104]

$$\varphi = \beta - \frac{1}{2}\beta'' y^2 + \frac{1}{24}\beta'''' y^4. \tag{2.79}$$

---

[102]Boussinesq kept the $\varepsilon'^2/\varepsilon$ terms, but neglected them when he integrated the equations.

[103]Boussinesq [1871c]; [1872b]. For a brief but accurate discussion of this memoir, cf. Miles [1981]. Miles notes that the memoir implicitly contains the Korteweg–de Vries (KdV) equation, but does not mention that Boussinesq [1877] explicity contains it (see later on pp. 83–4).

[104]The reader may wonder why Boussinesq now includes the fourth-order term, which he seems to have neglected in his earlier determination of the solitary profile. The reason is that the use of the differential condition (2.81) instead of the integral condition (2.69) requires a higher approximation of the potential.

Profil d'une onde solitaire.

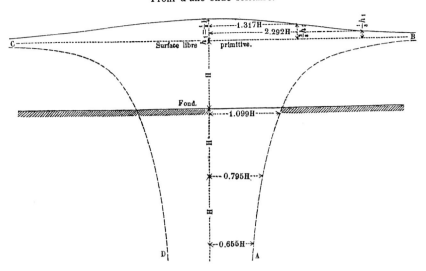

Fig. 2.23.   The profile of a solitary wave (curved solid line) according to Boussinesq ([1872b] p. 90).

The vanishing of the pressure at the free surface gives

$$g\sigma + \frac{\partial\varphi}{\partial t} + \frac{1}{2}(\nabla\varphi)^2 = 0 \quad \text{for } y = \sigma(x,t). \tag{2.80}$$

The condition that a particle originally on the surface should remain on the surface gives

$$\frac{\partial\varphi}{\partial y} = \frac{\partial\sigma}{\partial t} + \frac{\partial\varphi}{\partial x}\frac{\partial\sigma}{\partial x} \quad \text{if } y = \sigma(x,t). \tag{2.81}$$

At the lowest order of approximation, using dots for time derivatives and primes for derivatives with respect to x, these two conditions yield (in the reverse order)

$$\dot\sigma = -\beta'' h, \quad \dot\beta = -g\sigma. \tag{2.82}$$

The elimination of β gives Lagrange's wave equation

$$\ddot\sigma = gh\sigma''. \tag{2.83}$$

Consequently, at this order φ is the sum of a function of $x - c_0 t$ and a function of $x + c_0 t$, with $c_0 = \sqrt{gh}$. Boussinesq retained only the first component, which represents a perturbation traveling at the constant speed c_0 in the direction of increasing x.

At the next order of approximation, the two conditions give

$$\dot\sigma = -\beta'' h - \beta'' \sigma - \beta' \sigma' + \frac{1}{6}\beta'''' h^3, \quad \dot\beta = -g\sigma + \frac{1}{2}\dot\beta'' h^2 - \frac{1}{2}\beta'^2, \tag{2.84}$$

where $h + \sigma$ has been replaced by h in terms that have a derivative of third order or higher in factor. In order to eliminate β, Boussinesq derived the first equation with respect to time and the second equation twice with respect to x. This gives

$$\ddot{\sigma} = -\dot{\beta}''h - (\beta'\dot{\sigma})' - (\beta'\dot{\sigma})' + \frac{1}{6}\dot{\beta}''''h^3, \quad \dot{\beta}'' = -g\sigma'' + \frac{1}{2}\dot{\beta}''''h^2 - \frac{1}{2}(\beta'^2)''. \tag{2.85}$$

In the terms that follow the first, dominant term in each of these equations, $\dot{\sigma}$ and $\dot{\beta}$ can be replaced by their first approximation (2.82), and the operators $\partial/\partial t$ and $-c_0 \partial/\partial x$ are interchangeable. This gives

$$\sigma = -\dot{\beta}''h + g(\sigma^2)'' - \frac{1}{6}gh^3\sigma'''', \quad \dot{\beta}'' = -g\sigma - \frac{1}{2}gh^2\sigma'''' - \frac{1}{2}gh^{-1}(\sigma^2)''. \tag{2.86}$$

Hence follows Boussinesq's equation for the evolution of the perturbation:[105]

$$\ddot{\sigma} = gh\sigma'' + \frac{3}{2}g(\sigma^2)'' + \frac{1}{3}gh^3\sigma''''. \tag{2.87}$$

In order to ease the integration of this equation, Boussinesq imagined a series of fictitious vertical planes moving in such a manner that the volume of liquid between two consecutive planes remains constant. The velocity w of these planes is easily seen to depend on their abscissa x in such a way that

$$\dot{\sigma} = -(\sigma w)'. \tag{2.88}$$

With the notation

$$\gamma = \frac{3}{2}g\sigma^2 + \frac{1}{3}gh^3\sigma'', \tag{2.89}$$

eqn (2.87) leads to

$$\frac{\partial \sigma w}{\partial t} + c_0^2\sigma' + \gamma' = 0. \tag{2.90}$$

In terms of the auxiliary quantity

$$\chi = \sigma(w - c_0) - \frac{\gamma}{2c_0}, \tag{2.91}$$

this equation can be rewritten as

$$\dot{\chi} = c_0\chi', \tag{2.92}$$

provided that the operator $\partial/\partial t$ can be replaced by $-c_0\partial/\partial x$ when applied to the small quantity γ. This means that χ is a function of $x + c_0 t$ only. As it is also a combination of quantities that are functions of $x - c_0 t$ which vanish at infinity, it must vanish. This implies

$$w = c_0 + \frac{\gamma}{2c_0\sigma}, \tag{2.93}$$

---

[105]Boussinesq [1871c], [1872b] p. 74.

and, approximately,

$$w^2 = gh\left(1 + \frac{3}{2}\frac{\sigma}{h} + \frac{1}{3}\frac{h^2\sigma''}{\sigma}\right). \tag{2.94}$$

Boussinesq then substituted his expression for w into eqn (2.88) to get the convective variation of the height of the fluid slices as

$$\dot\sigma + w\sigma' = -c_0\left(\frac{3}{2}\frac{\sigma}{h} + \frac{1}{3}\frac{h^2\sigma''}{\sigma}\right)'. \tag{2.95}$$

He also verified that the volume, momentum, and energy of a swell evolving according to this equation were invariable. Most importantly, he identified a fourth invariant of the motion, namely the 'moment of instability'

$$M = \int\limits_{-\infty}^{+\infty}\left(\sigma'^2 - \frac{3\sigma^3}{h^3}\right)\mathrm{d}x. \tag{2.96}$$

He probably came to suspect its existence while studying the condition of permanent shape as follows.[106]

Remembering that w is the velocity of constant-volume slices of the swell, the shape of a swell is permanent if and only if w is a constant c which represents the celerity of the wave:

$$c^2 = gh\left(1 + \frac{3}{2}\frac{\sigma}{h} + \frac{1}{3}\frac{h^2\sigma''}{\sigma}\right).$$

This condition is identical to that reached earlier by Boussinesq using a more direct method (see eqn (2.74)). For anyone familiar with the calculus of variations, this equation obviously derives from the condition that the integral M should be a minimum for a fixed value of the integral $\int\limits_{-\infty}^{+\infty}\rho g\sigma^2 \mathrm{d}x$ that approximately gives the energy of the wave. This remarkable property of the quantity M presumably prompted Boussinesq to examine its time evolution for arbitrary swells. He found it to be a constant of motion. From the latter property, he inferred that M measured the departure of a swell from a solitary wave, or the speed at which its shape varied in time. This remark justified the name '*moment d'instabilité*'. It also explained the ease with which Russell and Bazin had produced solitary waves:[107]

> If the moment of instability of a wave slightly exceeds the minimum value, the shape of the swell will oscillate about that of a solitary wave with the same energy, without ever differing much from the latter wave: indeed a notable difference would imply an increase of the moment of instability, which is impossible, since this moment does not

[106]*Ibid.* pp. 76 (slices), 78 (eqn. 2.94), 79 (eqn. 2.95), 87 (moment). Equations (2.88) and (2.93) give $\dot\sigma = -(\sigma w)' = -c_0\sigma\sigma' - \gamma'/2c_0$ (KdV). The latter equation, time derivation under the integral sign of eqn (2.96), and eqn. (2.89) give $\dot M = -6\int\gamma\dot\sigma\mathrm{d}x = 6c_0\int\gamma\sigma'\mathrm{d}x + (3/c_0)\int\gamma\gamma'\mathrm{d}x = 0$, since the integral of the derivative of any function that vanishes at infinity is zero.

[107]*Ibid.* p. 100.

vary in time; or, rather, a solitary wave will soon be formed; because frictional forces, which we have neglected so far, damp the oscillations of the effective form of the swell about its limiting form.... And we may even conceive, in the absence of any stable form about which a wave might oscillate, that any swell susceptible, by its positive and moderate volume, to form a solitary wave with a height small enough not to break, should assume this form after a certain time. Thus is explained the ease with which solitary waves are produced.

2.4.6 *Torrents and tidal bores*

Lastly, Boussinesq used the expression (2.93) for the velocity w of constant-volume slices to determine the evolution of an arbitrary swell. Wherever the curvature σ'' is small compared to σ^2/h^3, this velocity is given by Airy's formula $w = \sqrt{g(h + \frac{3}{2}\sigma)}$. This applies, for instance, to the case of the flat horizontal part of a swell produced by the continuous injection of fluid at one end of a canal. Boussinesq's interest in this case was concerned with the distinction between river and torrents and with the theory of river tides.[108]

In 1870, by elementary reasoning based on momentum conservation, Saint-Venant had shown that a step-shaped swell propagated in a prismatic canal at the Lagrangian velocity $\sqrt{gh}$ in a first approximation, and at the velocity $\sqrt{g(h + \frac{3}{2}\sigma)}$ in a second approximation (σ being the height of the step). On this occasion, he proposed to call the velocity of wave propagation 'celerity' in order to distinguish it from the fluid velocity. Superposing a uniform flow at the velocity $-\sqrt{gh}$ onto this wave motion, he then synthesized a hydraulic jump (*ressaut*), that is, a sudden variation of the height of water on a constant stream. In a stream of velocity inferior to the critical value $\sqrt{gh}$, any such jump must drift in the downstream direction; in a stream of velocity superior to this critical value, jumps recede in the upstream direction. Therefore, when the water encounters an obstacle in the bed of the stream, it tends to accumulate upstream from the obstacle in the subcritical case (the accumulating water forms an upstream moving step); it tends to jump over the obstacle in the supracritical case. The former case defines a river, and the latter a torrent according to Saint-Venant.[109]

A few months later, the *Ponts et Chaussées* engineer Henri Partiot gave a theory of river tides based on Bazin's idea that the tidal flux entered the river through a succession of small step-swells propagating at the Lagrangian velocity for the height of the water they encountered during their progression.[110] Following Bazin, Partiot explained the tidal bore or *mascaret* by the fact that successive step-swells encountered higher and higher levels of water, and therefore propagated at higher and higher velocities. In this process, the later laminas of water catch up with the earlier ones, so that the front of the tidal wave becomes steeper and steeper. For strong tides or rapidly-narrowing beds, it can reach the vertical slope for which breaking occurs.[111]

---

[108]Boussinesq [1871*c*], [1872*b*] pp. 100–3.

[109]Saint-Venant [1870]. See also Chapter 6, pp. 227–9. The connection between gravity waves and the torrent/river distinction is roughly expressed in Darcy and Bazin [1865*a*] p. 34.

[110]Bazin was himself inspired by Théodore Brémontier, who, in 1809, analyzed river tides in terms of successive laminas of water (though without recourse to Lagrange's formula).

[111]Partiot [1871]; Bazin [1865] pp. 633–5.

After reading Partiot, Saint-Venant showed that the same evolution of the level of water along the river resulted from the general equation of non-permanent, gradually-varying flow that he had obtained by applying the momentum law.[112] For small step-swells, this equation retrieves the celerity formula $\sqrt{g(h + \frac{3}{2}\sigma)}$, in apparent contradiction with Russell's and Bazin's $\sqrt{g(h + \sigma)}$ formula. Whereas in his former communication Saint-Venant held friction responsible for the discrepancy, he now understood that the formula of Russell and Bazin applied to situations in which his approximation of gradually-varying flow was not allowed. For Russell, σ represented the height of a solitary wave. For Bazin, it represented the height of the surging head of a step-swell, which happened to be fifty per cent higher than the step itself.

When Boussinesq wrote on solitary waves, he made clear that Saint-Venant's formula only applied to a portion of a wave in which the curvature could be neglected. In the curved part of the swell, convexity implies a decrease of the velocity w, and concavity an increase. Through this simple remark, Boussinesq managed to justify the oscillatory shape of the front of Bazin's swell, as well as the oscillations behind Russell's negative waves. In the end, there was nothing in the multifarious wave phenomena observed by Bazin that Boussinesq could not explain through his powerful analysis. Saint-Venant applauded:[113]

> These numerous results of high analysis, founded on a detailed discussion and on judicious comparisons of quantities of various orders of smallness, sometimes to be kept, sometimes to be neglected or abstracted, and their constant conformity with the results obtained by the most careful experimenters and observers, appear most remarkable to me.

2.4.7 *Rayleigh on the solitary wave*

Five years after Boussinesq's note in the *Comptes rendus*, Lord Rayleigh independently reached the solitary wave equation and profile. With Lagrange and Boussinesq, he shared the idea of developing the fluid velocity in powers of the vertical coordinate y. His implementation of this idea was remarkably elegant, thanks to two subterfuges: he analyzed the fluid motion in a reference system bound to the wave; and he conjointly used Lagrange's potential φ and the stream function ψ such that $-v\,dx + u\,dy = d\psi$. The required power developments are

$$\varphi = \beta - \frac{y^2}{2!}\beta'' + \frac{y^4}{4!}\beta'''' - \ldots, \quad \psi = y\beta' - \frac{y^3}{3!}\beta''' + \ldots. \tag{2.97}$$

The stream line $\psi = 0$ forms the bottom of the channel. In Rayleigh's reference system, the motion is stationary, and the condition that a particle of the fluid surface should remain on this surface is replaced by the condition that this surface should be the stream line $\psi(x, y) = -ch$. The condition of uniform pressure at the free surface is

$$u^2 + v^2 = c^2 - 2g(y - h). \tag{2.98}$$

---

[112]Saint-Venant [1871*a*]. Saint-Venant was apparently unaware of Airy's earlier theory.

[113]Boussinesq [1872*b*] pp. 103–8; Saint-Venant [1873] p. XXI.

Rayleigh then inserted the power developments of φ and ψ into these two conditions, and neglected terms that involved orders of derivation higher than two. This led him to the differential equation

$$\frac{1}{y^2} + \frac{2}{3}\frac{y''}{y} - \frac{1}{3}\frac{y'^2}{y^2} = \frac{1}{h^2} - \frac{2g(y-h)}{c^2h^2} \tag{2.99}$$

for the function $y(x) = \sigma(x) + h$. The first integral of this equation is the same as Boussinesq's equation (2.76). Rayleigh discussed it and integrated it, and obtained results equivalent to those of Boussinesq.[114]

2.4.8 The so-called KdV equation

In a note to his monumental *Eaux courantes*, Boussinesq remarked that his second-order equation (see eqn (2.87))

$$\ddot{\sigma} - c_0^2\sigma'' = \gamma'' \ \left(\text{with } c_0^2 = gh \text{ and } \gamma = \frac{3}{2}g\sigma^2 + \frac{1}{3}gh^3\sigma'' \right)$$

for the deformation of a swell during its propagation could be integrated without recourse to the constant slice-motion, by rewriting it as

$$\left(\frac{\partial}{\partial t} - c_0\frac{\partial}{\partial x}\right)\left(\frac{\partial}{\partial t} + c_0\frac{\partial}{\partial x}\right)\sigma = \gamma'' \approx -\frac{1}{2c_0}\left(\frac{\partial}{\partial t} - c_0\frac{\partial}{\partial x}\right)\gamma'. \tag{2.100}$$

A reasoning similar to that given for the vanishing of the quantity χ of eqn (2.91) leads to the first-order equation

$$\left(\frac{\partial}{\partial t} + c_0\frac{\partial}{\partial x}\right)\sigma = -\frac{\gamma'}{2c_0}, \tag{2.101}$$

or

$$\dot{\sigma} = -\sqrt{gh}\left(\sigma + \frac{3}{4}\frac{\sigma^2}{h} + \frac{1}{6}h^2\sigma''\right)'. \tag{2.102}$$

This is the so-called KdV equation, which Boussinesq wrote some twenty years before its Dutch rediscovery. Rather than this equation, Boussinesq used the equivalent equations (2.93) for w and (2.95) for the convective variation of height, because they represented the deformation of the swell in a more direct manner.[115]

In 1895, the Dutch mathematician Diederik Johannes Korteweg and his doctoral student Gustav de Vries extended Rayleigh's method of 1876 to include oscillatory waves, arbitrary long waves of evolving shape, the effect of capillarity, and an investigation of higher-order terms in the Lagrange–Rayleigh expansion. They thus rediscovered the 'very important equation' that now bears their name, apparently unaware of Boussinesq's relevant study.[116]

---

[114]Rayleigh [1876a] pp. 256–61. Cf. Lamb [1932] pp. 424–6.   [115]Boussinesq [1877] p. 360n.

[116]Korteweg and de Vries [1895] p. 428. These authors gave the evolution of the wave in a reference system moving together with the wave. Hence their equation involved an undetermined constant depending on the celerity

Korteweg and de Vries also extended Rayleigh's derivation to periodic waves of permanent shape, not knowing that Boussinesq had already solved this problem in his *Eaux courantes*. In this case, the condition of constant pressure at the surface involves an undetermined constant, since the disturbance no longer vanishes at infinity. Consequently, the equation (2.76) for the slope of the wave is replaced by

$$\varepsilon'^2 = 3(\varepsilon - a)(\varepsilon - b)(k - \varepsilon), \tag{2.103}$$

where a, b, and k are three positive constants. The integral can be expressed in terms of the elliptic function 'cn', which prompted Korteweg and de Vries to call these periodic waves 'cnoidal'. As they showed, Stokes's finite oscillatory waves are large-depth approximations of the cnoidal waves. Solitary waves correspond to the limit of infinite period.[117]

Korteweg and de Vries believed that the permanence of the shape of their cnoidal waves was preserved at large orders, and gave a tentative proof of this long-debated fact. They thus sided with Stokes, who had believed since 1879 in the existence of waves of permanent type, both solitary and oscillatory. In 1891, Stokes identified the false step which had earlier led to the widespread belief in the impossibility of permanent solitary waves, namely, the assumption that, for a given height, a solitary wave could be so long that the horizontal velocity was the same on a vertical line. This was indeed the starting-point of Airy's theory of the nonlinear deformation of waves. The assumption is wrong, since the length of a solitary wave is determined by its height. Stokes could have added that an argument of his own, according to which, for a given wavelength, the height of a solitary wave could be so small as to undergo finite-depth dispersion, similarly fails. As modern soliton theorists know, the possibility of solitary waves rests on the exact compensation between a linear dispersive term and a nonlinear term in the equation of motion. For a given height of the wave, this compensation only occurs for a definite shape and length.[118]

2.5 The principle of interference

2.5.1 *Group velocity*

In his report on waves of 1844, Russell wrote:

> One observation which I have made is curious. It is that in the case of oscillating waves of the second order, I have found that the motion of propagation of the whole group is different from the apparent motion of wave translation along the surface.

The remark went largely unnoticed, until William Froude privately communicated a similar observation to Stokes and to Rayleigh in the early 1870s.[119]

---

of the wave. Strictly speaking, they did not write Boussinesq's equation (2.102), which is now called the KdV equation. On the precise connection between their equation and the KdV equation, cf. Miles [1981].

[117]Korteweg and de Vries [1895] p. 424; Boussinesq [1877] pp. 390–6. Cf. Lamb [1932] p. 426–7, Miles [1981] p. 137, who notes that Korteweg and de Vries's expression for the relation between cnoidal waves and Stokes's waves is not quite correct.

[118]Korteweg and de Vries [1895] pp. 438–43; Levi-Civita [1925]; Stokes [1880c], [1891].

[119]Russell [1845] p. 67.

At that time, Stokes was working on the measurement of sea waves for the Meteorological Council. In particular, he was asked to determine the origin of the strong swells sometimes observed in fine weather. Stokes immediately explained these swells by wave propagation from distant storms, and commented: 'It is curious to see that captains seem to have so little idea of the propagation of waves excited in a stormy region into a region where as regards the wind, it is comparatively calm.' According to the formula $c = \omega/k = \sqrt{g/k}$ of small deep-water waves, Stokes explained, the velocity c of a periodic wave is related to its time period $\tau = 2\pi/\omega$ through $c = g\tau/2\pi$. A measurement of τ would thus provide information on the location of the storm.[120]

In 1873, William Froude read the relevant section of Stokes's memorandum. He commented to the author:

> *Primâ facie*, the speed of such waves would determine the duration of their passage over a given distance. But this is not really so: because the foremost waves are perpetually dying out, as they invade the undisturbed water, and are undergoing metempsychosis in the ranks behind them.

For example, Froude went on, if the wheels of a paddle ship are stopped while its speed is kept constant by other means, the waves remain stationary with respect to the ship but their front moves away from the ship. From the perspective of an observer at rest, this means that the undulations within the train of waves advance faster than the front of the train. Froude had seen a lot of that in his towing tanks.[121]

In January 1876, Stokes reported to Airy:

> I have lately perceived a result of theory which I believe is new—that the velocity of propagation of roughness on water is, if the water be deep, only half of the velocity of propagation of the individual waves. This is of importance in connecting records of long swells which may be found in ships' logs with records like those of Ascension or St. Helena.

The following month he proposed the following problem for the Smith prize examination papers at Cambridge University:[122]

> Find the expression for the velocity of propagation of a series of simple periodic waves in water of uniform depth, the motion being small and in two dimensions.—If two such series, of equal amplitude and nearly equal wavelength, travel in the same direction, so as to form alternate lulls and roughness, prove that in deep water these are propagated with half the velocity of the waves; and that as the ratio of the depth to the wavelength decreases from ∞ to 0, the ratio of the two velocities increases from $\frac{1}{2}$ to 1.

Denoting by k and $k + \mathrm{d}k$ the wave numbers of the two superposed waves, and ω and $\omega + \mathrm{d}\omega$ the corresponding pulsations, the amplitude of the superposition varies as $\cos\frac{1}{2}(x\mathrm{d}k - t\mathrm{d}\omega)$. The resulting modulation travels with the velocity $\mathrm{d}\omega/\mathrm{d}k$. For small

[120]Stokes to Captain Toynbee, 5 Sept. 1878, in Stokes [1907] vol. 2, p. 141; Stokes to Colonel Sabine, 22 Sept. 1870, *ibid.* p. 136.

[121]Froude to Stokes, 17 Jan. 1873, *ibid.* pp. 156–7.

[122]Stokes to Airy, 5 Jan. 1876, *ibid.* p. 177; Stokes [1876].

waves in water of depth h, according to Kelland and Airy, $\omega^2 = gk \tanh kh$. The corresponding ratio between the group and phase velocities,

$$\frac{d\omega/dk}{\omega/k} = \frac{1}{2}(1 + kh(\tanh kh)^{-1} - kh \tanh kh), \qquad (2.104)$$

varies from $\frac{1}{2}$ to 1 when kh varies from ∞ to 0, as Stokes asked the Smith prize competitors to demonstrate.[123]

The following year, the Manchester engineering professor Osborne Reynolds reported his own observations of wave groups produced by throwing a stone into a pond, by the interference of sea waves, or by the motion of a ship. Like Russell and Froude, he noted that groups of waves in deep water traveled slower than the individual waves of which they were made. To explain this result, he first noted that the velocity of a wave group obviously represented the velocity of propagation of energy. He then showed that the latter velocity differed from the phase velocity. For instance, the waves produced by wind in a corn field obviously do not propagate any energy, since the motions of the individual corn stems are independent. In the more complex case of a sine wave on deep water, the particles of water move on circles with constant velocity, so that no kinetic energy is transmitted by the wave. In contrast, the potential energy is transmitted at the phase velocity. Since the potential energy of such waves is half their total energy, the speed of energy propagation is half the phase velocity. Therefore, the group velocity is half the phase velocity.[124]

In his influential *Theory of sound* of 1877, Rayleigh included Stokes's derivation of the group velocity, which he had independently obtained under Froude's stimulus. In a contemporary article, he proved Reynolds's equality between energy and group velocity in a precise mathematical manner. In the case of small waves on water of finite depth, he did this by computing the ratio between the work of pressure forces on a transverse section of the water and the energy density of the waves. In the general case of waves in an arbitrary dispersive medium, he astutely introduced a fictitious friction proportional to the absolute velocity of the parts of the medium. Assuming vibrational energy to be created at $x = 0$ and to propagate in the direction of increasing x, he computed the damping effect of the frictional force by noting that it turned the operator $\partial^2/\partial t^2$ into $\partial^2/\partial t^2 + \mu\partial/\partial t$, wherein μ is the friction coefficient divided by the fluid density ρ. This is nearly equivalent to changing the pulsation ω into $\omega - \frac{1}{2}i\mu$. The corresponding change of k is $-\frac{1}{2}i\mu dk/d\omega$. Consequently, the oscillating factor $e^{i(\omega t - kx)}$ of a forced oscillation at the pulsation ω is turned into $e^{-\frac{1}{2}\mu x dk/d\omega}e^{i(\omega t - kx)}$. The dissipated energy in the region $x > 0$ is the integral of $\mu\rho v^2$. It is therefore equal to 2μ times the kinetic energy, or else μ times the total energy in this region (according to a well-known theorem for harmonic oscillations). Denoting by E the energy per unit length near the source, this remark leads to the expression $\mu E \int_0^{+\infty} e^{-\mu x dk/d\omega} dx = E d\omega/dk$ for the dissipated energy. By energy conservation, this dissipation must be compensated for by the energy flux $E c_E$ through the section $x = 0$ of the water. Therefore, the velocity c_E of energy propagation must be identical to the group velocity $d\omega/dk$.[125]

---

[123]A more general argument with a continuous distribution of k is found in Rayleigh [1881].

[124]Reynolds [1877b].   [125]Rayleigh [1877], [1877–78].

The concept of group velocity could plausibly have emerged in the fields of physics where dispersion was first known, namely optics and acoustics.[126] In reality it did not. As we have just seen, observations made on deep-water waves played a crucial role. They motivated Stokes's and Rayleigh's theoretical considerations, although these assumed familiarity with interference and beat phenomena in optics and acoustics. As for Froude's understanding of group velocity, it derived from his engineering concern with the energy carried by the waves.

2.5.2   Thomson's fishing line

In early 1871, the catastrophic sinking of the HMS *Captain* prompted the British Admiralty to name a 'Committee on designs for Ships of War'. On behalf of this committee, William Thomson asked his friend Stokes a few questions about waves: 'The longest waves that have been observed?—by whom?—their length from crest to crest?—and height from hollow to crest?' The following summer, while sailing on his personal yacht the *Lalla Rookh*, he observed a gentler but no less interesting phenomenon: a fishing line hanging from the slowly-cruising yacht caused very short waves or 'ripples' directly in front of the line, and much longer waves in its wake. The whole pattern was steady with respect to the line, so that the celerity of both kinds of waves was equal to the velocity of the line's progression through the water. Unknown to Thomson, the French military engineer–mathematician Jean Victor Poncelet had already described this phenomenon with his colleague Joseph Aimé Lesbros, and Scott Russell had already identified capillarity as the cause of the ripples. Thomson was the first, however, to solve the hydrodynamic equations in this case.[127]

In a similar manner to Poisson, Thomson sought solutions of the form $\cos(kx - \omega t)$ for the linearized equations of motion. The only difference with the Lagrange–Poisson conditions for the velocity potential is the substitution of $g\sigma - T\sigma''$ for $g\sigma$ in the pressure equation at the free surface, where T denotes the superficial tension per unit density. Consequently, g must be replaced by $g + k^2 T$ in the dispersion formula $\omega^2 = gk$ for waves on deep water. The corresponding celerity is

$$c = \sqrt{\frac{g}{k} + Tk}. \tag{2.105}$$

Hence, for a given value of the celerity there are two possible values of the wavelength $2\pi/k$, as observed at the front and rear of the fishing line. When c^2 is large compared to $\sqrt{gT}$, the smaller waves approximately obey $c = \sqrt{Tk}$ as capillarity waves would exactly do, and the larger waves approximately obey $c = \sqrt{g/k}$ as gravity waves would exactly do. The formula (2.105) further indicates the existence of a minimum velocity, $\sqrt{gT}$, below which the waves can no longer be formed. Thomson verified this last point on his yacht with the help of an eminent guest, Hermann Helmholtz.[128]

---

[126]On a possible anticipation in William Rowan Hamilton's optics, cf. Lamb [1932] p. 381n.

[127]Thomson to Stokes, 3 March 1871, *ST*; Thomson [1871*b*]; Poncelet [1831]. In the same papers, Thomson treated the wave-generating instability of a water surface under wind, see Chapter 5, pp. 188–90.

[128]Thomson [1871*a*], [1871*c*] p. 88 (Helmholtz).

2.5.3   Rayleigh's solution

Thomson only reasoned on free waves and did not try to analyze the process through which the fishing line caused the waves. Rayleigh accomplished this much more difficult task in 1883. Using a favorite stratagem, he first turned the problem of progressive waves into a steady-wave problem by selecting the reference system bound to the perturbing cause (the fishing line). Then he computed the distribution of surface pressure that corresponds to a sine wave in the restricted two-dimensional problem. Although he only treated the case of infinite depth, the finite-depth formulas are given here to allow a parallel discussion of later related works.[129]

The assumed expressions of the potential φ and the stream function ψ are

$$\frac{\varphi}{c} = x + \alpha \cosh ky \, e^{ikx}, \quad \frac{\psi}{c} = y + i\alpha \sinh ky \, e^{ikx}, \tag{2.106}$$

where α is a small constant (the extraction of the real part of complex expressions is understood). The unperturbed motion ($\alpha = 0$) is a uniform flow at the velocity c in the direction of increasing x. The stream line $\psi = 0$ corresponds to the bottom $y = 0$ of the water. The free surface fits the stream line $\psi = ch$. The corresponding surface deformation is

$$\sigma = -i\alpha \sinh kh \, e^{ikx}. \tag{2.107}$$

The pressure $\overline{P}$ applied on the free surface differs from the fluid pressure by the capillary force $\rho T \sigma''$. As the latter pressure obeys Bernoulli's law, we have

$$\frac{\overline{P}}{\rho} = -g\sigma + T\sigma'' - \frac{1}{2}(u^2 + v^2 - c^2). \tag{2.108}$$

To first order in the small quantity α, this gives

$$\frac{\overline{P}}{\rho} = i\alpha[(g + Tk^2)\sinh kh - c^2 k \cosh kh)]e^{ikx}. \tag{2.109}$$

Consequently, the surface deformation that corresponds to the pressure point

$$\overline{P} = F\delta(x) = \frac{F}{2\pi} \int_{-\infty}^{+\infty} e^{ikx} dx \tag{2.110}$$

of intensity F is

$$\sigma = \frac{F}{2\pi\rho g} \int_{-\infty}^{+\infty} \frac{e^{ikx}}{(1 + Tk^2/g)(c^2/c_k^2 - 1)} dk, \tag{2.111}$$

with

$$c_k^2 = \left(\frac{g}{k} + Tk\right)\tanh kh. \tag{2.112}$$

---

[129]Rayleigh [1883b].

When the wave number k is such that the velocity of the corresponding free wave is equal to the velocity c of the stream, this integral is ill-defined. In order to circumvent this difficulty, Rayleigh introduced a small, fictitious frictional force $\mu(\mathbf{c} - \mathbf{v})$ that damped any free oscillation of the uniform stream. As he had already shown in his *Theory of sound*, Lagrange's theorem for the existence of the potential remains true in the presence of this force. Its only effect on the previous calculation is an additional term $\mu(cx - \varphi)$ in the pressure equation.[130]

From a formal point of view, Rayleigh thus anticipated the adiabatic turning on of the perturbing force that is commonly used in modern scattering theory. Indeed, a slow variation of the coefficient α implies an additional term $-\partial\varphi/\partial t = (\dot\alpha/\alpha)(cx - \varphi)$ in the pressure equation (2.108). This term has exactly the same form as Rayleigh's frictional term.

Taking into account the frictional term, Rayleigh replaced eqn (2.111) by

$$\sigma = \frac{F}{2\pi\rho g} \int_{-\infty}^{+\infty} \frac{e^{ikx}}{(1 + Tk^2/g)(c^2/c_k^2 - 1 - i\varepsilon_k)}\,dk, \qquad (2.113)$$

where ε_k is a small quantity that has the same sign as k. In the case of infinite depth, Rayleigh expressed this integral in terms of elementary or already tabulated functions (the sine integral 'Si'). It is more convenient, however, to retain a large but finite depth (for the integrand to be meromorphic) and to make use of Cauchy's theorem of residues.[131] For positive x, the integration path can be closed in the complex k-plane by the upper half of an infinite circle centered on the origin, as shown in Fig. 2. 24. Hence the integral is given by the sum of the residues in the upper half of the complex k-plane. Symmetrically, for

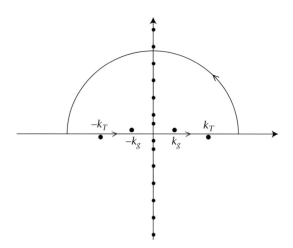

Fig. 2.24.   Integration curve and poles in the complex k-plane for evaluating a certain integral.

---

[130]Rayleigh [1877/78] par. 239.   [131]Cf. Lamb [1895] pp. 396–7; [1932] pp. 406–10.

negative x the integral is given by the sum of the residues in the lower half of the complex k-plane. The poles of the integrand are represented in the figure. The four poles close to the real axis correspond to the two wavelengths for which the celerity of free waves is equal to the velocity of the stream. Two of the poles marked on the imaginary axis correspond to the wavelength for which the free waves have minimum celerity in deep water. The remaining poles on the imaginary axis correspond to the infinite sequence of imaginary wavelengths for which the celerity of free waves is equal to the velocity of the stream. Their distance $|k|$ from the origin is approximately given by the successive zeros of the function $\tan|k|h - (c^2/gh)|k|h$.

The contribution of the imaginary poles is a series of terms that decrease exponentially with x. The physically important terms are the oscillatory terms given by the quasi-real poles. For positive x, the two symmetric poles of larger wavelength contribute an oscillation at this wavelength; for negative x, the contributing poles are those of smaller wavelength. Concretely, the pressure point induces shorter capillary waves upstream, and longer gravity waves downstream, in conformance with Thomson's observations.

A fuller analysis of the wave pattern created by a fishing line requires a three-dimensional analysis. For this purpose, Rayleigh superposed the disturbances produced by pressures constantly applied on straight horizontal lines passing through a fixed point of the water surface, the direction of the line being uniformly distributed. The individual wave patterns are those of the two-dimensional problem. Their wavelengths $2\pi/k$ are such that the corresponding celerity c_k is equal to the projection $c\cos\psi$ of the velocity of the stream on their wave normal. The crests of the various component waves thus form continuous families of straight lines whose distance from the origin is a given function of their orientation. Presumably inspired by an analogy with caustic surfaces in optics, Rayleigh obtained the crests of the combined disturbance as the envelopes of the successive families of straight lines (see Fig. 2.25).[132]

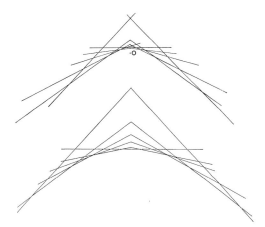

Fig. 2.25.   Rayleigh's construction of the waves created by a drifting fishing-line ([1883b] p. 267). The plane of the figure represents the water surface, the point O the upwards drifting trace of the line. The two families of straight lines represent the first crest of the capillarity (in front of O) and gravity (behind O) waves caused by straight lines of pressure passing through O. Their curved envelope represents the first waves created by the line.

2.5.4 *Houston's paradox solved*

Three years later, Thomson studied the similar problem of the waves produced by a uniformly-moving boat. As he eloquently argued:[133]

> Of all the beautiful forms of water waves that of Ship Waves is perhaps most beautiful, if you can compare the beauty of such beautiful things. The subject of ship waves is certainly one of the most interesting in mathematical science. It possesses a special and intense interest, partly from the difficulty of the problem, and partly from the peculiar complexity of the circumstances concerned in the configuration of the waves.

In the two-dimensional canal case, Thomson pushed the analysis far enough to explain Houston's old towing paradox. He enthusiastically reported to Stokes:[134]

> I have been getting out some very curious things about waves (water), among them complete confirmation of Scott Russell's doctrine of sudden diminution of force, in towing a *boat* in a canal, when the velocity is got to exceed $\sqrt{gh}$. I find (which is now quite obvious) that if water were inviscid, zero force would suffice to keep a boat moving at any constant speed $> \sqrt{gh}$, whether in a canal or in open water.

As Thomson explained in an evening lecture for a popular audience, his theory relied on the group-velocity concept, and on balancing the work produced by the towing force and the energy emitted by the boat in the form of waves. The procession of waves behind a boat, he began, is known to be steady with respect to the boat. Therefore, the phase velocity of this procession must be equal to the velocity of the boat. According to the Kelland–Airy formula (2.52), the former velocity cannot be larger than the velocity $\sqrt{gh}$ of infinitely-long waves. Therefore, the procession can only exist if the boat moves slower than this critical velocity, in conformity with Russell's 'accurate observations and well-devised experiments.'[135]

As the boat must have started from rest, the wave procession necessarily has a finite length. Its end moves with the group velocity, which is smaller than the phase velocity. Therefore, the length and the global energy of the procession increase in time, and an equivalent work must be spent to propel the boat. If the boat moves faster, the procession lengthens at a slower rate but the waves are much higher, so that the resistance grows. A crisis occurs when the velocity of the boat approaches that of infinitely-long waves. 'Once that crisis has been reached,' Thomson asserted, 'away the boat goes merrily.' Thomson then recalled how 'the discovery [had been] made by a horse' and had permitted for a few years a system of fly-boats between Edinburgh and Glasgow on the Forth and Clyde Canal, until, in the early 1840s, the development of railways had rendered this poetical notion of speed obsolete.[136]

In the previous year Thomson had published abundant, complex calculations that justified this theory. The basic mathematical problem was to determine the disturbance

---

[132]Rayleigh [1883*b*].   [133]Thomson [1887*f*] p. 410.   [134]Thomson to Stokes, 8 Nov. 1886, *ST*.
[135]Thomson [1887*f*] pp. 415–20.   [136]*Ibid*. pp. 418–19.

of a uniform flow caused by a local pressure on the water surface of a canal. Thomson first solved the similar problem of the waves produced by a bump at the bottom of the canal when water flows at constant velocity. These two problems resemble Rayleigh's fishing-line problem, except that capillarity is now neglected and the depth is finite. Like Rayleigh, Thomson obtained the desired solution by the superposition of sinusoidal solutions. His execution of this plan seems awkward to a modern reader. Instead of generating the pressure peak by the direct superposition of sine functions, he used the mathematical intermediate of a periodic succession of Lorentzian peaks, and then let the distance between two successive peaks tend to infinity. He encountered enormous difficulties in evaluating the resulting integral for the surface disturbance. His results were, nonetheless, the same as those of the following calculation based on the method of residues.[137]

For the two-dimensional problem of a local pressure disturbing a uniform flow, the disturbance is given by the vanishing-capillarity limit of eqn (2.113):

$$\sigma = \frac{F}{2\pi\rho g} \int\limits_{-\infty}^{+\infty} \frac{e^{ikx}}{c^2/c_k^2 - 1 - i\varepsilon_k}\, dk, \tag{2.114}$$

in which $c_k^2 = (g/k)\tanh kh$. The lowest possible value of c_k^2 is its infinite-wavelength limit gh. Therefore, when c exceeds $\sqrt{gh}$, the integrand only has imaginary poles, and the integral is an exponentially-decreasing function of x. There is no wave production, and the boat can 'travel merrily'. In the opposite case, the integrand has two symmetric, quasi-real poles $\pm k_g + i\varepsilon$ in the upper half of the complex k-plane that yield a downstream undulation of the water surface, with a period equal to the length $2\pi/k_g$ of free waves traveling at the speed c. This explains why waves are produced by a boat at subcritical speed, and why these waves always *follow* the boat. When the speed c is slightly below the critical velocity $\sqrt{gh}$, the two residues are $(3/2k_g h^2)e^{\pm ik_g}$. The amplitude of the resulting oscillations diverges together with their period $2\pi/k_g$ when the pressure point reaches the critical velocity, in conformance with the 'crisis' described in Thomson's popular lecture. In the limit of infinite depth, the two residues are $k_g e^{\pm ik_g x}$, so that the amplitude of the oscillations is inversely proportional to the wavelength. Lastly, Thomson computed the necessary propelling force by balancing the energy flux of the waves with the work done by this force.

2.5.5 *Echelon waves*

Thomson's greatest achievement in this area was to derive the ship-wave pattern in the three-dimensional case.[138] Like Rayleigh, Thomson superposed the disturbances produced by pressures constantly applied on straight horizontal lines passing through a fixed point O of the water surface, to be identified with the location of the boat. Had he followed Rayleigh even further, he could have obtained the wave pattern geometrically, by

---

[137]Thomson [1886].

[138]Thomson [1887*f*], [1906]. In 1887, Thomson only gave the formulas for the configuration of the wave crests, which he claimed to have obtained by Stokes's principle of group velocity ([1887*f*] p. 423). In the following it is assumed that the relevant calculations were similar to those of Thomson [1906]. One could speculate that Thomson reasoned in the more elementary manner given at the end of this chapter. That manner, however, does not seem to yield the height of the waves, which Thomson claimed to have computed in 1887.

tracing the envelopes of the component wave crests.[139] However, he preferred a more analytical method that also yielded the intensity of the waves.

Denote by r and θ the polar coordinates of the point P of the wake with respect to the origin O and to the trajectory of the boat, and by ψ the angle that the rearward normal of one of the pressure lines makes with the axis $\theta = 0$. The distance δ of the point P from this line is $r \cos(\psi - \theta)$. The wave number k of the resulting wave component must be such that the corresponding wave velocity $\sqrt{g/k}$ is equal to the projection $V \cos \psi$ of the velocity V of the boat on the wave normal. Hence the phase ϕ of this component at the point P is

$$\phi = k\delta = \frac{gr}{V^2} \frac{\cos(\psi - \theta)}{\cos^2 \psi}. \tag{2.115}$$

Its amplitude is proportional to $k = g/V^2 \cos^2 \psi$. The angle ψ is uniformly distributed between $\theta - \pi/2$ and $\theta + \pi/2$, since P must belong to the wake of the ψ line. The resultant disturbance has the form[140]

$$\sigma \propto \int_{\theta-\pi/2}^{\theta+\pi/2} \frac{\cos \phi}{\cos^2 \psi} \, d\psi. \tag{2.116}$$

In order to evaluate this integral, Thomson appealed to 'the principle of interference, as set forth by Prof. Stokes and Lord Rayleigh in their theory of group-velocity and wave-velocity.'[141] At a distance from the boat much larger than the characteristic wavelength $\lambda = g/V^2$, the phase ϕ is very large and therefore $\cos \phi$ oscillates very quickly between positive and negative values when ψ varies. This oscillation implies destructive interference, unless there are particular values of ψ for which the phase is stationary, that is, $d\phi/d\psi = 0$.

If x and y denote the Cartesian coordinates of the point P, and τ is the tangent of the angle ψ, we have

$$\phi = \frac{2\pi}{\lambda}(x + y\tau)(1 + \tau^2)^{1/2}. \tag{2.117}$$

The condition of stationary phase then gives

$$x\tau + y(1 + 2\tau^2) = 0. \tag{2.118}$$

This quadratic equation has real roots only if $(y/x)^2 < 1/8$. Hence the disturbance is confined between the two half-lines that originate in the (point-like) boat and make an angle of $\tan^{-1}\sqrt{1/8} \approx 19°28'$ with the mid-wake of the boat. The curves of constant phase obey the parametric equations[142]

$$x = a(1 + 2\tau^2)(1 + \tau^2)^{-3/2}, \quad y = -a\tau(1 + \tau^2)^{-3/2}, \tag{2.119}$$

---

[139]Lamb did so in the 1895 edition of his treatise. The equation of the envelope is easily seen to be identical to the condition of stationary phase that Thomson presumably used in 1887.

[140]Thomson [1906] p. 409.   [141]Thomson [1887g] p. 303.

[142]The formulas of Thomson [1887f] have a different parameter, $w = (1 - 2\tau^2)/(1 + 2\tau^2)$, but represent the same curves, despite Larmor's contrary statement ([1907] p. 413n).

SHIP WAVES.

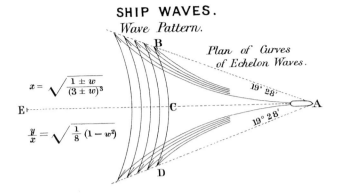

Wave Pattern.

Plan of Curves of Echelon Waves.

$$x = \sqrt{\frac{1 \pm w}{(3 \pm w)^3}}$$

$$\frac{y}{x} = \sqrt{\frac{1}{8}(1-w^2)}$$

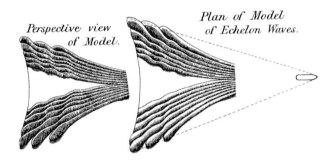

Perspective view of Model.

Plan of Model of Echelon Waves.

Perspective view of Echelon Waves.

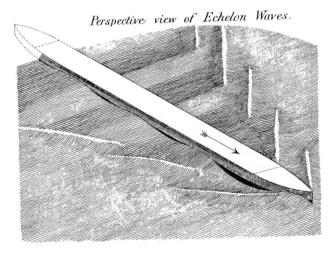

Fig. 2.26.   Kelvin's ship-waves (Thomson [1887*f*] plate; perspective view borrowed by Kelvin from R.E. Froude: Thomson must have considered that the impulse of the prow of the long barge approximately determined the wave pattern).

where $a = (\lambda/2\pi r)\phi$. They have the 'beautiful' shape represented in Fig. 2.26. Thomson further determined the amplitude of the waves by summing the contributions of the two roots of eqn (2.118) to the integral (2.116). He even had a clay model made to represent the wave pattern.[143]

2.5.6 *The stationary-phase method*

Thomson's success in completing the theoretical analysis of ship waves crucially depended on the stationary-phase method. An anticipation of this method is found in a mathematical paper of 1850 by Stokes, in the context of Airy's spurious rainbows.[144] Stokes did not explain why the procedure worked, and did not provide a clear criterion for judging which integral was amenable to it. In contrast, Thomson imagined the destructive interference of the rapidly-oscillating integrand, and showed that the method applied to the integral of any rapidly-oscillating function. In 1887, he published a striking application of this method to the Poisson–Cauchy integral given in eqn (2.30), namely

$$\sigma = \frac{A}{\pi} \int_0^{+\infty} dk \cos kx \cos \omega_k t,$$

that represents, in two dimensions, the water-surface disturbance caused by a local deformation around $x = 0$.[145]

The progressive part of this disturbance is the real part of the integral

$$\sigma_+ = \frac{A}{2\pi} \int_0^{+\infty} dk e^{i(kx - \omega_k t)}. \tag{2.120}$$

The phase is stationary when $x\, dk - t\, d\omega_k = 0$, that is, when the group velocity $d\omega_k/dk$ is equal to x/t. For gravity waves on infinitely-deep water, $\omega_k = \sqrt{gk}$. The phase is stationary for $k = \kappa = gt^2/4x^2$. Its value around this stationary point is

$$\phi \approx -\frac{gt^2}{4x} + \beta(k - \kappa)^2, \tag{2.121}$$

where $\beta = x^3/gt^2$ is the value of $\frac{1}{2}d^2\phi/dk^2$ for $k = \kappa$. The resulting approximation of the integral for large values of $gt^2/4x$ is

$$\sigma_+ = \frac{A}{2\pi} e^{-igt^2/4x} \int_{-\infty}^{+\infty} dk e^{i\beta(k-\kappa)^2} = \frac{A}{2\pi} e^{-igt^2/4x} \sqrt{\frac{\pi}{\beta}} e^{i\pi/4} = \frac{At}{2}\sqrt{\frac{g}{\pi x^3}}\, e^{-i(gt^2/4x - \pi/4)}, \tag{2.122}$$

in conformance with Poisson's equation (2.36).

In the same spirit, Horace Lamb later noted that, in the large-phase approximation and for relatively small variations of the distance x from the origin, the disturbance at a given

---

[143]Thomson [1906] p. 413; [1887*f*] p. 424.

[144]Stokes [1850*a*]; Airy [1838]. Cf. Darrigol [2003] pp. 85–6.

[145]Thomson [1887*g*].

time differs very little from a sine wave with the wave number $k = gt^2/4x^2$. Therefore, the disturbance created by a non-local deformation with the profile $f(a)$ results from the interference of a system of sine waves with an amplitude proportional to $f(a)$ and a phase shifted by $-ka$. As in diffraction theory, this picture leads to replacing the coefficient A by the Fourier transform of the profile in the expression (2.122) of the disturbance created by a point-like perturbation. Lamb thus short-circuited Poisson's delicate and lengthy derivation of eqn (2.38).[146]

In summary, Thomson and Lamb substituted a physico-mathematical analysis for the purely formal developments of Poisson, Cauchy, and Stokes. The gain was enormous: whereas the anterior treatment rested on formal tricks that required much ingenuity and worked only in particular cases, the stationary-phase method offered an intuitive strategy that automatically gave the asymptotic behavior of large-phase integrals. Under Thomson's magic wand, much of the enormous memoirs of Cauchy and Poisson collapsed into a few lines of physico-mathematical common sense. Moreover, the formidable problem of ship waves received a strikingly simple solution.

2.5.7   After-math

This solution was not quite definitive. Although Thomson gave the result and the method of stationary phase in 1887, he only published the calculations in 1906, at age 83. As he knew, the pressure obtained by isotropic superposition of pressures localized on straight lines passing through a fixed point varies as the inverse of the distance from this point. In modern notation, this results from the identity $\int_0^{2\pi} \delta(r\cos\theta)\mathrm{d}\theta = 2/r$. Such a slowly-decreasing function cannot realistically represent the pressure exerted by a boat on the water surface. In the year of his death, Thomson was still working on an improved version of his theory in which the perturbation was more sharply localized.[147]

In his paper of 1887, Thomson suggested a more direct approach. The disturbance produced by the ship, he noted, may be regarded as the superposition of the disturbances produced by a succession of impulses along its path. A Newcastle lecturer in applied mathematics, Thomas Havelock, managed to do the corresponding calculations in 1908, thanks to a repeated application of the method of stationary phase. At a point P in the wake of the ship A (see Fig. 2.27), the wave created by an individual impulse is the superposition of monochromatic, circular waves with the phase $\omega\tau - kd$ (up to a constant), where k is the wave number, d is the distance EP between the impulse and the point P, ω is the deep-water pulsation $\sqrt{gk}$, and t is the time that has elapsed since the ship was at E.[148]

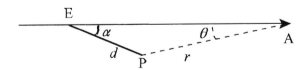

Fig. 2.27.   Diagram for Havelock's calculation of Kelvin's ship-wave pattern.

---

[146]Lamb [1932] pp. 392–4.   [147]Thomson [1907].

[148]Thomson [1887f]; Havelock [1908]. See also Lamb [1916] and Lamb [1932] pp. 433–7.

In order to avoid destructive interference between the waves created by successive impulses, this phase must be stationary with respect to a variation of the time t. Hence the phase velocity ω/k must be equal to $\dot{d} = V \cos\alpha$, where V is the velocity of the boat and α is the angle that EP makes with the direction of motion of the ship. Furthermore, the phase must be stationary with respect to a variation of k. This implies that the group velocity $d\omega/dk$ must be equal to the ratio d/t.

The first condition of stationarity and some trigonometry lead to the expression

$$\phi = \frac{gr}{V^2} \frac{\cos(\alpha - \theta)}{\cos^2\alpha} \tag{2.123}$$

for the phase, where r is the distance from P to the ship's present location A, and θ is the angle that AP makes with the direction of motion of the ship. This expression is the same as Thomson's equation (2.115), although the angles α and ψ have different interpretations. As the variation with respect to k is equivalent to a variation with respect to α, Havelock's calculation yields the same lines of constant phase as Kelvin's. Only the height of the wave crests differs, because the amplitude of Havelock's spherical component waves differs from the amplitude of Kelvin's straight-line component waves.[149] This correction does not really improve the comparison with the experimental pattern, because the latter depends on the form of the ship, and because at the cusps of the curves of constant phase the theory leads to a divergent amplitude that is incompatible with the original small-wave assumption.[150]

In Kelvin's and Havelock's derivations of the echelon shape of ship waves, there seems to be a disproportion between the simplicity of the results and the complexity of the calculations. In his popular lecture of 1887, Thomson hinted at a more elementary derivation. After noting that the disturbance produced by the ship could be regarded as the superposition of the disturbances produced by a succession of impulses along its path, he declared that the point E in Fig. 2.26(top) (such that EC = CA) represented the position of the ship at the time when it caused the impulse responsible for the disturbance around C. His justification holds in one sentence: 'Calculate out the result from the law that the group-velocity is half the wave-velocity—the velocity of a group of waves at sea is half the velocity of the individual waves.' Indeed, if the disturbance travels from E to C at the group velocity, and if the phase velocity along the x-axis is equal to the velocity of the ship, this law implies that the ship must move twice as fast as the disturbance. Thomson seems to have grasped these two conditions intuitively, through the picture of a train of waves made of individual waves that are steady with respect to the ship. As we have just seen, they can be justified through the method of stationary phase.[151]

Thomson's consideration may now be extended to the disturbance around a point P that is no longer on the x-axis. This disturbance has traveled from E with the group velocity in the direction EP. For steadiness with respect to the ship, the phase velocity in this direction must be equal to the projection of the ship velocity on this direction. Hence the angle EPC

---

[149]This agreement should be expected, because the disturbance created by a diffuse pressure is the superposition of geometrically-similar echelon patterns created by symmetrically-distributed pressure points.

[150]On more realistic theories of ship waves, cf. Lamb [1932] pp. 437–9.

[151]Thomson [1887f] p. 426.

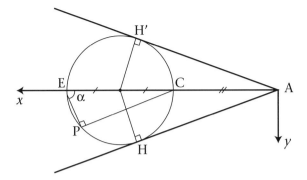

Fig. 2.28.   Diagram for elementary calculation of Kelvin's ship-wave pattern.

is a right angle, and the point P must be located on the circle of diameter EC (see Fig. 2.28). The significantly disturbed part of the water surface is therefore confined between the tangents AH and AH′, which make the angle $\sin^{-1}(\mathrm{OH/OA}) = \sin^{-1} 1/3 = \tan^{-1}\sqrt{1/8}$ with the axis.[152]

Denoting by X the diameter EC and α the angle that the emission line EP makes with the axis, the Cartesian coordinates of the point P are (see Fig. 2.28)

$$x = X(2 - \cos^2 \alpha), \quad y = X \sin \alpha \cos \alpha \qquad (2.124)$$

On a given curve of constant phase, y is a function of x; or, equivalently, X is a function of α. This function can be determined through the condition $\mathrm{d}y/\mathrm{d}x = \cot\alpha$, which means that the curve of constant phase that goes through P is normal to the direction EP of propagation. Computing $\mathrm{d}y/\mathrm{d}x$ from the previous expressions for x and y, we obtain

$$\frac{\mathrm{d}X}{\mathrm{d}\alpha} = -X \tan \alpha \qquad (2.125)$$

The integral $X = a\cos\alpha$ of this equation then gives

$$x = a\cos\alpha(2 - \cos^2 \alpha), \quad y = a\sin\alpha\cos^2\alpha, \qquad (2.126)$$

which are the same as eqns (2.119) with $\tau = \tan\alpha$.[153]

The extreme simplicity of this derivation strikingly illustrates the transformation of mathematical physics announced in the introduction to this chapter. In 1775, Laplace already knew the equations of hydrodynamics that are needed to formulate the ship-wave problem mathematically. Had he dared to approach this problem, he would probably have

---

[152]This reasoning is from Lighthill [1957] pp. [21–2, [1978] pp. 269–79. See also Billingham and King [2000] pp. 99–105. Thomson ([1887f] pp. 425–7) gives this geometrical construction of the characteristic angle, without the physical interpretation.

[153]The form $X = a\cos\alpha$ of the constant-phase condition also derives from $\phi = \omega t - kd$ (with $t = 2X/V$, $d = X\cos\alpha$, $V\cos\alpha = \omega/k = \sqrt{g/k}$), which leads to $\phi = gX/V^2 \cos\alpha$.

fallen into the same error as in the waves-by-emersion problem, for he did not know how to synthesize local perturbations from sinusoidal ones. Some forty years later, Poisson and Cauchy could have written the multiple integral that yields the water disturbance behind the ship. But they lacked efficient means to evaluate this integral. Ninety years later, Thomson succeeded in this task thanks to 'the principle of interference'. Through the related intuition of wave groups, he even suggested a way to circumvent the integral and reason in geometric terms.

This story exemplifies a symbiotic evolution of mathematical analysis and physical interpretation in the nineteenth century. The need to solve the differential equations of physics problems such as the propagation of heat inspired new mathematical tools such as Fourier analysis. In turn, the application of these tools to a broad array of physical phenomena provided them with physical interpretations that suggested more efficient ways of handling them. From raw, algebraic procedures for combining and transforming mathematical expressions, they became genuine physico-mathematical tools. Whereas in their more primitive guise they often generated impenetrable integrals, in their mature form they revealed the behavior of the integrals.

This evolution largely explains the success of nineteenth-century theorists in dealing with complex wave patterns in the linear approximation. That Stokes, Boussinesq, and Rayleigh could also solve an important class of nonlinear problems depended on another quality, namely, their ability to develop methods of approximation that combined two different small parameters, the slope and the elevation of the waves. In both cases, a century elapsed between the basic formulation of the problem in Lagrange's memoir of 1781 and a fairly complete mastery of the observed wave behaviors. Although this may seem a long time, it is less than what was needed for a fragmentary answer to other hydrodynamic questions.

With hindsight, there are three peculiarities of water-wave motion that make it more easily amenable to mathematical analysis than other forms of fluid motion. Firstly, it can be studied with reasonable accuracy without taking into account the small viscosity of water. Secondly, in the same approximation it can be regarded as irrotational (except for Gerstner's waves) and therefore admits a harmonic velocity potential. Thirdly, it is stable and non-turbulent, except in the limit of breaking waves. We will now leave this relatively simple domain and enter more troubled waters.

3

VISCOSITY

> M. Navier himself only gives his starting principle as a hypothesis that can be verified solely by experiment. If, however, the ordinary formulas of hydro-dynamics resist analysis so strongly, what should we expect from new, far more complicated formulas?[1] (Antoine Cournot, 1828)
>
> As far as I can see, there is today no reason not to regard the hydrodynamic equations [of Navier and Stokes] as the exact expression of the laws that rule the motions of real fluids.[2] (Hermann Helmholtz, 1873)

In the early nineteenth century, the rational fluid mechanics of d'Alembert, Euler, and Lagrange remained irrelevant to the mundane problems of pipe flow and ship resistance. Engineers had their own empirical formulas, and mathematicians their own paper theory of perfectly unresisted flow. A similar contrast existed in the case of elasticity: the formulas established by mathematicians for the flexion of prisms were of little help in evaluating the limits of rupture in physical constructions. In the 1820s and 1830s, a new breed of French engineer–mathematicians trained at the Ecole Polytechnique, mainly Navier, Cauchy, and Saint-Venant, struggled to fill this gap between theory and practice. As a preliminary step toward a more realistic theory of elasticity, in 1821 Navier announced the general equations of equilibrium and motion for an (isotropic, one-constant) elastic body. Transposing his reasoning to fluids, he soon obtained a new hydrodynamic equation for viscous flow, namely the Navier–Stokes equation.

Navier's latter theory received little contemporary attention. The Navier–Stokes equation was rediscovered or rederived at least four times, by Cauchy in 1823, by Poisson in 1829, by Saint-Venant in 1837, and by Stokes in 1845. Each new discoverer either ignored or denigrated his predecessors' contribution. Each had his own way to justify the equation, although they all exploited the analogy between elasticity and viscous flow. Each judged differently the kind of motion and the nature of the system to which it applied. The comparison between the various derivations of this equation—or of the equations of motion of an elastic body—brings forth important characteristics of mathematical physics in the period 1820–1850.

A basic methodological and ontological issue was the recourse to molecular reasoning. Historians have often perceived an opposition between Laplacian molecular physics on the one hand, and macroscopic continuum physics on the other, with Poisson being the champion of the former physics, and Fourier the champion of the latter. Closer studies of Fourier's heat theory have shown that the opposition pertains more to the British reading of this work than to its actual content. Fourier actually combined molecular intuitions

---

[1] Cournot [1828] p. 13.

[2] Helmholtz [1873] p. 158.

with more phenomenological reasoning. Viscous-fluid and elastic-body theorists similarly hybridized molecular and continuum physics. Be they engineers or mathematicians, they all agreed that the properties of real, concrete bodies required the existence of non-contiguous molecules. However, they differed considerably over the extent to which their derivations materially involved molecular assumptions.

At one extreme was Poisson, who insisted on the necessity of discrete sums over molecules. At the other extreme was Cauchy, who combined infinitesimal geometry and spatial symmetry arguments to define strains and stresses and to derive equations of motion without referring to molecules. Yet the opposition was not radical. Poisson relied on Cauchy's stress concept, and Cauchy eventually provided his own molecular derivations. Others compromised between the molecular and the molar approach. Navier started with molecular forces, but quickly jumped to the macroscopic level by considering virtual works. Saint-Venant insisted that a clear definition of the concept of stress could only be molecular, but nevertheless provided a purely macroscopic derivation of the Navier–Stokes equation. Stokes obtained the general form of the stresses in a fluid by a Cauchy type of argument, but he justified the linearity of the stresses with respect to deformations by reasoning on hard-sphere molecules.

These methodological differences largely explain why Navier's successors ignored or criticized his derivation of the Navier–Stokes equation. His short cuts from the molecular to the macroscopic levels seemed arbitrary or even contradictory. Cauchy and Poisson simply ignored Navier's contribution to fluid dynamics. Saint-Venant and Stokes both gave credit to Navier for the equation, but believed an alternative derivation to be necessary. To this day, Navier's contribution has been constantly belittled, even though his approach was far more consistent than a superficial reading may suggest.

This wide spectrum of methodological attitudes, both in fluid mechanics and in elasticity theory, corresponds to different views of mathematical rigor and different degrees of concern with engineering problems. Navier's way of injecting physical intuition into mathematical derivations was alien to Cauchy and Poisson, who were the least involved in engineering and the most versed in higher mathematics. Yet many engineers judged Navier's approach too mathematical and too idealized. Personal ambitions and priority controversies enhanced, and at times even determined, the disagreements. Acutely aware of these tensions, Saint-Venant developed innovative strategies that combined the demands of mathematical rigor and practical usefulness.

The many fathers of the Navier–Stokes equation also differed in the types of application they envisioned. Navier and Saint-Venant had pipe and channel flow in mind. Cauchy's and Poisson's interests were more philosophical than practical. Cauchy did not even intend the equation to be applied to real fluids; he derived it for 'perfectly inelastic solids', and noted its identity with Fourier's heat equation in the limiting case of slow motion. Stokes was motivated by British geodesic measurements that required aerodynamic corrections to pendulum oscillations.

To Navier's disappointment, his equation worked well only for slow, regular motions, as occurs around pendulums and within capillary tubes. In most hydraulic cases, there seemed to be no alternative to the empirical approach of engineers. It was not even clear whether the Navier–Stokes equation could be maintained. Many years elapsed before this equation acquired the fundamental status that we now ascribe to it.

The first section of this chapter is devoted to the hydraulic failure of Euler's hydrodynamics, and to Girard's study of flow in capillary tubes, on which Navier relied. Section 3.2 describes Navier's achievements in the theory of elasticity, their transposition to fluids, and the application to Girard's tubes. Section 3.3 discusses Cauchy's stress–strain approach to elasticity and its adaptation to a 'perfectly inelastic solid'. Section 3.4 recounts Poisson's struggle for rigor in the molecular approach, Cauchy's own implementation of the same approach, and Navier's response to Poisson's attacks. Section 3.5 concerns Saint-Venant's unique brand of applied mechanics, and his contributions to elasticity and hydraulics. The Final Section, Section 3.6, deals with Hagen's and Poiseuille's experiments on narrow-pipe discharge, and their long-delayed explanation by the Navier–Stokes equation.

3.1 Mathematicians' versus engineers' fluids

3.1.1 *Resistance and retardation*

As we saw in Chapter 1, d'Alembert regarded the vanishing of fluid resistance in his theory as a challenge for future geometers. As a possible clue to this paradox, he evoked an asymmetry of the fluid motion (around a rear–front symmetric body) owing to 'the tenacity and the adherence of the fluid particles'. However, he did not try to formalize this effect, presumably because he regarded the molecular interactions as too complex to yield well-defined mathematical laws at the macroscopic level.[3]

Well before the fundamental equations of fluid motion were known, Euler had already shown that momentum balance, when applied to the tubes of flow around the immersed body, led to a vanishing resistance. In a modernized version of his argument, the momentum gained by the immersed body (whatever its shape may be) in a unit of time should be equal to the difference of momentum fluxes across normal plane surfaces situated far ahead and far behind the body; this difference vanishes because of the equality of velocity and mass flux on the two surfaces.[4]

Yet Euler, unlike d'Alembert, did not conclude that the theoretical resistance necessarily vanished. For obscure reasons, he believed that only the front part of the tubes of flow contributed to the momentum balance. This assumption, which he still defended in 1760, yields results similar to Edme Mariotte's and Issac Newton's old theories of fluid resistance, according to which the impact of fluid particles on the front of the immersed body completely determines the resistance. The true form of the flow and the shape of the rear of the body do not matter in such theories. Although their experimental inexactitude and their *ad hoc* character were already recognized in Euler's day, they remained popular until the beginning of the nineteenth century for the lack of any better theory.[5]

---

[3]D'Alembert [1780] p. 211. Cf. Saint-Venant [1887b] p. 10. The fluid resistance data used in 1877 by the Academic Commission for the Picardie Canal, to which d'Alembert belonged, were purely empirical, cf. Redondi [1997].

[4]Euler [1745] chap. 2, prop. 1, rem. 3 (French transl. pp. 316–17). Cf. Saint-Venant [1887b] pp. 29–31. A more rigorous reasoning would have required a cylindrical wall to limit the flow laterally, together with a proof that the works of pressure forces on the two plane faces of the cylinder are equal and opposite. This cancellation results from the equality of pressures on the two faces, which itself derives from Bernoulli's theorem or from the conservation of live force. Compare with Saint-Venant's proof of 1837, discussed on p. 134.

[5]Euler [1760b]. Cf. Saint-Venant [1887b] pp. 34–6. On Newtons' theory, cf. *ibid.* pp. 15–29, G. Smith [1998], Chapter 7, pp. 265–6.

The great geometers of the eighteenth century were even less concerned with the hydraulic problems of pipe and channel flow than with fluid resistance. Available knowledge in this field was mostly empirical. Since Mariotte's *Traité du mouvement des eaux* (1686), hydraulic engineers assumed a friction between running water and walls, proportional to the wetted perimeter, and increasing faster than the velocity of the water. This velocity was taken to be roughly uniform in a given cross-section of the pipe or channel, in conformance with common observation. Claude Couplet, the engineer who designed the elaborate water system of the Versailles castle, performed the first measurements of the loss of head in long pipes of various sections. Some fifty years later, Charles Bossut, a Jesuit who taught mathematics at the engineering school of Mézières, performed more precise and extensive measurements of the same kind.[6]

So did his contemporary Pierre Du Buat, an engineer with much experience in canal and harbor development, and the author of a very influential hydraulic treatise. Du Buat's superiority rested on a sound mechanical interpretation of his measurements. He was the first, in print, to give the condition for steady flow by balancing the pressure gradient (in the case of a horizontal pipe) or the parallel component of fluid weight (in the case of an open channel) with the retarding frictional force. He took into account the loss of head at the entrance of pipes (due to the sudden increase in velocity), whose neglect had flawed his predecessors' results for short pipes. Lastly, he proved that fluid friction, unlike solid friction, did not depend on pressure.[7]

Bossut found the retarding force to be proportional to the square of the velocity, and Du Buat found it to increase somewhat slower than that with velocity. Until the mid-nineteenth century, German and French retardation formulas were usually based on the data accumulated by Couplet, Bossut, and Du Buat. In 1804, the Directeur of the Ecole des Ponts et Chaussées, Gaspard de Prony, provided the most popular formula, which made the friction proportional to the sum of a quadratic and a small linear term. The inspiration for this form came from Coulomb's study of fluid coherence, to be discussed shortly.[8]

3.1.2   *Fluid coherence*

For Du Buat's predecessors, the relevant friction occurred between the fluid and the walls of the tube or channel. In contrast, Du Buat mentioned that viscosity was needed to check the acceleration of internal fluid filaments. He observed that the average fluid velocity used in the retardation formulas was only imaginary, that the real flow velocity increased with the distance from the walls, and even vanished at the walls in the case of a very small flux. The molecular mechanism he suggested for the resistance implied the adherence of fluid molecules to the walls, so that the retardation truly depended on internal fluid processes. Specifically, Du Buat imagined that the adhering fluid layer impeded the motion of the rest of the fluid, partly as a consequence of molecular cohesion, and mostly

[6]Mariotte [1686] part 5, discourse 1. Cf. Saint-Venant [1887*b*] pp. 39–40, Rouse and Ince [1957] pp. 114 (Couplet), 126–8 (Bossut).

[7]Du Buat [1786], vol. 1, pp. xvii, 14–15, 40. Cf. Saint-Venant [1866], Rouse and Ince [1957] pp. 129–34. In 1775, Antoine Chézy had already given the condition of steady motion in an unpublished report for the Yvette Canal (cf. *ibid.* pp. 117–20). More will be said on Bossut and Du Buat in Chapter 6, pp. 221–2.

[8]Cf. Rouse and Ince [1957] pp. 141–43. In 1803, Girard had used a non-homogenous $v + v^2$ formula, also inspired by Coulomb.

because of the granular structure of this layer. This structure implied a 'gearing' of traveling-along molecules (*engrenage des molécules*), through which they lost a fraction of their momentum proportional to their average velocity, at a rate itself proportional to this velocity. Whence came the quadratic behavior of the resistance.[9]

In 1800, the military engineer Charles Coulomb used his celebrated torsional-balance technique to study the 'coherence of fluids and the laws of their resistance in very slow motion'. The experiments consisted of measuring the damping of the torsional oscillations of a disk suspended by a wire through its center and immersed in various fluids. Their interpretation depended on Coulomb's intuition that the coherence of fluid molecules implied a friction proportional to the velocity, and that surface irregularities implied an inertial retardation proportional to the square of the velocity. In conformance with this view, Coulomb found that the quadratic component depended only on density and that the total friction became linear for small velocities. From his further observation that greasing or sanding the disk did not alter the linear component, he concluded:[10]

> The part of the resistance which we found to be proportional to the velocity is due to the mutual adherence of the molecules, not to the adherence of these molecules with the surface of the body. Indeed, whatever be the nature of the plane, it is strewn with an infinite number of irregularities wherein fluid molecules take permanent residence.

Although Du Buat's and Coulomb's emphasis on internal fluid friction or viscosity was exceptional in their day, the notion was far from new. Newton had made it the cause of the vortices induced by the rotation of an immersed cylinder, and he had even provided a derivation (later considered to be flawed) of the velocity field around the cylinder. He assumed (in conformance with later views) that the friction between two consecutive, coaxial layers of the fluid was proportional to their velocity difference. After a century during which this issue was virtually ignored, in 1799 the Italian hydraulic engineer Giovanni Battista Venturi offered experiments that displayed important effects of internal fluid friction.[11]

Venturi intended to prove 'the lateral communication of motion in fluids' and to show its consequences for various kinds of flow. Some of the effects he described, such as the increase of efflux obtained by adding a divergent conical end to the discharging pipe, were purely inertial effects already known to Daniel Bernoulli. Others, such as the formation of eddies, genuinely depended on internal friction. The eddies that Leonardo da Vinci had beautifully drawn for the flow past immersed bodies, those evoked by Daniel Bernoulli for sudden pipe enlargement, or those commonly seen in the smoke from chimneys or in rivers behind bridge pillars, were all due, Venturi explained, to 'motion communicated from the more rapid parts of the stream to less rapidly moving lateral parts' (see Fig. 3.1).

---

[9]Du Buat [1786] vol. 1, pp. 22, 39–41, 58–59, 89–90. Du Buat's notion of fluid viscosity or cohesion was not quite identical with internal friction as we now understand it. Du Buat meant an 'adhesion' of the molecules that needed to be overcome to separate them, the resistance to this separation being proportional to its suddenness. He believed (*ibid.* p. 41) that the microscopic structure of the surface of the pipe or channel had no effect on the retardation, for it was hidden by the adhering layer of fluid.

[10]Coulomb [1800] pp. 261 (two kinds of resistance), 287 (quote). Cf. Gillmor [1971] pp. 165–74.

[11]Newton [1687] book 2, prop. 51; Venturi [1797]. Cf. Saint-Venant [1887b] pp. 41–4 and Dobson [1999] (Newton), Rouse and Ince [1957] pp. 134–37 (Venturi).

(a)

(b)

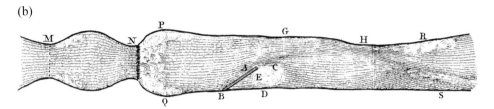

Fig. 3.1.   Eddy formation according to (a) da Vinci and (b) Venturi (from Rouse and Ince [1957] p. 46 and
Venturi [1797] plate).

Accordingly, Venturi made eddy formation one of the principal causes of retardation in
rivers, which current wisdom attributed to friction against banks and the bottom.[12]

Venturi prudently avoided deciding whether the lateral communication of motion was
occasioned 'by the viscidity or mutual adhesion of the parts of the fluids, or their mutual
engagement or intermixture, or the divergence of those parts which are in motion.' Nor did
he venture to suggest new equations of fluid motion. As he explained in his introduction,

> The wisest philosophers have their doubts with regard to every abstract theory
> concerning the motion of fluids: and even the greatest geometers avow that those
> methods which have afforded them such surprising advances in the mechanics of
> solid bodies, do not afford any conclusions with regards to hydraulics, but such as are
> too general and uncertain for the greater number of particular cases.

Venturi's memoir enjoyed a favorable review by the French Academicians Bossut, Coulomb, and Prony. Together with Du Buat's and Coulomb's works on fluid friction, it
contributed to revive the old Newtonian notion of friction between two contiguous layers
of fluid.[13]

3.1.3   Girard's capillary tubes

In 1816, the Paris water commissioner and freshly-elected Academician Pierre-Simon
Girard applied Newton's notion to a six-month-long study of the motion of fluids in
capillary tubes. While his prominent role in the construction of the *Canal de l'Ourcq* and
his contribution to several hydraulic projects amply justified his interest in flow retardation, Girard had the more philosophical ambition of participating in Laplace's novel
molecular physics. He believed the same molecular cohesion forces to be responsible for
the capillarity phenomena analyzed by Laplace and for retardation in pipe flow. By

---

[12]Venturi [1797] transl. in Tredgold [1826] p. 165.

[13]*Ibid.* pp. 132–33, 129; Prony, Bossut, and Coulomb [1799].

experimenting on fluid discharge through capillary tubes, he hoped to contribute both to the theory of molecular forces and to the improvement of hydraulic practice.[14]

In conformance with Du Buat's observations of reduced flows, Girard assumed that a layer of fluid adhered to the walls of the tube, and that the rest of the fluid moved with a roughly uniform velocity. Flow retardation then resulted from friction between the moving column of fluid and the adherent layer. Girard favored experiments on capillary tubes, no doubt because measurements were easier in this case, but also because he believed (incorrectly by later views) that the uniformity of the velocity of the central column would apply better to narrower tubes (because of a presumably higher cohesion of the fluid). He operated with copper tubes of two different diameters (D) of around 2 mm and 3 mm and lengths (L) varying between 20 cm and 2.20 m. The tubes were horizontal and fed by a large water vessel under a constant height H (see Fig. 3.2). Girard took the pressure gradient in the tube to be equal to $\rho g H/L$, where g is the acceleration of gravity and ρ is the density of water. Following Coulomb and Prony, he assumed the form $av + bv^2$ for the retarding force on the unit surface of the tube, where v is the flow velocity and a and b are two tentative constants. The balance of the forces acting on a cylindrical slice of fluid then gives[15]

$$\frac{\rho g D H}{4L} = av + bv^2. \tag{3.1}$$

Girard measured the rate of discharge $\pi D^2 v/4$ for various lengths and charges, at a temperature varying with the season or controlled artificially. His first conclusion was that the quadratic friction term disappeared for tubes of sufficient length. Consequently, he

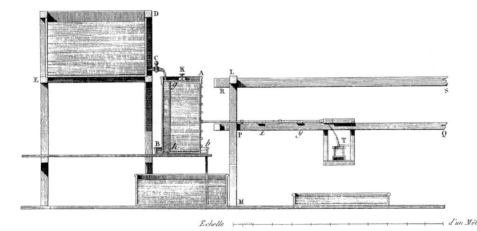

Fig. 3.2. Girard's apparatus for measuring discharge through narrow tubes (from Girard [1816] plate). The water from the tank D is maintained at a constant level in the tank A and flows through the horizontal tube (lying on xy) into the bucket T.

---

[14]Girard [1816]. Cf. Grattan-Guinness [1990] vol. 1, pp. 563–65.

[15]Girard [1816] pp. 257–58, 265.

assumed the friction to be fundamentally linear, and the quadratic contribution to be due to the lack of (recti)linearity of the flow near the entrance of the tube (involving Newton's *vena contracta* and its subsequent oscillations). He then focused on the linear behavior, apparently forgetting the engineer's interest in the quadratic contribution (which dominates in the case of large pipes of any length). He found that the 'constant' *a* significantly decreased when the temperature rose, and that it varied with the diameter of the tube.[16]

Girard produced a nice molecular explanation for these effects. A temperature increase, he reasoned, implies a dilation of the fluid and therefore a decrease in the mutual adhesion of the fluid molecules expressed in the constant *a*. As for the dependence of *a* on the tube's diameter, Girard evoked the finite thickness *e* of the adherent layer of fluid, which implies the substitution of $D - 2e$ for D in eqn (3.1). For high temperatures the thickness *e* should be negligible since there is little adhesion between the fluid and wall molecules. Then the original formula (3.1) (with $b = 0$) and the proportionality of the discharge to the cube of the diameter hold approximately, as Girard's measurements with heated water seemed to confirm. In a sequel to this memoir, Girard used glass tubes instead of copper and various liquids instead of water, meaning to confirm his view that the thickness of the adhering layer depended on molecular forces between the layer and the wall.[17]

As he had little to offer to the hydraulic engineer, Girard wrote something for the physiologist. The capillary dimensions of vessels and the wetting of their walls, he noted, was essential to explain blood or sap circulation in animals and plants. Otherwise, body temperature could not control the circulation, and friction would wear the vessels. Girard expressed his amazement at the 'simplicity of the means of Nature and the perfection of her works' when seen in the light of his own research. His self-confident tone and his professional authority easily convinced his contemporaries, including the Academicians who welcomed him. Yet his experimental method and his theoretical reasoning falter when compared with those of the best French experimenters of the day.[18]

In the absence of contemporary criticism, we may only imagine what flaws a more careful contemporary could have detected in Girard's work. While estimating the charge *H* of the tube, Girard did not include the loss of head due to the entrance in the tube, even though Du Buat had noted the importance of this correction for short pipes. In considering the variation of the discharge rate with the diameter of the tube, he used only two different diameters and did not indicate how he had measured them. Judging from Gotthilf Hagen's later measurements, the numbers provided by the manufacturer or a simple external measurement could not be trusted.

These circumstances may in part explain why Girard did not obtain the D^4 law for the discharge, which we know to be quite accurate, why he found glass to provide a stronger discharge than copper, and why he believed that retardation would be linear for any diameter and velocity if the tube were long enough. On the theoretical side, he conflated

---

[16]Girard [1816] p. 285. Girard insisted (*ibid.* p. 287) that, contrary to Coulomb's case, the velocity did not need to be small for the quadratic term to disappear. Girard borrowed the expression for the accelerating force and the expression 'linear motion' from Euler (*ibid.* p. 307).

[17]*Ibid.* pp. 315–21, 328–29; Girard [1817] p. 235. In this second memoir, Girard used and praised the graphic method that Prony had used for channel flow; he found that the linear term did not exist for mercury, as he expected from the fact that mercury does not wet glass.

[18]Girard [1817] p. 259.

adhesion with friction, and therefore did not appreciate the circumstances that determine the velocity profile. Girard nevertheless obtained the linear behavior in H/L for the discharge through narrow tubes, which is as well known today as it was surprising to contemporary hydraulicians.

3.1.4 *The rational and the practical*

In summary, at the beginning of the nineteenth century no one expected rational fluid dynamics to explain the practically important phenomena of fluid resistance and flow retardation. Most knowledge of these phenomena was empirical and derived from the observations and measurements accumulated by hydraulic engineers. Although some notion of internal friction had been available since Newton, and although Du Buat, Venturi, Coulomb, and Girard somewhat revived it at the turn of the century, there was no attempt to apply this insight to the mathematical determination of fluid motion.

It may seem surprising that no one before Navier tried to insert new terms into Euler's hydrodynamic equations. A first explanation is that the new hydrodynamics was part of a rational mechanics that valued clarity, formal generality, and rigor above empirical adequacy. Another is that Euler's equations were complex enough to saturate contemporary mathematical capability. They were among the first partial differential equations ever written, and they involved the nonlinearity that has troubled mathematical physicists to this day. Even if someone had been willing to modify Euler's equations, he would have lacked empirical clues about the structure of the new terms, because the concept of internal friction was as yet immature.

Last, but perhaps most important, the French mathematicians who were the most competent at inventing new partial differential equations all accepted d'Alembert's fundamental principle of dynamics, according to which the equations of motion of a mechanical system can be obtained from the equilibrium condition between impressed forces and inertial forces. From this point of view, the hydrodynamic equations should result directly from the laws of hydrostatics. Since the latter were solidly established, Euler's equations seemed unavoidable.[19]

3.2 Navier: molecular mechanics of solids and fluids

3.2.1 *X+Ponts*

In the jargon of the *Grandes Ecoles*, Claude-Louis Navier was an 'X+Ponts', that is, an engineer trained first at the Ecole Polytechnique and then at the Ecole des Ponts et Chaussées. He embodied a new style of engineering that combined the analytical skills acquired at the Polytechnique with the practical bent of the *Ecoles d'application*. Through his theoretical research and his teaching he contributed to a renewal of the science of mechanics that made it fit much better to the needs of engineers. Navier famously promoted considerations of 'live force' (kinetic energy) and 'quantity of action' (work) in the theory of machines, thus following Lazare Carnot's pioneering treatise and facilitating Gaspard Coriolis's and Jean-Victor Poncelet's later developments.[20]

---

[19]Cournot expressed this view in his comment on Navier's equation, discussed later on p. 118.

[20]Cf. McKeon [1974] pp. 2–5. On the new style of engineering, cf. Belhoste [1994], Picon [1992] chaps 8–10. On the concept of work, cf. Grattan-Guinness [1984].

Orphaned at fourteen, Navier was educated by his uncle Emiland Gauthey, a renowned engineer of bridges and canals. He later expressed his gratitude through a careful edition of Gauthey's works, published in 1809–1816. His competence in hydraulic architecture then led him to edit Bernard Forest de Bélidor's voluminous treatise, which had been a canonical reference on this subject since its first publication in 1737. In this new edition, published in 1819, Navier left Bélidor's text intact but denounced numerous theoretical misconceptions that still affected engineering practice in France and elsewhere. His footnotes and appendices constituted a virtual book within the book, including a new presentation of mechanics and a theory of machines based on live forces.[21]

Navier found Bélidor's treatment of hydraulic problems most defective, as appears from his judgment of the included theory of efflux:

> The preceding theory, with which the author seems so pleased, now appears to be one of the most defective of his work. In truth one had not, at the time he was writing, gathered a sufficient amount of experiments so as to establish the exact measure of phenomena; but this does not justify the totally vicious theory that he gives of it, nor the trust with which he presents it.

In order to correct Bélidor on this subject, Navier only had to return to Daniel Bernoulli and to refer to Venturi's relevant experiments.[22]

Fluid resistance was harder to rectify. As Navier well knew, numerous experiments by Jean-Charles de Borda in the 1760s and by Bossut and Du Buat in the 1780s and 1790s had disproved the old impact theory recalled by Bélidor. In his notes, Navier could only deplore that contemporary hydrodynamics did not permit a definitive solution to this problem. He agreed with Euler that momentum balance applied to the tubes of flow around the immersed body should yield the value of the resistance. No more than Euler, however, could he justify the truncation of the tubes that allowed for a nonzero resistance proportional to the squared velocity. Nor could he account for the negative pressure that Du Buat had found to exist at the rear of the body.[23]

From Coulomb, Navier also knew that the resistance became proportional to the velocity for very slow motion. He agreed with Coulomb that in this case the retarding force resulted from 'the mutual adhesion of the fluid molecules among themselves or at the surface of the immersed bodies.' In summary, he considered two causes of fluid resistance, namely, a non-balanced distribution of pressure around the immersed body owing to some particularity in the shape of the lines of flow around the body, and friction occurring between the body and the successive layers of fluid owing to 'molecular adhesion'. He respected Bélidor's omission of pipe flow.[24]

3.2.2 *Laplacian physics*

Another novelty of Navier's edition was the respect he paid to Laplace's new molecular physics. Imitating Newton's gravitation theory and some of his queries, the French

---

[21]Cf. McKeon [1974], Prony [1864], Grattan-Guinness [1990] vol. 2, pp. 969–74.

[22]Navier, note to Bélidor [1819] p. 285n.

[23]*Ibid.* pp. 339n–356n. On the fluid-resistance experiments by Borda, Bossut, and Du Buat, cf. Dugas [1950] pp. 297–305, Rouse and Ince [1957] pp. 124, 128, 133–4, Nemenyi [1962] pp. 77–9.

[24]Navier, note to Bélidor [1819] p. 345n. Navier briefly mentioned (*ibid.* p. 292n) 'friction of the fluid on the [pipe] walls' (but not the internal adhesion in this case).

astronomer sought to explain the properties of matter by central forces acting between molecules. His first successful attempt in this direction was a theory of capillarity published in 1805/06. In the third edition of his *Système du monde*, published in 1808, he also indicated how optical refraction, elasticity, hardness, and viscosity could all be reduced to short-range forces between molecules. In an appendix to the fifth volume of the *Mécanique céleste*, published in 1821, he gave a detailed molecular theory of sound propagation, based on his and Claude-Louis Berthollet's idea that molecular repulsion depended on the compression of elastic atmospheres of caloric.[25]

In the foreword to his edition of Bélidor, Navier approved Laplace's idea of the constitution of solids:

> Even though the intimate constitution of bodies is unknown, the phenomena which they show allow us to clearly perceive a few features of this constitution. From the faculty that solid bodies have to dilate under heating, to contract under cooling, and to change their figure under effort, it cannot be doubted that they are made of parts which do not touch each other and which are maintained in equilibrium at very small distances from each other by the opposite actions of two forces, one of which is an attraction inherent in the nature of matter, and the other a repulsion due to the principle of heat.

At that time, Navier used this conception of solids only to banish the ideally-hard bodies of rational mechanics from collision theory. He referred to Laplace's theory of capillarity in a footnote. The conditions of equilibrium of fluids, he emphasized, could not be rigorously established without the molecular viewpoint. *A fortiori*, fluid motion had to depend on molecular processes, as he argued in his discussion of Coulomb's fluid-friction experiments.[26]

3.2.3   *Elastic beams and plates*

In his engineering role, Navier acted mostly as an expert on bridge construction. In the 1810s, he designed three new bridges on the River Seine, and oversaw an important bridge and embankment project in Rome. This work, as well as his edition of Gauthey's works, brought to his attention the empirical inadequacies of the existing theoretical treatments of the elasticity of solid bodies. Previous calculations of the compression, extension, and flexion of beams had assumed the existence of mutually-independent longitudinal fibers that resisted extension or compression by a proportional tension or pressure; otherwise, they relied on an even cruder idealization in which the beam was replaced by a line or blade with a curvature-driven elastic response. Navier worked to improve the fiber-based reasoning in order to address the practically essential question of rupture. He still taught this point of view in the course he began to teach at the Ponts et Chaussées in 1819, although he also told his students that the true foundation of elasticity should be molecular.[27]

[25]Cf. Heilbron [1993] pp. 1–16, Fox [1971], [1974], Crosland [1967], Grattan-Guinness [1990] chap. 7.

[26]Navier, in Bélidor [1819] pp. x–xi, 208n. *Ibid.* on p. 215n Navier rejected Daniel Bernoulli's and Bélidor's kinetic interpretation of pressure.

[27]Cf. Prony [1864] pp. xliii–xliv, Saint-Venant [1864*b*] pp. civ–cix. On the history of elasticity, see also Truesdell [1960], Todhunter and Pearson [1886–1893], Timoshenko [1953], Benvenuto [1991]. On Navier's course, cf. Picon [1992] pp. 482–495.

In August 1820, Navier submitted to the Academy of Sciences a memoir on vibrating plates in which he still reasoned in terms of continuous deformations. The problem of vibrating plates had occupied several excellent minds since the German acoustician Ernst Chladni, with fine sand and a violin bow, had revealed their nodal lines to French Academicians in 1808. Whereas Sophie Germain and Lagrange still reasoned on the basis of presumptive relations between curvature and restoring force, in 1814 Laplace's close disciple Siméon Denis Poisson offered a first molecular theory. He considered a two-dimensional array of molecules, and computed the restoring force acting on a given molecule by summing the forces exerted by the surrounding displaced molecules. To perform this sum, Poisson assumed, as Laplace had done in his theory of capillarity, that the sphere of action of a molecule was very small compared to a macroscopic deformation and nevertheless contained a very large number of molecules. Consequently, he replaced the molecular sums with integrals and retained only low-order terms in the Taylor expansion of the deformation.[28]

Poisson's analysis confirmed the differential equation used by Lagrange and Germain. Navier, nonetheless, sought another derivation, because he believed that the boundary conditions were still in doubt. In his memoir of 1820, he applied Lagrange's method of moments, which has the advantage of simultaneously yielding the equation of motion and the boundary conditions. As we saw in Chapter 1, in his seminal memoir on fluid motion, Lagrange had used this method to derive Euler's equations and the appropriate boundary conditions. In his *Méchanique analitique*, he had also introduced the moment (virtual work) $\iint F\delta S$ of the elastic tension F that arises in response to the stretching of an elastic membrane.[29]

For simplicity, Navier assumed that the local deformation of a plate could be decomposed into flexion and isotropic stretching. To Lagrange's expression for the moment of the tension caused by the stretching, he added the moment of the elastic tensions and pressures that arise in response to the flexion of a plate of finite thickness (the fibers on one side of the plate are compressed while those on the other side are extended). Lastly, he obtained the equation of equilibrium and the boundary conditions balancing the total moment of a virtual deformation with the moment of the external forces.[30]

3.2.4   *The general equations of elasticity*

On the one hand, Navier admired Lagrange's method for its power to yield the boundary conditions. On the other, he approved of Laplace's and Poisson's molecular program. A few months after submitting his memoir on elastic plates, he managed to combine these two approaches. Presumably, he first rederived the moments for the elastic plate by summing molecular moments. Having done so, he realized that this procedure could easily be extended to an arbitrary, small deformation of a three-dimensional body. He thereby obtained the general equations of elasticity for an isotropic body (with one elastic constant only). In the memoir he read on 14 May 1821, he gave two different derivations of these

---

[28]Poisson [1814]. Cf. Saint-Venant [1864b] pp. ccliii–cclviii, Dahan [1992] chap. 4, Grattan-Guinness [1990] vol. 1, pp. 462–5.

[29]Navier [1820], [1823a]; Lagrange [1788] pp. 139–45, 158–62 (membrane), 438–41. Cf. Dahan [1992] pp. 50–1.

[30]Cf. Saint-Venant [1864b] pp. cclix–cclx, Grattan-Guinness, [1990] vol. 2, pp. 977–83.

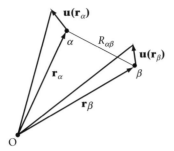

Fig. 3.3.   Diagram for displacements in an elastic body.

equations. The first derivation was by a direct summation of the forces acting on the given molecules, and the second was by the balance of virtual moments. This second route, Navier's favorite, goes as follows.[31]

For a solid in its natural state of equilibrium, the moment of molecular forces vanishes. After a macroscopic deformation such that a particle (i.e., a small portion of the solid) originally located at $\mathbf{r}$ goes to the point $\mathbf{r} + \mathbf{u}(\mathbf{r})$, the vector $\mathbf{R}_{\alpha\beta}$ joining the two molecules α and β alters by $\delta\mathbf{R}_{\alpha\beta} = \mathbf{u}(\mathbf{r}_{\beta}) - \mathbf{u}(\mathbf{r}_{\alpha})$ (see Fig. 3.3). To first order in $\mathbf{u}$, the corresponding change of distance $\delta R_{\alpha\beta}$ is given by the projection u_R of the vector $\mathbf{u}(\mathbf{r}_{\beta}) - \mathbf{u}(\mathbf{r}_{\alpha})$ onto the line joining the two molecules. Navier assumed that, for small deformations, the force between two molecules varied by an amount proportional to the change in their distance, the proportionality coefficient being a rapidly-decreasing function $\phi(R_{\alpha\beta})$ of their distance. This restoring force must be attractive for an increase of distance, and repulsive for a decrease of distance.

Now consider a virtual displacement $\mathbf{w}(\mathbf{r})$ of the particles of the solid. To first order in $\mathbf{u}$, the deformation $\mathbf{u}$ implies a change of moment (virtual work) $-\phi w_R \delta R$ for the forces between the molecules α and β, where w_R is the projection of the difference $\mathbf{w}(\mathbf{r}_{\beta}) - \mathbf{w}(\mathbf{r}_{\alpha})$ onto the line joining these two molecules (the indices α and β affecting R are dropped to simplify the notation; an attraction is understood to be positive). Consequently, the total moment of molecular forces after the deformation is

$$M = -\frac{1}{2}\sum_{\alpha\beta} \phi(R)u_R w_R. \qquad (3.2)$$

Exploiting the rapid decrease of the function $\phi(R)$, Navier replaced u_R with its first-order Taylor approximation $R^{-1}x_i x_j \partial_i u_j(\mathbf{r}_{\alpha})$. In this modernized notation, x_i denotes the ith coordinate of $\mathbf{R}$, ∂_i is the partial derivation with respect to the ith coordinate of $\mathbf{r}$, and summation over repeated indices is understood. With a similar substitution for w_R, we have

$$\phi(R)u_R w_R \approx R^{-2}\phi(R)x_i x_j x_k x_l \partial_i u_j \partial_k w_l. \qquad (3.3)$$

[31]Navier [1827] (read on 14 May 1821), [1823b]. Cf. Saint-Venant [1864b] pp. cxlvii–cxlix, Dahan [1992] chap. 8, Grattan-Guinness [1990] pp. 983–5. In the extract of his memoir on elastic plates (Navier [1823a]), Navier assumed a molecular foundation for the flexion moment.

Navier then replaced the sum over β in eqn (3.2) by a volume integral weighted by the number N of molecules per unit volume (since his calculation of the moment M was limited to first order in $\mathbf{u}$, he could neglect the variation of N caused by the deformation). Separating the integration over R and that over angular variables yields

$$\sum_{\beta} \phi u_R w_R = 2NK(\partial_i u_j \partial_i w_j + \partial_i u_i \partial_j w_j + \partial_i u_j \partial_j w_i),$$ (3.4)

with

$$K = \frac{2\pi}{15} \int \phi(R) R^4 \mathrm{d}R.$$ (3.5)

In order to obtain the total molecular moment M, Navier then performed the sum over α, which he also replaced by an integral. The result can be put in the form

$$M = \int \tau_{ij} \partial_i w_j \mathrm{d}\tau,$$ (3.6)

with

$$\tau_{ij} = -KN^2(\delta_{ij}\partial_k u_k + \partial_i u_j + \partial_j u_i),$$ (3.7)

where δ_{ij} is the unit tensor.

By analogy with Lagrange's hydrodynamic reasoning, Navier then integrated by parts to obtain

$$M = \oint \tau_{ij} w_j \mathrm{d}S_i - \int (\partial_i \tau_{ij}) w_j \mathrm{d}\tau.$$ (3.8)

The deformed solid is in equilibrium if and only if this moment is balanced by the moment of the applied forces, which may include an internal force density $\mathbf{f}$ (such as gravity) and an *oblique* pressure $\overline{\mathbf{P}}$ on the surface of the solid. For virtual displacements that occur entirely within the body, the balance requires that $f_j - \partial_i \tau_{ij} = 0$ or, in vector notation,

$$\mathbf{f} + KN^2[\Delta\mathbf{u} + 2\nabla(\nabla \cdot \mathbf{u})] = \mathbf{0}.$$ (3.9)

The second term represents the restoring force that acts on a volume element of the deformed solid. According to d'Alembert's principle, the equations of motion of the elastic solid are simply obtained by equating this force to the acceleration times the mass of the element. For virtual displacements at the surface of the body, the balance of the surface term of eqn (3.8) with the moment $\int -\overline{\mathbf{P}} \cdot \mathbf{w} \, \mathrm{d}S$ of the oblique external pressure gives the boundary condition

$$\tau_{ij} \, \mathrm{d}S_j = \overline{P}_i \, \mathrm{d}S.$$ (3.10)

Navier, of course, used Cartesian notation, which gives a forbidding appearance to his calculation. However, the basic structure of his reasoning was as simple as the above rendering suggests. The only step in the tensor calculation that may imply more than Navier had in mind is the introduction of the tensor τ_{ij} to prepare for the partial integration of eqn (3.6). Navier treated each term of this equation separately. He none-theless wrote the following Cartesian version of eqn (3.10):

$$X' = \varepsilon \left[\cos l \left(3\frac{dx'}{da'} + \frac{dy'}{db'} + \frac{dz'}{dc'} \right) + \cos m \left(\frac{dx'}{db'} + \frac{dy'}{da'} \right) + \cos n \left(\frac{dx'}{dc'} + \frac{dz'}{da'} \right) \right],$$

$$Y' = \varepsilon \left[\cos l \left(\frac{dx'}{db'} + \frac{dy'}{da'} \right) + \cos m \left(\frac{dx'}{da'} + 3\frac{dy'}{db'} + \frac{dz'}{dc'} \right) + \cos n \left(\frac{dy'}{dc'} + \frac{dz'}{db'} \right) \right], \quad (3.11)$$

$$Z' = \varepsilon \left[\cos l \left(\frac{dx'}{dc'} + \frac{dz'}{da'} \right) + \cos m \left(\frac{dy'}{dc'} + \frac{dz'}{db'} \right) + \cos n \left(\frac{dx'}{da'} + \frac{dy'}{db'} + 3\frac{dz'}{dc'} \right) \right],$$

which gives the local response of the solid to an oblique external pressure.[32]

3.2.5   A new hydrodynamic equation

Soon after presenting this memoir on elasticity, Navier thought of adapting his new molecular technique to fluid mechanics. First considering a fluid in equilibrium, he assumed a force $f(R)$ that acted between every pair of molecules and which decreased rapidly with the distance R (an attraction being understood as positive). Denoting by $\mathbf{w}(\mathbf{r})$ a virtual displacement of the particles of the fluid, and using the notation of the previous section, the corresponding moment is

$$M = -\frac{1}{2} \sum_{\alpha\beta} f(R) w_R \approx -\frac{1}{2} \sum_{\alpha\beta} R^{-1} f(R) x_i x_j \partial_i w_j. \quad (3.12)$$

Replacing the sums by integrals, and separating angular variables in the first integration yields

$$M = -\int N^2 \varpi \nabla \cdot \mathbf{w} \, d\tau, \quad (3.13)$$

with

$$\varpi = \frac{2\pi}{3} \int_0^\infty R^3 f(R) \, dR. \quad (3.14)$$

When the fluid is subjected to an internal force density $\mathbf{f}$ and to an external pressure $\overline{P}$, the equilibrium condition reads:

$$\int (\mathbf{f} \cdot \mathbf{w} + \varpi N^2 \nabla \cdot \mathbf{w}) \, d\tau - \int \mathbf{w} \cdot \overline{P} \, d\mathbf{S} = 0, \quad (3.15)$$

which has the same form as eqn (1.44) that Lagrange gave for the equilibrium of an incompressible fluid.

In conformance with this analogy, Navier took the density N to be nearly constant (he gave it the value *one*) but made the parameter ϖ vary from one particle of the fluid to another. This odd assumption (it seems incompatible with the expression for ϖ), of which more will be said later, brought him back to the Euler–Lagrange conditions of

---

[32]Navier [1827] p. 390.

equilibrium, namely $\mathbf{f} + \nabla(N^2\varpi) = \mathbf{0}$ within the fluid, and $\overline{P} = -N^2\varpi$ on its surface, so that $P = -N^2\varpi$ plays the role of internal pressure. In Navier's words, ϖ 'measures the resistance opposed to the pressure that tends to bring the fluid parts closer to each other.'[33]

Navier then turned to the case of a fluid moving with a velocity $\mathbf{v}(\mathbf{r})$, and he assumed that 'the repulsive actions of the molecules increased or diminished by a quantity proportional to the velocity with which the distance of the molecules decreased or increased.' Denoting by $\psi(R)$ the proportionality coefficient, this intuition implies a new contribution of the form

$$M' = -\frac{1}{2}\sum_{\alpha\beta}\psi v_R w_R \tag{3.16}$$

to the moment of the molecular forces. By analogy with the corresponding formula (3.2) for elastic solids, this leads to an additional force $\mu\Delta\mathbf{v}$ in the equation of motion of an incompressible fluid, with

$$\mu = \frac{2\pi}{15}\int N^2\psi(R)R^4\mathrm{d}R. \tag{3.17}$$

The new equation of motion reads

$$\rho\left[\frac{\partial\mathbf{v}}{\partial t} + (\mathbf{v}\cdot\nabla)\mathbf{v}\right] = \mathbf{f} - \nabla P + \mu\Delta\mathbf{v}, \tag{3.18}$$

which is now known as the 'Navier–Stokes equation' (for an incompressible fluid).[34]

3.2.6   Boundary conditions

Navier gave this equation in a memoir read on 18 March 1822 at the Academy of Sciences and published it in summary form in the *Annales de chimie et de physique*. There he assumed, as Girard had, that the velocity $\mathbf{v}$ vanished at the wall, in which case the balance of moments gives no additional boundary condition.[35] Under this hypothesis, Navier calculated the uniform flow in a pipe of rectangular section and found a discharge proportional to the pressure gradient, as Girard had observed for 'linear motions' (that is, laminar flow). According to the same calculation, the average fluid velocity in a square tube should be proportional to the square of its perimeter (as it is according to Poiseuille's later law for circular tubes). Navier (wrongly) believed this result to agree with Girard's observation of a departure from the expected proportionality to the perimeter (in the case of circular tubes).[36]

At the same time, Navier deplored a contradiction with another of Girard's results, namely, the difference between the discharge in glass and copper tubes. He now faced the following dilemma: either he maintained the boundary condition $\mathbf{v} = \mathbf{0}$ and thus contradicted Girard's experimental finding, or he gave up this condition and contradicted the most

---

[33]Navier [1823c] p. 395. Cf. Saint-Venant [1864b] pp. lxii–lxiv, Dugas [1950] pp. 393–401, Grattan-Guinness [1990] pp. 986–92 (with questionable chronology), Belhoste [1997].

[34]Navier [1823c] p. 414.

[35]However, the tangential stress must vanish at the *free* surface of the fluid.

[36]Navier [1822] p. 259.

essential assumption of Girard's theory. As he indicated toward the end of his memoir, he preferred the second alternative. On 16 December 1822, he read a second memoir in which he proposed a new boundary condition based on an evaluation of the moment of the forces between the molecules of the fluid and those of the wall. The form of this moment is

$$M'' = E \int \mathbf{v} \cdot \mathbf{w} \, \mathrm{d}S, \qquad (3.19)$$

where E is a molecular constant. This is to be cancelled by the surface term

$$\oint \mu(\partial_i v_j + \partial_j v_i) w_i \, \mathrm{d}S_j \qquad (3.20)$$

of the moment M', for any displacement $\mathbf{w}$ that is parallel to the wall. The resulting condition is

$$E\mathbf{v} + \mu \partial_\perp \mathbf{v}_{/\!/} = \mathbf{0}, \qquad (3.21)$$

where $\partial_\perp$ is the normal derivative and $\mathbf{v}_{/\!/}$ is the component of the fluid velocity parallel to the surface.[37]

With this new boundary condition, Navier redid his calculation of uniform square-pipe flow, and also treated the circular pipe by Fourier series. Taking the limit of narrow tubes, he found the average flow velocity to be proportional to the surface coefficient E, to the pressure gradient, and to the diameter of the tube, in rough agreement with Girard's data. Note that he no longer believed Girard's data to support a quadratic dependence of the velocity on the diameter. In fact, Girard's theoretical formula assumed a linear dependence, and his experimental results indicated an even slower increase with diameter. As he had no reason to distrust Girard's experiments on the differences between glass and copper tubes, Navier built the old idea of fluid–solid slip into the theory of a viscous fluid.[38]

3.2.7 A useless equation

For large pipes, Navier's theory no longer implies a significant surface-slip effect, but still makes the loss of head proportional to the average fluid velocity. Since Navier knew that in most practical cases the loss of head was nearly quadratic, he did not bother taking the large-section limit of his resistance formulas. He only noted that, in this limit, the flow obviously did not have the (recti)linearity assumed in his calculations. Probably discouraged by this circumstance, he never returned to his theory of fluid motion. In the hydraulic section of his course at the Ponts et Chaussées, he only mentioned his formula for capillary tubes, which agreed with 'M. Girard's very curious experiments'. The theory on which this formula is based, he immediately noted, 'cannot suit the ordinary cases of application. Since the more complicated motion that the fluid takes in these cases has not been submitted to calculation, the results of experience are our only guide.'[39]

---

[37]Navier [1823c]. In Cauchy's stress language, the condition means that the tangential stress is parallel and proportional to the sliding velocity.

[38]Navier [1823c] pp. 432–40.   [39]Ibid. p. 439; Navier [1838] pp. 88–9.

The two commissioners for Navier's first memoir, Poisson and Joseph Fourier, and the three for his second memoir, Girard, Fourier, and Charles Dupin, never wrote their reports, perhaps because Navier was elected to the Academy in 1824, well before the publication of his second memoir. However, the mathematician Antoine Cournot wrote a review for Férussac's *Bulletin* that may reflect the general impression that Navier's memoir made at the French Academy. Being Laplace's admirer and Poisson's protégé, Cournot welcomed Navier's theory as a new contribution to the then prosperous molecular physics. Yet he suspected a few inconsistencies in Navier's basic assumptions.[40]

In his derivation of hydrostatic pressure, Cournot noted, Navier assumed incompressibility, which seemed incompatible with the molecular interpretation of pressure as a reaction to a closer packing of the molecules. In fact, according to Navier's formula (3.14) the coefficient ϖ should be a constant, which excludes a variable pressure if the density N is also a constant. Upon closer inspection, Navier's procedure is more coherent than Cournot believed. Here and elsewhere, Navier's formulas did not quite reflect his basic intuition. In his mind, the distance R in the force function $f(R)$ did not represent the distance of the molecules in the actual state of the fluid, but their distance before compression. For a real substance, which can only be approximately incompressible, the difference between those two distances is extremely small but finite, so that Navier's f function could vary with the local state of the fluid.[41]

Another worry of Cournot's was that Navier admitted the same equations of equilibrium of a fluid as Euler and Lagrange, and yet obtained different equations of motion, against d'Alembert's principle. 'The matter', Cournot deplored, 'does not seem to be free from obscurity.' Today we would solve this apparent paradox by noting that dissipative forces, such as those expressing fluid viscosity or the viscous friction between two solids, are to be treated, in the application of d'Alembert's principle, as additional, motion-dependent forces that are impressed on the system. At the molecular level, where Navier reasoned, the difficulty is that his calculation seems to rely on velocity-dependent forces unknown to Laplacian physics.[42]

Even here Navier's formulas did not directly reflect his intentions. As a close reading of his text shows, he meant that the macroscopic motion of the fluid modified the distribution of intermolecular distances: 'If the fluid is moving', he wrote, 'which implies, in general that the neighboring molecules come closer to or further from one another, it seems natural to assume that the [intermolecular] repulsions are modified by this circumstance.' This occurs in the Laplacian conception of fluids, because the trajectory of an individual molecule undulates around the path that is imposed overall by the macroscopic motion. At any instant, the molecules of a fluid are in positions that slightly deviate from an equilibrium configuration that continually changes over time. Thus, the molecular force function ψv_R in Navier's moment formula (3.16) does not refer to the actual distance of the molecules, but to the distance that they have in the nearest equilibrium configuration; and the

---

[40]Cournot [1828] pp. 11–14.

[41]*Ibid.* pp. 11–12; Navier [1823c] p. 392: 'La force répulsive qui s'établit entre les deux molécules dépend de la situation du point M [lieu de la première molécule], puisqu'elle doit balancer la pression, qui peut varier dans les diverses parties du fluide.'

[42]Cournot [1828] p. 12.

difference between these two distances obviously depends on the macroscopically-impressed motion. This is how the fluid velocity enters the expression of Navier's molecular forces, even though the true forces depend only on the distances between the molecules.[43]

Unfortunately, Navier never explained as much—so that none of his successors (except Saint-Venant) could make sense of his calculation. In Cournot's eyes, the premises of Navier's equation seemed as arbitrary as its applicability to concrete problems was difficult to judge:[44]

> M. Navier himself only gives his starting principle as a hypothesis that can be solely
> verified by experiment. However, if the ordinary formulas of hydrodynamics resist
> analysis so strongly, what should we expect from new, far more complicated formu-
> las? The author can only arrive at numerical applications after a large number of
> simplifications and particular suppositions. The applications no doubt show great
> analytical skill; but can we judge a physical theory and the truth of a principle after
> accumulating so many approximations? In one word, will the new theory of M.
> Navier make the science of the distribution and expense of waters less empirical? I
> do not feel able to answer such a question. I can only recommend the reading of this
> memoir to all who are interested in this kind of application.

3.3 Cauchy: stress and strain

3.3.1 *The stress system*

Like Navier, Augustin Cauchy was an 'X+Ponts' with superior mathematical training and with engineering experience. However, his poor health and mathematical genius soon confined him to purely academic activities. In 1822, his study of Navier's memoir on elastic plates led him to new considerations that still constitute the basis of elasticity theory. If we are to trust Cauchy's own account, then what triggered his main inspiration was Navier's appeal to two kinds of restoring forces produced by extension and flexion.[45]

The second kind of force, Cauchy surmised, could be avoided if forces of the first kind were no longer assumed to be perpendicular to the sections on which they acted. With this insight, he then imitated Euler's hydrodynamics and reduced all elastic actions to pressures acting on the surface of portions of the body. The only difference was the non-normality of the pressure. Previous students of elasticity, in particular Coulomb and Young, had already considered tangential pressures (our shearing stresses) in specific problems such as the rupture of beams. In his memoir of May 1821 on a molecular derivation of the general equations of elasticity, Navier had introduced oblique external pressures and boundary conditions that entailed the Cauchy stress system. Whether or not Cauchy relied on such anticipations, he was the first to base the theory of elasticity on a general definition of internal stresses.[46]

---

[43]Navier [1823c] p. 390.   [44]Cournot [1828] pp. 13–14.

[45]Cauchy [1823]. Cf. Belhoste [1991] pp. 93–102; Grattan-Guinness [1990] pp. 1005–13.

[46]Cauchy [1823], [1827b]. Cf. Truesdell [1968], Dahan [1992] chap. 9. In his memoir on elastic plates [1820], Navier noted that in general the pressures would not be parallel to the faces of the element. Fresnel's theory of light was perhaps another source of Cauchy's inspiration, cf. Belhoste [1991] pp. 94–5. The stress–strain terminology is William Rankine's. Cauchy and contemporary French writers used the words *pression/tension* and *condensation/dilatation*.

As in hydrodynamics, Cauchy introduced the pressures (or tensions) that act on a volume element by recourse to the forces that would act on its surface after an imaginary solidification of the element. For a unit surface element normal to the jth axis, denote by τ_{ij} the ith component of the force acting on the negative side ($x_j < 0$) of the element. Note that, with this convention, adopted by most of Cauchy's followers, a tension is understood to be positive. Cauchy proved three basic theorems in a manner that is still used in modern texts on elasticity.

The first theorem stipulates that the pressure on an arbitrary surface element dS is given by the sum $\tau_{ij}dS_j$. In modern words, the stress system τ_{ij} is a tensor of second rank. This results from the fact that the resultant of the pressures acting on the pyramidal volume element $\{x, y, z, > 0; \alpha x + \beta y + \gamma z < \varepsilon\}$ would be of second order in the small quantity ε, and therefore could not be balanced by the resultant of a volume force (which is of third order) if the theorem were not true. Cauchy's second theorem states the symmetry of the stress system, namely $\tau_{ij} = \tau_{ji}$, without which the resultant of the pressure torques on a cubic element of the solid would be of third order and therefore could not be balanced by the torque of any volume force, which is of fourth order. Thirdly, and most obviously, the resultant of the pressures acting on a (cubic) volume element is $\partial_j\tau_{ij}$ per unit volume.[47]

3.3.2 *Strain and motion*

As Cauchy knew from the theory of quadratic forms (which he had recently applied to inertial moments), the symmetry of the pressure system implies the existence of three principal axes for which the pressures are normal (in modern terms, the stress tensor is then diagonal). Cauchy used this property to relate the pressure system to the local deformations of the system. If $\mathbf{u}(\mathbf{r})$ is the displacement of a solid particle at the point of space $\mathbf{r}$, he showed, then the first-order variation in the distance between two points whose coordinate differences have the very small values dx_i is given by $dx_i\, dx_j\, \partial_i u_j$. In modern terms, this quadratic form is associated with the symmetric tensor $e_{ij} = \partial_i u_j + \partial_j u_i$. This tensor has three principal axes, which means that the local deformation is reducible to three dilations or contractions along three orthogonal axes.[48]

Cauchy then argued that, for an isotropic body, the principal axes of the tensors τ_{ij} and e_{ij} were necessarily identical. He further assumed that the pressure ratios between two such axes were equal to the dilation ratios. This implies that the two tensors are proportional. Lastly, Cauchy assumed that the proportionality coefficient was a constant independent of the deformation, which is a generalization of Hooke's law. He thus obtained an equation of equilibrium similar to Navier's equation (3.9), though without the factor 2 in the $\nabla(\nabla \cdot \mathbf{u})$ term. The boundary conditions immediately result from the balance of internal and external pressures.[49]

---

[47]Cauchy [1827*b*], [1827*d*].   [48]Cauchy [1823], [1827*c*].

[49]Cauchy [1823], [1828*a*]. Cauchy introduced the word 'isotrope' in 1839/40, for example in Cauchy [1840].

3.3.3   *The perfectly inelastic body*

In a last section, Cauchy considered the case of a 'non-elastic body', defined as a body for which the stresses at a given instant only depend on the change of form experienced by the body in a very small time interval preceding this instant. He found it natural to assume that the stress tensor was proportional to the tensor representing the velocity of deformation (again reasoning with respect to principal axes). For an incompressible body, the resulting equation of motion is the one Navier had given for viscous fluids, save for the pressure term. Cauchy, however, mentioned neither Navier's result nor any similarity between a real fluid and his 'non-elastic body'. Instead, he noted that, for very slow motion, the linearized equation of motion was identical to Fourier's equation for the motion of heat, and claimed 'a remarkable analogy between the propagation of caloric and the propagation of vibrations in a body entirely deprived of elasticity.'[50]

3.3.4   *Final foundations?*

Cauchy announced his theory of elasticity on 30 September 1822, and published it in summary form the following year. He waited six more years before complete publication in his own, self-serving journal, the *Exercices de mathématiques*. The reason for this delay may have been the courtesy of waiting for Navier's memoir of 1821 to be published. In the final version of his theory, Cauchy proposed the more general, two-constant relation

$$\tau_{ij} = K'(\partial_i u_j + \partial_j u_i) + K'' \delta_{ij} \partial_k u_k \qquad (3.22)$$

between stress and deformation. This allowed him to retrieve Navier's equation of equilibrium as the particular case for which $K' = K''$. The two-constant theory is the one now accepted for isotropic elasticity.[51]

Cauchy's memoirs on elasticity were written with incomparable elegance and rigor. For this reason, and also because of their strikingly modern appearance, they have often been regarded as the first and final foundation of this part of physics. Cauchy's contemporaries thought differently. In the years following his publication, theorists of elasticity were not satisfied with this purely macroscopic-continuum approach, even though they all adopted Cauchy's stress. In their eyes, the true foundation of elasticity had to remain molecular, as it should be in Laplace's grand unification of physics.[52]

It would also be wrong to regard Cauchy's stress–strain approach as an indication that he supported a continuist view of matter. For theological reasons, he was a finitist in mathematics and an atomist in physics. That he first derived the equations of elasticity without reference to the molecular level only proves that he possessed the geometrical and algebraic skills that make this route natural and easy. He in fact provided the most complete and rigorous molecular theory of elasticity, even before his first theory of elasticity was published. On this ground, he found himself again in competition with Poisson, undoubtedly the most aggressive supporter of the molecular approach.[53]

---

[50]Cauchy [1823], [1828*a*] par. 3.   [51]Cauchy [1828*a*].   [52]Cf. Saint-Venant [1864*b*], pp. cliv–clv.

[53]In his Torino lectures of 1833, Cauchy argued that extended molecules would be indefinitely divisible, against the principle that 'only God is infinite, everything is finite except him' (Cauchy [1833] pp. 36–7). However, he never used molecular considerations in print before his molecular theory of elasticity (I thank Bruno Belhoste for this information).

3.4   Poisson: the rigors of discontinuity

3.4.1   *Laplacian motivations*

Unlike Navier and Cauchy, Poisson did not have engineering training and experience, for he settled at the Polytechnique as a *répétiteur* and then as a professor. His interest in elasticity came from his enthusiastic embrace of Laplace's molecular program. His 1814 theory of elastic plates was already molecular. Presumably stimulated by Navier's publications of 1820/21, he returned to this subject in the late 1820s. His memoir read on 14 April 1828 contains his famous plea for a *mécanique physique*:[54]

> It would be desirable that geometers reconsider the main questions of mechanics under this physical point of view which better agrees with nature. In order to discover the general laws of equilibrium and motion, one had to treat these questions in a quite abstract manner; in this kind of generality and abstraction, Lagrange went as far as can be conceived when he replaced the physical connections of bodies with equations between the coordinates of their various points: this is what *analytical mechanics* is about; but next to this admirable conception, one could now erect a *physical mechanics*, whose unique principle would be to reduce everything to molecular actions that transmit from one point to another the given action of forces and mediate their equilibrium.

Poisson's memoir of 1828 can, to some extent, be seen as a reworking of Navier's memoir of 1821 on the molecular derivation of the general equations of elasticity. Both authors aimed at a derivation of the general equations and boundary conditions of elasticity by the superposition of short-range molecular actions. However, there were significant differences in their assumptions and methods. Whereas the only molecular forces in Navier's calculations were those produced by the deformation of the solid, Poisson retained the total force $f(R)$ between two molecules. Also, Poisson avoided Navier's method of moments and instead directly summed the molecular forces acting on a given molecule.

Cauchy worked on a similar molecular theory in the same period. Competition was so intense that Cauchy decided to deposit a draft of his calculation as a *pli cacheté* at the Academy, and Poisson decided to read his memoir in a still unripe form. Cauchy's assumptions and methods were essentially the same as Poisson's; this should not surprise us as they were both following Laplacian precepts without Navier's personal touch. Yet Cauchy's execution surpassed Poisson's in rigor, elegance, and compactness.[55]

By summation of the forces acting from one side of a given surface element to the other side, the molecular theory leads to the stress system

$$\tau_{ij} = N(\lambda_{ij} - \lambda_{ij}\partial_k u_k + \lambda_{ik}\partial_k u_j + \lambda_{jk}\partial_k u_i + \lambda'_{ijkl}\partial_k u_l), \tag{3.23}$$

where N is the original number of molecules per unit volume,

---

[54]Poisson [1829a] p. 361. Cf. Arnold [1983], Grattan-Guinness [1990] pp. 1015–25, Dahan [1992] chap. 10.

[55]Cf. Belhoste [1991] pp. 99–100.

$$\lambda_{ij} = \sum_{\beta} \frac{1}{2} f R^{-1} x_i x_j, \tag{3.24}$$

and

$$\lambda'_{ijkl} = \sum_{\beta} \frac{1}{2} R^{-1} \frac{\mathrm{d} f R^{-1}}{\mathrm{d} R} x_i x_j x_k x_l \tag{3.25}$$

(R is the length, and x_i is the ith coordinate of the vector joining a molecule α situated at the point at which the stress is computed and an arbitrary molecule β). For an isotropic solid, there are only two independent constants λ and λ', in terms of which the stress system reads

$$\tau_{ij} = N[\lambda \delta_{ij} + (\lambda' - \lambda)\delta_{ij}\partial_k u_k + (\lambda + \lambda')(\partial_i u_j + \partial_j u_i)]. \tag{3.26}$$

This result agrees with Cauchy's earlier macroscopic theory, except for the pressure λ in the original state.[56]

3.4.2   Sums versus integrals

Poisson and Cauchy both investigated the limiting case of a continuous medium, in which the sums (3.24) and (3.25) expressing the coefficients λ_{ij} and λ'_{ijkl} can be rigorously replaced by integrals. As Cauchy (but not Poisson) saw, isotropy follows without further assumption, and the coefficients λ and λ' are given by

$$\lambda = \frac{2\pi}{3} N \int_0^\infty f R^3 \mathrm{d}R \tag{3.27}$$

and

$$\lambda' = \frac{2\pi}{15} N \int_0^\infty R^5 \frac{\mathrm{d} f R^{-1}}{\mathrm{d}R} \mathrm{d}R. \tag{3.28}$$

Integrating the latter expression by parts yields the relation

$$\lambda + \lambda' = \lim_{R \to 0} R^4 f(R). \tag{3.29}$$

Poisson and Cauchy both assumed the limit to be zero. Then the medium loses its rigidity since the transverse pressures disappear. As Cauchy further observed, the continuous limit of the stress has the form

$$\tau_{ij} \propto \rho^2 \delta_{ij}, \tag{3.30}$$

---

[56]Cauchy [1828b], [1829]; Poisson [1829a]. Cf. Saint-Venant [1864b] pp. clv–clxi, [1868a], Dahan [1992] chap. 11, Darrigol [2002a] pp. 121–4. Regarding the molecular definition of stress, see Saint-Venant's intervention mentioned later on p. 130.

where ρ is the density in the deformed state. This means that the body is an elastic fluid whose pressure varies as the square of the density.[57]

In his own manner, Poisson also obtained the absence of transverse pressures in the continuum limit. He used this conclusion to dismiss Navier's theory and to denounce the general impossibility of substituting integrals for molecular sums in the new physical mechanics. He also claimed to be the first to have offered a genuinely molecular theory of elasticity.[58]

3.4.3 *Navier's defense*

There followed a long, bitter polemic in the *Annales de chimie et de physique*. Navier first recalled that Poisson and Laplace had had no qualms replacing sums with integrals in their past works. The newer emphasis on a supposed rigor could only betray a desire to belittle his own achievement. It was he, Navier, who in 1821 'conceived the idea of a new question, one necessary to the computation of numerous phenomena that interest artists and physicists.' It was he who 'recognized the principle on which this solution had to rest.' This principle, however, was not what Poisson thought it should be; it made the variation of intermolecular forces during a deformation of a solid body depend linearly on the variation of molecular distances, but did not require that the molecules should interact through central forces only. Consequently, Navier believed that his theory was immune to Poisson's arguments on sums versus integrals.[59]

Navier then counter-attacked Poisson for failing to provide a description of the force function $f(R)$ that would account for the stability and elastic behavior of solids. For example, in order that the internal pressure vanishes in an unstrained solid Poisson required the vanishing of the sum $\sum Rf(R)$, without exhibiting a choice for f that met this condition. If Poisson were willing to presuppose so much about the function f, Navier argued, why did he not consider a nonzero value of the limit of $R^4 f$ when R reaches zero? This would avoid the fatal $\lambda + \lambda' = 0$, and allow the use of integrals instead of sums.[60]

From this extract of Navier's defense, one may judge that he was hesitating between two strategies. The first option was to deny the general applicability of the Laplacian doctrine of central forces, and to deal only with the forces that arise when an equilibrium *of unknown nature* is disturbed. This option agreed with Navier's positivist sympathies and with the style of applied physics that he embodied at the Ponts et Chaussées; and it could accommodate later, unforeseen changes in molecular theory.[61]

The second option was to admit the Laplacian reduction to central forces and to show that appropriate results could nevertheless be obtained by substituting integrals for sums. Here Navier erred, because a Laplacian continuum, that is, a continuous set of material

[57]Cauchy [1828*b*] p. 266.   [58]Poisson [1829*a*] pp. 397–8, 403–4.

[59]Navier [1828*a*], [1828*b*], [1829*a*], [1829*b*]; Poisson [1828*a*], [1828*b*]. Cf. Saint-Venant [1864*b*] pp. clxi–clxvii, Arnold [1983] parts 6 and 8.

[60]Navier [1829*a*], [1829*b*]. Poisson also objected to Navier's occasional assumption that in the natural state of the body the forces between any two molecules vanished. Navier, however, did not regard this assumption as necessary to his derivations.

[61]Physicists today regard the existence of the equilibrium state of a solid as a quantum property, but they nevertheless allow a classical treatment of small perturbations of this state.

points subjected to central forces acting in pairs, cannot have rigidity. The subterfuge of a nonzero limit of $R^4 f$ is unavailable, because that would imply the divergence of the integral $\int R^3 f \, dR$. More fundamentally, the lack of rigidity is an immediate consequence of the symmetry properties of a central-force continuum. Neither Cauchy nor Poisson saw this fact, which is only evident to modern physicists trained to exploit symmetries. It was Saint-Venant who first remarked that the lack of shear stress in a perfectly continuous body resulted from the perfect invariance of the central forces acting in such a body for a large class of internal, shearing deformations. As a simple example, take a global shift of the half of an (infinite) body situated on one side of a fixed plane.[62]

Another of Poisson's objections to Navier was that the method of moments, which Lagrange had successfully used for continuous media, did not apply to molecular systems. This is a surprising statement, since the principle of virtual velocities does not presuppose the continuity of the material system to which it is applied. Poisson probably meant that Navier's estimate of the total moment did not properly include the contribution of molecules whose sphere of action intersects the surface of the body. Indeed, the moments of the forces between such a molecule and all other molecules of the body do not sum up to the full value given in eqn (3.4). Nevertheless, the contribution of these bordering molecules is negligible, because their moment is to the total moment what the radius of action is to the average radius of the body. Although Navier never gave this justification, his intuitive estimate of the total moment was correct.[63]

Navier's methods were more coherent than Poisson believed, and they had considerable advantages. They minimized assumptions concerning the nature of molecular forces, and they provided a direct link between these assumptions and macroscopic properties. For this reason, several modern commentators have seen in Navier's theory an anticipation of George Green's potential-based theory of elasticity of 1837. Regarding the necessity of preserving discrete sums, Poisson was essentially correct. However, he exaggerated the difficulty; in the isotropic case the substitution of integrals for sums does not affect the structure of the equations of motion as long as the integration over distance is not explicitly performed.[64]

3.4.4 *Fluids as temporary solids*

In 1829, Poisson, the champion of molecular rigor, had to correct several flaws in his 1828 memoir that Cauchy's memoir had made apparent. He took this opportunity to offer a theory of fluid motion based on the following assumption: a fluid, like a solid, experiences stresses during its motion, but these stresses spontaneously relax in a very short time. In this picture, the liquid goes through a rapid alternation of stressed and relaxed states.

---

[62]Saint-Venant [1834] sect. 2, [1844]. The remark on the limit of $R^4 f$ is mine. By varying Poisson's central forces around equilibrium, Navier's elastic force ϕ is easily seen to be related to Poisson's f (in my notation) by $\phi = R^{-1} f + R \, d(R^{-1} f)/dR$, which implies that the integral of $R^4 \phi$ and Navier's elastic constant vanish. Saint-Venant's argument may have been inspired by Fresnel's remark, in his molecular ether-model of 1821, that resistance to the shift of a slice of ether required molecular constitution with intermolecular distances much smaller than this shift (Fresnel [1821] pp. 630–2).

[63]Poisson [1829a] p. 400.

[64]Reference to Green is found, e.g., in Dahan [1992]. One way to save Navier's procedure is to introduce a finite lower limit in his integrals, see Clausius [1849] pp. 56–8.

Poisson further assumed that the average stress system of the fluid was to the fluid's rate of deformation what the stress system of an isotropic solid was to its strain. This hypothesis leads to the Navier–Stokes equation, with some additions to the pressure gradient term that depend on the compressibility of the fluid.[65]

Poisson did not refer to Navier's memoir on fluid motion, which he must have judged incompatible with sound Laplacian reasoning. Nor did he mention Cauchy's 'perfectly inelastic solid', despite the similarity between his and Cauchy's ways of relating the fluid stresses to those in an isotropic elastic solid.

3.5   Saint-Venant: slides and shears

3.5.1   *Le Pont des Invalides*

Navier's and other Polytechnicians' efforts to reconcile theoretical and applied mechanics had no clear effect on French engineering practice. Industry prospered much faster in Britain, despite the lesser mathematical training of its engineers. Some of Navier's colleagues saw this and ridiculed the use of transcendental mathematics in concrete problems of construction.[66] In the mid-1820s, a spectacular incident apparently justified their disdain. Navier's *chef-d'oeuvre*, a magnificent suspended bridge at the Invalides, had to be dismantled in the final stage of its construction.

Navier had learnt the newer technique of suspension during official missions to England and Scotland in 1820 and 1823. At the end of his ministerial report, he argued in favor of a new suspended bridge of unprecedented scale across the River Seine and facing the Invalides (see Fig. 3.4). In Prony's and Saint-Venant's well-informed opinion, Navier's innovative design was based on sound experience and calculation. Yet, as the bridge was nearly finished, an accidental flood caused the displacement of one of the rounded stones on which the suspending chains changed direction before anchoring (see Fig. 3.4(b)). As Saint-Venant later explained, Navier had mis-estimated the direction of the force exerted by the chain on the stone—a kind of oversight that frequently occurs in engineering construction and that is easily corrected on the spot. Hostile municipal authorities nevertheless obtained the dismantlement of Navier's bridge.[67]

According to Saint-Venant, the incident meant more than a local administrative deficiency:

> At that time there already was a surge of the spirit of denigration, not only of the *savants*, but also of science, disparaged under the name of *theory* opposed to *practice*;

[65]Poisson [1831a] pp. 139–74. Stokes showed that, for small compressions, Poisson's additional gradient term is $(\mu/3)\nabla(\nabla \cdot \mathbf{v})$, as in Stokes's own molecular fluid model.

[66]Cf. Belhoste [1994] pp. 24–5. Belhoste explains how this state of affairs prompted reforms at the Ecole Polytechnique and at the Ecoles d'applications.

[67]Navier [1823d], [1830]. Cf. Prony [1864] pp. xlv–xlvii, Saint-Venant [1864a] pp. lxv–lxix, Grattan-Guinness [1990] pp. 994–1000, Picon [1992] pp. 372–84, Kranakis [1997], Cannone and Friedlander [2003]. The popular perception of this event differed from Saint-Venant's, as shown by this extract from Honoré de Balzac's *Le curé de village*: 'All France knew of the disaster which happened in the heart of Paris to the first suspension bridge built by an engineer, a member of the Academy of Sciences; a melancholy collapse caused by blunders such as none of the ancient engineers—the man who cut the canal at Briare in Henri IV's time, or the monk who built the Pont Royal—would have made; but our administration consoled its engineer for his blunder by making him a member of the Council-general' (transl. by K. P. Wormeley, quoted in Cannone and Frielander [2003] p. 7).

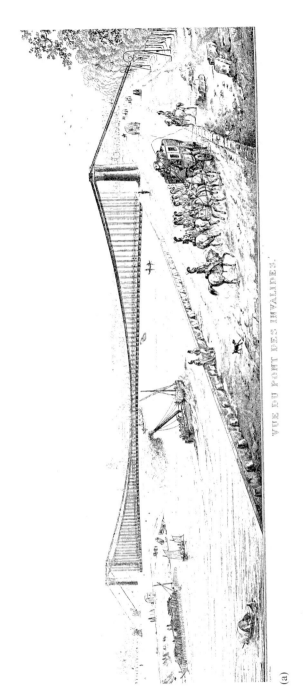

VUE DU PONT DES INVALIDES.

(a)

(b)

Fig. 3.4. (a) Navier's projected Pont des Invalides on the River Seine and (b), the anchoring system for the chains. From Navier [1830] plates.

one henceforth exalted practice in its most material aspects, and pretended that higher mathematics could not help, as if, when it comes to results, it made sense to distinguish between the more or less elementary or transcendent procedures that led to them in an equally logical manner. Some *savants* supported or echoed these unfounded criticisms.

Some engineers were indeed fighting the theoretical approach that Navier embodied. In 1833, the *Ingénieur en chef des Ponts et Chaussées*, Louis Vicat, already acclaimed for his improvement of hydraulic limes, cements, and mortars, performed a number of experiments on the rupture of solids. His declared aim was 'to determine the causes of the imperfection of known theories, and to point out the dangers of these theories to the constructors who, having had no opportunity to verify them, would be inclined to lend them some confidence.' He measured the deformations and the critical charge for various kinds of loading, and observed the shapes of the broken parts. He believed to have refuted Coulomb's and Navier's formulas for the collapse of pillars, as well as Navier's formulas for the flexion and the torsion of prisms. Moreover, he charged Coulomb and Navier with erroneous conceptions of the mode of rupture.[68]

3.5.2   *Vicat's ruptures*

Vicat distinguished three ways in which the aggregation of a solid could be destroyed: pull (*tirage*), pressure (*pression*), and sliding (*glissement*). He called the corresponding forces pulling force (*force tirante*), sustaining force (*force portante*), and transverse force (*force transverse*). This last force (our shearing stress) he defined as 'the effort which tends to divide a body by making one of its parts slide on the other (so to say), without exerting any pressure nor pull outside the face of rupture.' The usual theories of sustaining beams, Vicat deplored, ignored the slides and transverse forces, even though they controlled the rupture of short beams under transvers load. A important exception was Charles Augustin Coulomb, of whose theory Vicat however disapproved.[69]

Vicat published his memoir in the *Annales des Ponts et Chaussées*, but also ventured to send a copy for review to the Academy of Sciences. The reviewers, Prony and Girard, defended their friends Coulomb and Navier, arguing that Vicat had used granular, inflexible materials and short beams for which the incriminated formulas were not intended. They judged that Vicat's measurements otherwise confirmed existing theories. They also emphasized that only Coulomb's theory could justify the use of reduced-scale models, on which Vicat's conclusions partly depended.[70]

In his response, Vicat compared the two Academicians to geometers who would declare the law 'surface equals half-product of two side lengths' to apply to any triangle because they had found it to hold for rectangular triangles. In a less ironic tone, he showed that some of his measurements did contradict the existing theories in their alleged domain of validity. Navier himself did not respond to Vicat's aggression. However, some modifications in his course at the Ponts et Chaussées suggest that he took Vicat's conclusions on the

---

[68]Saint-Venant [1864*a*] p. lxviii; Vicat [1833] p. 202. On Vicat, his work on limes, cements, and mortars, and his implicit criticism of Navier's conception of suspended bridges, cf. Picon [1992] pp. 364–71, 384–5.

[69]Vicat [1833] p. 201. Cf. Benvenuto [1998] pp. 18–19.

[70]Prony and Girard [1834].

importance of slides and transverse forces seriously. His former student Saint-Venant certainly did.[71]

3.5.3 *Molecules, slides, and approximations*

Adhémar Barré de Saint-Venant had an 'X+Ponts' training, and an exceptional determination to reconcile engineering with academic science. His mathematical fluency and his religious dedication to the improvement of his fellow citizens' material life determined this attitude. He rejected both the narrow empiricism of Vicat and the arbitrary idealizations of French rational mechanics. His own sophisticated strategy may be summarized in the following five steps.[72]

(i) Start with the general mechanics of bodies as they are in nature, which is to be based on the molecular conceptions of Laplace, Poisson, and Navier.

(ii) Determine the macroscopic kinematics of the system, and seek molecular definitions for the corresponding macroscopic dynamics.

(iii) Find macroscopic equations of motion, if possible, by summing over molecules, or else by macroscopic symmetry arguments. The molecular level is thus black-boxed in adjustable parameters.

(iv) Develop analytical techniques and methods of approximation to solve these equations in concrete situations.

(v) Test consequences and specify adjustable parameters by experimental means.

Saint-Venant developed this methodology while working on elasticity and trying to improve on Navier's methods. He regarded the first, molecular step as essential for a clear definition of the basic concepts of mechanics and for an understanding of the concrete properties of matter. In his mind, the most elementary interaction was the direct attraction or repulsion of two mass points. Consequently, there could be no continuous solid (as Poisson and Cauchy had proved in 1828). Matter had to be discontinuous, and all physics had to be reduced to central forces acting between non-contiguous point-atoms.[73]

In the second, kinematic step Saint-Venant characterized the macroscopic deformations of a quasi-continuum in harmony with Vicat's analysis of rupture. Cauchy had introduced the quantities $e_{ij} = \partial_i u_j + \partial_j u_i$, but only to determine the dilation or contraction $(1/2)e_{ij}dx_i dx_j$ of a segment $d\mathbf{r}$ of the body. While studying a carpentry bridge on the River Creuze in 1823, and later in his lectures at the Ponts et Chaussées, Saint-Venant gave a precise geometrical definition of Vicat's slides and took them into account in a computation of the flexion of beams. According to this definition, the jth component of slide (*glissement*) in a plane perpendicular to the ith axis is, at a given point of the body, the cosine of the angle that two concrete lines of the body intersecting at this point and originally parallel to the ith and jth axes make after the deformation (see Fig. 3.5). To first order in $\mathbf{u}$, this is the same as Cauchy's e_{ij}. Saint-Venant used the slides not only to

[71]Vicat [1834].   [72]Cf. Boussinesq and Flamant [1886], Melucci [1996], Darrigol [2001].

[73]Saint-Venant [1834], [1844].

e_{ij}

Fig. 3.5.   Geometrical meaning of Saint-Venant's slide e_{ij} with respects to the orthogonal axes i and j (in the plane of the figure).

investigate the limits of rupture, but also to develop a better intuition of the internal deformations in a bent or twisted prism.[74]

Saint-Venant defined the pressure on a surface element dS as the resultant of the forces between any two molecules such that the line joining them crosses the surface element. This pressure has the form $\tau_{ij}\,dS_j$, which defines the stress system τ_{ij}. Saint-Venant then determined the relation between these stresses and the strains e_{ij}, in one of Cauchy's two manners. Although the first manner, based on symmetry only, was simpler, he believed that only the second, molecular manner could give the correct number of independent elastic constants (one constant instead of two in the isotropic case).[75]

On this very theoretical basis, Saint-Venant struggled to solve concrete problems of engineering. He was aware of the great variety of available strategies of approximation that could help in this task:

> Between mere groping and pure analysis, there are many intermediaries: the methods of false position, the variation of arbitrary constants, the solutions by series or continuous fractions, the methods of successive approximations, integration by the computation of areas or by the formulas of Legendre and Thomas Simpson, the reduction of the equations to more easily soluble ones by the choice of an unknown of which one may neglect a few powers or some function in a first approximation, graphical procedures, figurative curves drawn on squared paper, the use of curvilinear coordinates, etc. etc.

None of these methods, however, sufficed to solve the outstanding problem of the engineer of wood and iron structures, namely the flexion and torsion of prisms. For some twenty years, Saint-Venant worked hard to avoid the simplifications used in previous solutions, such as the absence of slides, small deformation, perpendicularity of longitudinal fibers and transverse sections, flatness of transverse sections, etc.[76]

[74]Saint-Venant [1837], [1843a] p. 943: 'Je fais entrer dans le calcul les effets de glissement latéral dus à ces composantes transversales dont l'omission a été l'objet principal d'une sorte d'accusation portée par M. Vicat contre toute la théorie de la résistance des solides.' Cf. Boussinesq and Flamant [1886] p. 560 (bridge on the River Creuze), Todhunter and Pearson [1886–1893] vol. 1, pp. 834–6, 843, vol. 2, pp. 394–5, Benvenuto [1998] pp. 20–4.

[75]Saint-Venant [1843b], [1834/35]. In their molecular theories, Cauchy and Poisson used a less consistent definition of pressure that makes it the resultant of the forces between all the molecules on one side of the plane of the surface element and the molecules belonging to a straight cylinder based on the other side of the element. Cauchy [1845] approved Saint-Venant's definition. Cf. Darrigol [2002a] pp. 122–3.

[76]Saint-Venant [1834/35]. For the successive steps of Saint-Venant's work on the flexion and torsion of prisms, see Saint-Venant [1864c].

His most impressive achievement was the 'semi-inverse' method he developed in the 1830s. The 'direct' problem of elasticity, which is the determination of impressed forces knowing the deformation, is easily solved by applying the stress–strain relation. In contrast, the practically important 'inverse' problem, which is the determination of deformations under given impressed forces, leads to differential equations whose integration in finite terms is usually impossible. Saint-Venant's important idea was to replace the inverse problem with a solvable, mixed problem in which the deformation and the impressed forces were both partly given. He then showed that the exact solutions of the latter problem did not significantly differ from the practically needed solution of the inverse problem.[77]

3.5.4 *On fluid motion*

Although Saint-Venant is best known for his work on elasticity, he also had a constant interest in hydraulics. Early in his career, he reflected on waterwheels and the channels and weirs that fed them. He also began to think about the scientific control of waters in rural areas, which he later called *hydraulique agricole*. In this field, as for elasticity, Saint-Venant avoided narrow empiricism. He wanted to base the determination of channel and pipe flow on fundamental hydrodynamics. Navier's failed attempt in this direction no doubt stimulated him.[78]

In 1834, Saint-Venant submitted to the Academy of Sciences a substantial, though never published, memoir on the dynamics of fluids. To start with, he expressed his approbation of the *mécanique physique* by citing Poisson: 'It is important for the progress of sciences that rational mechanics should no longer be an abstract science, founded on definitions referring to an imaginary state of bodies.' He rejected ideal solids, argued for central forces and point-atoms, and proved the discontinuity of matter in the earlier-mentioned manner. He defined the average 'translatory' motion observed in hydraulic experiments and the invisible 'non-translatory' motion that molecular interactions necessarily implied. Then he gave his molecular definition of internal pressures (which he called 'impulsions'), and showed the existence of transverse pressures in moving fluids by a detailed consideration of the perturbation of the translatory motion by molecular encounters. In harmony with his kinematics of elastic bodies, he characterized the transverse pressure as being opposed to the sliding of successive layers of the fluid on one another.[79]

This transverse pressure depends on the microscopic non-translatory motion of the molecules, which propagates through the whole fluid mass 'and gets lost to the outside by producing, in the walls and in the exterior air foreign agitation and other effects foreign to the translatory motion of the fluid.' The live force of the macroscopic motion thus diminishes at the price of hidden microscopic motion. Later, in the 1840s, Saint-Venant identified the non-translatory motions with heat.[80]

---

[77]Saint-Venant introduced this method in 1847 and 1853. His fullest study of the torsion and flexion of prisms is Saint-Venant [1855].

[78]Cf. Melucci [1996], Darrigol [2001].

[79]Saint-Venant [1834] sects 1 (molecular mechanics), 2 (no continuous matter), 4 (undulated motion of molecules), 5 (definition of impulsions), 6–7 (transverse pressures); Poisson [1831a] p. 130.

[80]Saint-Venant [1834] sect. 7. For the identification with heat, cf. Saint-Venant [1887b] p. 73 n.

Deterred by the complexity he saw in the friction-related molecular motions, Saint-Venant renounced a purely molecular derivation of the pressure system. Instead, he appealed to a symmetry argument in the spirit of Cauchy's first theory of elasticity. He assumed that the transverse pressure on a face was parallel to the fluid slide on this face, and (erroneously, he later realized) took the slide itself to be parallel to the projection of the fluid velocity on the face. This led him to an equation of motion far more complicated than Navier's, with five parameters instead of one, and with variations of these parameters depending on the internal, microscopic commotions of the fluid. Saint-Venant applied this equation to flow in rectangular or semi-circular open channels and described a new method of fluid-velocity measurement. He thus wanted to prepare the experimental determination of the unknown functions that entered his equations.[81]

3.5.5   A first-class burial

The commissioners Ampère, Navier, and Félix Savary approved Saint-Venant's memoir. Yet Savary, who was supposed to write the report, never did so and instead expressed disagreements in letters to the author. From Saint-Venant's extant replies, we may infer that Savary did not know of the contradiction between Du Buat's measurements and Navier's equation and that he condemned the recourse to adjustable parameters in fundamental questions of hydrodynamics. In his defense, Saint-Venant clarified the purpose of his memoir: 'My principal goal is all practical: it is the solution of the open-channel problem for a bed of variable and arbitrary figure.' He then formulated an interesting plea for a semi-inductive method:[82]

> My equations contain indeterminate quantities and even indeterminate functions; but is it not good to show how far, in fluid dynamics, we may proceed with a theory that is free of hypotheses (save for *continuity*, at least on *average*), that brings forth the unknown and prepares its experimental determination? A bolder march may sometimes quickly lead to the truth.... However, you will no doubt admit that in such an important matter it may be advantageous to consider things from another point of view, to avoid every supposition and to appeal to experimenters to fix the values of indeterminate quantities by means of *special* experiments prepared so as to isolate the effects that the theory will later try to explain with much more assurance and to represent by expressions that are as free of empiricism as possible.

3.5.6   Re-founding Navier's equation

Three years later, Saint-Venant discovered his error about the direction of slides, and ceased to request a report from Savary. Instead, he inserted a more cogent argument in the manuscript deposited at the Academy. He still assumed that the transverse pressure on a face was parallel to the slide on this face, or, equivalently and even more naturally, that the transverse pressure was zero in the direction of the face for which the slide vanished. However, he now used the correct expression $\partial_i v_j + \partial_j v_i$ for the slides (per unit time) corresponding to the fluid velocity $\mathbf{v}$ and the orthogonal directions i and j. He further

---

[81]Saint-Venant [1834] sects 11 (hypothesis), 15 (equation), 18–24 (consequences), 25–8 (suggested experiments).

[82]Saint-Venant to Savary, 25 Aug. 1834, Bibliothèque de l'Institut de France, MS 4226; see also the letters of 27 July and 10 Sept. 1834, *ibid.*

noted that $\tau_{ii} - \tau_{jj}$ represented twice the transverse pressure along the line bisecting the angle $\hat{ij}$, and $\partial_i v_i - \partial_j v_j$ represented the slide along the same line. Granted that the components of slide must be proportional to the components of transverse pressure, the ratios $\tau_{ij}/(\partial_i v_j + \partial_j v_i)$ and $(\tau_{ii} - \tau_{jj})/2(\partial_i v_i - \partial_j v_j)$ are all equal for every choice of i and j. Denoting by ε their common value at a given point of the fluid and ϖ an undetermined isotropic pressure, this implies that

$$\tau_{ij} = \varepsilon(\partial_i v_j + \partial_j v_i) + \varpi\delta_{ij}. \tag{3.31}$$

As Saint-Venant noted, this stress system yields the Navier–Cauchy–Poisson equation in the special case of a constant ε, with a gradient term contributing to the normal pressure.[83]

For a modern reader familiar with tensor calculus, Saint-Venant's reasoning looks like another proof of the fact that the expression (3.31) is the most general symmetrical second-rank tensor that depends linearly and isotropically on the tensor e_{ij}. Yet this is not the case, because Saint-Venant did not assume the linearity. Admittedly, his hypothesis of the parallelism of slides and tangential pressures implies more than mere isotropy; for instance, it excludes terms proportional to $e_{ik}e_{kj}$. However, it allows for an ε that varies from one particle of the fluid to another, and from one case of motion to another.[84]

Saint-Venant believed that Du Buat's and others' experiments on pipe and channel flow required a variable ε, which expressed the effects of local 'irregularities of motion' on internal friction. The velocity $\mathbf{v}$ in his reasoning referred to the average, smooth, large-scale motion. Smaller-scale motions only entered the final equation as a contribution to tangential pressures defined at the larger scale. Whether or not Saint-Venant regarded Navier's equation with constant ε as valid at a sufficiently small scale is not clear. In any case, he believed that the value of ε should be determined experimentally without prejudging its constancy from place to place or from one case of motion to another.[85]

In the mid-1840s, the military engineer Pierre Boileau undertook a series of experiments on channel and pipe flow. Unlike most hydraulicians, who were only interested in the global discharge, Boileau planned measurements of the velocity profile of the flow. Saint-Venant congratulated him for this intention, because such knowledge was necessary to estimate the friction between successive fluid filaments, or the variable ε of his equation of fluid motion. He advised Boileau on the most suitable channel and pipe shapes and on the technique of velocity measurement. As we will see in Chapter 6, this sort of experiment and the correlative idea of an effective, eddy-related viscosity had a future.[86]

[83]Saint-Venant to Savary, 13 Jan. 1837, *ibid.*; Saint-Venant [1834] new version (later than 1837) of sect. 15; Saint-Venant [1843c].

[84]Saint-Venant [1843c] p. 1243 for variable ε. *Ibid.* p. 1242n, Saint-Venant noted that Cauchy's pressure theorems were valid to second order in the dimensions of the volume elements, 'which allows us to extend their application to the case when partial irregularities of the fluid motion forces us to take faces of a certain extension so as to have regularly varying averages.'

[85]Saint-Venant [1843c]; Saint-Venant to Savary (ref. 82), 27 July 1834 (on Du Buat); Saint-Venant [1834] new sect. 15: 'It is experiment that should determine whether ε is constant or variable.' Perhaps Saint-Venant did not believe in a constant ε, even at the small scale, because, for the tumultuous flows observed in rivers channels and occurring in pipes of not too small diameter, Saint-Venant believed that any irregularity of motion cascaded to a smaller and smaller scale by 'molecular gearing'.

[86]Boileau [1847], [1854]. Saint-Venant to Boileau, 29 Mar. 1846, Fond Saint-Venant, reproduced and discussed in Melucci [1996] pp. 65–71.

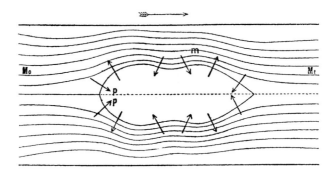

Fig. 3.6.   Drawing for Saint-Venant's proof of d'Alembert's paradox (from Saint-Venant [1887*b*] p. 50).

3.5.7   *Fluid resistance*

In 1846, Saint-Venant tackled the old, difficult problem of fluid resistance. He first showed that the introduction of internal friction solved d'Alembert's paradox. For this purpose, he borrowed from d'Alembert, Du Buat, and Poncelet the idea of placing the immersed body inside a cylindrical pipe (see Fig. 3.6), from Euler the balance of momentum, and from Borda the balance of live forces. If the body is sufficiently far from the walls of the pipe, the action of the fluid on the body should be the same as for an unlimited flow. If the body is fixed, the flow is steady, and the fluid is incompressible, then the momentum which the fluid conveys to the body in unit time is equal to the difference $P_0S - P_1S$ between the pressures on the faces of a column of fluid extending far before and after the body, because the momentum of the fluid column remains unchanged. For an ideal fluid, the work $(P_0S - P_1S)v_0$ of these pressures in unit time must vanish, because the live force of the fluid column is also unchanged. Hence the two pressures are equal, and the fluid resistance vanishes. This is d'Alembert's paradox, as proved by Saint-Venant.[87]

In a molecular fluid, the (negative) work of internal friction must be added to the work of the pressures P_0 and P_1, or, equivalently, the live force of non-translatory motions must be taken into account. Hence the pressure falls when the fluid passes the body, and the resistance no longer vanishes. The larger the amount of non-translatory motion induced by the body, then the higher is the resistance. When tumultuous, whirling motion occurs at the rear of the body, the resistance largely exceeds the value it would have for a perfectly smooth flow. After drawing these conclusions, Saint-Venant improved on a method invented by Poncelet to estimate the magnitude of the resistance and based on the assumption that the pressure P_1 at the rear of the body does not differ much from the value that Bernoulli's law gives in the most contracted section of the flow (see Fig. 3.7).[88]

In summary, Saint-Venant did not accept the dichotomy between a hydrodynamic equation for ideally smooth flow on the one hand, and completely empirical retardation and resistance formulas for hydraulic engineers on the other. He sought a *via media* that

---

[87]Saint-Venant [1846*b*], [1887*b*] pp. 45–9. In Borda [1766] p. 605, the Chevalier de Borda had derived the paradox in an even simpler manner, by applying the conservation of live forces to a body pulled uniformly through a calm fluid. For a modern derivation, see Appendix A.

[88]Saint-Venant [1846*b*] pp. 28, 72–8, 120–1, [1887*b*] pp. 56–192.

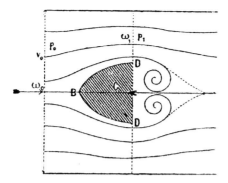

Fig. 3.7.   Drawing for Poncelet's and Saint-Venant's evaluation of fluid resistance (from Saint-Venant [1887*b*]
p. 89).

would bring theoretical constraints to bear on practical flows and yet would allow for
some experimental input. One of his strategies, later pursued by Joseph Boussinesq and
successfully applied to turbulent flow to this day, consisted of reinterpreting Navier's
hydrodynamic equation as controlling the average, smoothed-out flow with a variable
viscosity coefficient. Another was the astute combination of momentum and energy
balances with empirically-known features of the investigated flow. For hydraulics, as for
elasticity, Saint-Venant was a most persevering and imaginative conciliator of fundamental and practical aims.

3.6   Stokes: the pendulum

3.6.1   *A swimming mathematician*

Until the 1830s at least, the production of advanced mathematical physics in an engineering context remained a uniquely French phenomenon, largely depending on the creation
of the Ecole Polytechnique. The main British contributors to elasticity theory and hydrodynamics in this period had little or no connection with engineering. Typically, they were
astronomers like Airy and Challis, or mathematicians like Green and Kelland. Their work
on elasticity was subordinate to their interest in the new wave optics, and the aspects
of hydrodynamics that captured their attention tended to be wave and tide theory.
A Cambridge-trained mathematician, and the first Wrangler and Smith prize winner
(1841), George Gabriel Stokes was not much closer to the world of engineers. He nonetheless was a keen observer of nature, a first-rate swimmer, and a naturally gifted
experimenter. He was quick to note the gaps between idealized theories and real processes,
and sometimes eager to fill them.[89]

During the two decades preceding Stokes's student years, British mathematical physics
had undergone deep reforms that eliminated archaic Newtonian methods in favor of the
newer French ones. While Fourier's theory of heat and Fresnel's theory of light were most

---

[89]Cf. Stokes [1846*a*], Parkinson [1976], Wilson [1987].

admired for their daring novelty, the hydrodynamics of Euler and Lagrange provided the simplest illustration of the necessary mathematics of partial differential equations. The famous Cambridge coach William Hopkins made it a basic part of the Tripos examination, and persuaded Stokes to choose it as his first research topic. In his first papers, published in the early 1840s, Stokes already noted discrepancies between real and ideal flows and suggested a few remedies, including the introduction of viscosity.[90]

3.6.2 *The pendulum*

Stokes's interest in imperfect fluidity derived from the pendulum experiments performed by Edward Sabine in 1829. This artillery officer had led a number of geodesic projects, one of which, in 1821/22, dealt with the pendulum determination of the figure of the Earth. In 1828, the German astronomer Friedrich Bessel published a memoir on the seconds' pendulum that brought pendulum studies, and quantitative experiment in general, to an unprecedented level of sophistication. Bessel not only improved experimental procedures and data analysis, but he also brought new theoretical insights into the various effects that altered the ideal pendulum motion. Most importantly, he was the first to take into account the inertia of the air moved by the pendulum. His study played a paradigmatic role in defining a Königsberg style of physics. It also induced further experimental and theoretical pendulum studies in Britain and France.[91]

   While investigating Bessel's inertial effect, Captain Sabine found that the mass correction of a pendulum oscillating in hydrogen was much higher than the density ratio between hydrogen and air would imply. Sabine suggested that gas viscosity could be responsible for this anomaly. The remark prompted Stokes to study the way viscosity affected fluid motion. His first strategy, implemented in a memoir of 1843, was to study special cases of perfect-fluid motion in order to appreciate departures from reality:[92]

> The only way by which to estimate the extent to which the imperfect fluidity of fluids may modify the laws of their motion, without making any hypothesis on the molecular constitution of fluids, appears to be, to calculate according to the hypothesis of perfect fluidity some cases of fluid motion, which are of such a nature as to be capable of being accurately compared with experiment.

[90]On the transformation of British physics, cf. Smith and Wise [1989] chap. 6. On Hopkins's role, cf. Wilson [1987] p. 132. Henry Moseley's hydrodynamic treatise [1830], written for the students of Cambridge University under Challis's advice, marked a transition between older Newtonian methods and Euler's hydrodynamics: it only introduced the fundamental equations (in integral form) at a very late stage, and based most reasoning on pre-Eulerian techniques such as Bernoulli's law or d'Alembert's principle; it gave a Newtonian treatment of fluid resistance, ignored d'Alembert's paradox, and failed to mention Navier's equations of fluid motion.

[91]On Stokes and pendulums, cf. Stokes [1850b] pp. 1–7. On Sabine, cf. Reingold [1975] pp. 49–53. On Bessel's work, cf. Olesko [1991] pp. 67–73. On pendulum studies in general, cf. Wolf [1889]. Bessel's inertial effect was already known to d'Alembert [1768] and to Du Buat [1786] vol. 3, in a hydraulic context. Poisson [1832] gave the theoretical value of the inertial mass correction for a sphere as half of the mass of the displaced fluid, in conformance with Du Buat's result for water.

[92]Sabine [1829], commentary to his eighth experiment; Stokes [1850b] p. 2 (Sabine); Stokes [1843] pp. 17–18 (quote). Stokes assumed that the motion started from rest, which implies the existence of a velocity potential for a perfect liquid. Stokes hoped that this property would still hold approximately for the small oscillations of a real fluid (*ibid.* p. 30; this turned out to be wrong in the pendulum case).

Among his cases of motion Stokes included oscillating spheres and cylinders that could represent the bulb and the suspending thread of a pendulum. In the spherical case, he found the mass correction to be equal to half the mass of the fluid expelled by the sphere, in conformance with a calculation made by Poisson in 1831, but only five-ninths of Bessel's experimental result of 1828. Stokes also transposed Thomson's method of electrical images to show that a rigid wall placed near the oscillating sphere modified the mass correction. Lastly, he addressed the most evident contradiction with observation, namely, that a perfect fluid does not have any more damping effect on oscillatory motion than it would have on a uniform translational motion.[93]

Stokes evoked three possible causes of the observed resistance, namely, fluid friction, discontinuous flow, and instability leading to a turbulent wake. As he did not yet feel ready to explore any of these options by means of theory, he looked for further experimental results. He was not himself planning pendulum measurements, presumably because the required apparatus and protocol were too complex for his taste; he usually favored experiments that could be performed with the minimum equipment and time consumption. For testing the departure of real fluids from perfect ones, he judged that the moments of inertia of water-filled boxes offered a better opportunity. Unfortunately, the experiments he soon performed with suspended water boxes could only confirm the perfect-fluid theory. They were not accurate enough to show any effect of imperfect fluidity.[94]

3.6.3 *Fluid friction*

Having exhausted the possibilities of his first strategy for studying the imperfection of fluids, Stokes tried another approach. In 1845, he sought to include internal fluid friction in the fundamental equations of hydrodynamics. To Du Buat's arguments for the existence of internal friction he added pendulum damping and a typically British observation: 'The subsidence of the motion in a cup of tea which has been stirred may be mentioned as a familiar instance of friction, or, which is the same, of a deviation from the law of normal pressure.' From Cauchy he borrowed the notion of transverse pressure, as well as the general idea of combining symmetry arguments and the geometry of infinitesimal deformations.[95]

Stokes's first step was the decomposition of the rate of change $\partial_i v_j dx_i$ of an infinitesimal fluid segment **dr** into a symmetric and an antisymmetric part:

$$\partial_i v_j dx_i = \frac{1}{2}(\partial_i v_j + \partial_j v_i)dx_i + \frac{1}{2}(\partial_i v_j - \partial_j v_i)dx_i. \qquad (3.32)$$

Then he showed that the antisymmetric part corresponded to a rotation of the vector **dr**, and the symmetric part to the superposition of three dilations (or contractions) along three orthogonal axes. That $\partial_i u_j - \partial_j u_i$ represents the rotation of an element of a continuum for a small deformation **u** was known to Cauchy. No one, however, had explicitly given

---

[93]Stokes [1843] pp. 36, 38–49, 53; Poisson [1832]. Stokes made his calculation in the incompressible case, knowing from Poisson that the effects of compressibility were negligible in the pendulum problem.

[94]*Ibid.* pp. 60–8; Stokes [1846b] p. 196. On Stokes' experimental style, cf. Liveing [1907].

[95]Stokes [1849a] pp. 75–6; [1848a] p. 3 (cup of tea). Stokes ([1849a] p. 118) refers to Cauchy as follows: 'The method which I have employed is different from [Cauchy's], although in some respects it much resembles it.'

Stokes's decomposition and its geometrical interpretation. Cauchy and other theorists of elasticity directly studied the quadratic form $(1/2)e_{ij}\mathrm{d}x_i\mathrm{d}x_j$ that gives the change in the squared length of the segment $\mathbf{dr}$.[96]

Stokes then required, as Cauchy had done, the principal axes of pressure to be identical with those of deformation. He decomposed the three principal dilations into an isotropic dilation and three 'shifting motions' along the diagonals of these axes. To the isotropic dilation he associated an isotropic normal pressure, and to each shift a parallel transverse pressure. In order to get the complete pressure system, he superposed these four components and transformed the result back to the original system of axes. So far, Stokes's procedure was similar to Saint-Venant's, except that Saint-Venant dealt directly with slides in the original system of axes and did not require any superposition of principal pressures nor any transformation of axes.[97]

The analogy with Saint-Venant—whose communication Stokes was probably unaware of—ends here. Stokes wanted the pressures to depend linearly on the instantaneous deformations. He justified this linearity (including the above-mentioned superposition), as well as the zero value he chose for the pressure implied by an isotropic compression, by means of a somewhat obscure model of 'smooth molecules acting by contact'. His previous approach to the imperfect fluid had been deliberately non-molecular. The new, internal-friction approach was explicitly molecular. Undoubtedly Stokes grew to be an overcautious physicist who avoided microphysical speculation as much as he could. Yet, no more than his French predecessors could he conceive of internal friction without transverse molecular actions.[98]

3.6.4 *Elastic bodies, ether, and pipes*

Stokes's reasoning of course led to the Navier–Stokes equation, since this is the only hydrodynamic equation that is compatible with local isotropy and a linear dependence between stress and distortion rate. After reading Poisson's memoir of 1829, which proceeded from the equations of elastic bodies to those of real fluids, Stokes tried the reverse course and transposed his hydrodynamic reasoning to elastic bodies. From the 'principle of superposition of small quantities', he derived the linearity of the stress–strain relation. He then exploited isotropy in the principal-axis system to introduce two elastic constants, one for the shifts, and the other for isotropic compression.

Stokes thus retrieved the two-constant stress system that Cauchy had obtained for isotropic elastic bodies in his non-molecular theory of 1828. He imputed Poisson's single-

---

[96]Stokes [1849a] pp. 80–4; Cauchy [1841] p. 321 (cf. Dugas [1950] pp. 402–6). Stokes's reasoning did not seem too clear to Saint-Venant; see his letter to Stokes, 22 Jan. 1862, in Larmor [1907] vol. 1, pp. 156–159. Larmor's comment, 'The practical British method of development in mathematical physics, by fusing analysis with direct physical perception or intuition, still occasionally present similar difficulties to minds trained in a more formal mathematical discipline', does not seem to apply well to Saint-Venant, although it certainly applies to the continental perception of Larmor's own work.

[97]Stokes [1849a] pp. 83–4.

[98]*Ibid.* pp. 84–6. Cf. Yamalidou [1998]. Stokes mentioned Saint-Venant's proof in his [1846a] pp. 183–4, with the observation: 'This method does not require the consideration of ultimate molecules at all.' Stokes's model implies a zero trace for the viscous stress tensor, so that his equation includes the term $(\mu/3)\nabla(\nabla \cdot \mathbf{v})$ (besides the $\mu\Delta\mathbf{v}$ term) in the case of a compressible fluid.

constant result to the assumption that the sphere of action of a given molecule contained many other molecules—which only shows that he had not read the memoir in which Cauchy proved this assumption to be unnecessary. More pertinently, Stokes argued that soft solids such as India rubber or jelly required two elastic constants, for they had a much smaller resistance to shifts than to compression. He also suggested that the optical ether might correspond to the case of infinite resistance to compression, for which longitudinal waves no longer exist. In summary, Stokes had both down-to-earth and ethereal reasons to require two elastic constants instead of one. With George Green, whose works he praised, he inaugurated the British preference for the multi-constant theory.[99]

Stokes's immediate purpose was, however, a study of the role of internal friction in fluid resistance and flow retardation. Here boundary conditions are essential. When, in 1845, Stokes read his memoir on fluid friction, he was already inclined to assume a vanishing relative velocity at a rigid wall. He worried, however, about the resulting pipe retardation law, which contradicted Bossut's and Du Buat's results. Navier's and Poisson's condition that the tangential pressure at the wall should be proportional to the slip did not work any better, except for a very small velocity, in which case the measured retardation became proportional to the velocity. Girard's measurements, as interpreted by Navier, seemed to require a finite slip in this case, although Du Buat had found a zero velocity near the walls of a very reduced flow. In this perplexing situation, Stokes refrained from publishing discharge calculations. He only gave the parabolic velocity profile for cylindrical pipes with zero velocity at the walls.[100]

3.6.5 *Back to the pendulum*

In the pendulum case Stokes knew the retardation to be proportional to velocity, in conformance with both the Navier–Poisson boundary condition and the zero-slip condition. He also knew, from a certain James South, that a tiny piece of gold leaf attached perpendicularly to the surface of a pendulum's globe remained perpendicular during oscillation. This observation, together with Du Buat's and Coulomb's small-velocity results, brought him to try the analytically simpler zero-slip condition. The success of this choice required justification. In his major memoir of 1850 on the pendulum, Stokes argued that it was 'extremely improbable' that the forces called into play by an infinitesimal internal shear and by a finite wall shear would be of the same order of magnitude, as they should be for the dynamical equilibrium of the layer of fluid next to the wall.[101]

Neglecting the quadratic $(\mathbf{v} \cdot \nabla)\mathbf{v}$ terms in the Navier–Stokes equation, Stokes found an exact analytical solution for an oscillating sphere representing the globe of the pendulum, and a power-series solution for an oscillating cylinder representing the suspending thread of the pendulum. The results explained Sabine's mass-correction anomaly, and permitted a close fit with Francis Baily's extensive experiments of 1832. Ironically, Stokes obtained this impressive agreement with a wrong value for the viscosity coefficient. The explanation of this oddity is that his data analysis depended on the assumption that viscosity is

[99]Stokes [1849*a*] sects 3–4.

[100]*Ibid.* pp. [93–9]; Stokes [1846*a*] 186. For large pipes, Stokes assumed a tangential pressure proportional to the velocity squared at the walls, justified in Du Buat's and Coulomb's manner by surface irregularities.

[101]Stokes [1850*b*] pp. 7, 14–15.

proportional to density, which is at variance with the approximate constancy later proved by James Clerk Maxwell.[102]

Stokes also considered the case of uniform translation, still in the linear approximation of the Navier–Stokes equation. For a sphere of radius R moving at the velocity V, he obtained the expression $-6\pi\mu RV$ for the resistance, now called 'Stokes' formula'. *En passant*, he explained the suspension of clouds: according to his formula, the resistance experienced by a falling droplet decreases much more slowly with its radius than its weight does. In the case of a cylinder, he found that no steady solution existed, because the quantity of dragged fluid increased indefinitely. He speculated that this accumulation implied instability, a trail of eddies, and nonlinear resistance.[103]

At that time, Stokes did not discuss other cases of nonlinear resistance, such as the swiftly-moving sphere. In later writings, he adopted the view that the Navier–Stokes condition with the zero-shift boundary condition applied generally, and that the non-linearity of the resistance observed beyond a certain velocity corresponded to an instability of the regular solution of the equation, leading to energy dissipation through a trail of eddies. This is essentially the modern viewpoint.[104]

3.7   The Hagen–Poiseuille law

3.7.1   *Hagen's pipes*

Stokes's pendulum memoir contains the first successful application of the Navier–Stokes equation with the boundary condition which is now regarded as correct. For narrow-pipe flow, Stokes (and previous discoverers of the Navier–Stokes equation) knew only of Girard's results, which seemed to confirm the Navier–Poisson boundary condition. Yet a different law of discharge through narrow tubes had been published twice before Stokes' study, in 1839 and in 1841.

The German hydraulic engineer Gotthilf Hagen was the first to discover this law, without knowledge of Girard's incompatible results. Hagen had learned precision measurement under Bessel and had traveled through Europe to study hydraulic constructions. As he had doubts about Prony's and Johann Eytelwein's widely-used formulas for pipe retardation, he performed his own experiments on this subject in 1839. In order to best appreciate the effect of friction, he selected pipes of small diameter, between 1 mm and 3 mm. Although the principle of the experiment was similar to Girard's, Hagen eliminated important sources of error that had escaped Girard's attention. For example, he carefully measured the internal diameter of his pipes by weighing their water content. Also, he avoided the irregularities of open-air efflux by having the pipe end in a small tank with a constant water level (see Fig. 3.8).[105]

To his surprise, Hagen observed that, beyond a critical pipe-flow velocity of order $(2gh)^{1/2}$, with h being the pressure head, the flow became highly irregular. For better

[102]Stokes [1850b] sects 2–3. On the wrong value of the viscosity coefficient, cf. Stokes, note appended to his [1850b], *SMPP* 3, pp. 137–41; Stokes to Wolf, undated (*c.* 1991), in Larmor [1907], vol. 2 pp. 323–4.

[103]Stokes [1850b] 59, pp. 66–7. More on the cylinder case will be said in Chapter 5, pp. 186–7.

[104]Cf. Stokes's letters of the 1870s and 1880s in Larmor [1907].

[105]Hagen [1839]. Cf. Schiller [1933] pp. 83–4, Rouse and Ince [1957] pp. 157–61.

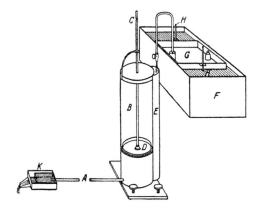

Fig. 3.8.   Hagen's apparatus for measuring fluid discharge (from Hagen [1839]). The tank F feeds the cylinder
B through the regulating device H. The water level in the cylinder is determined by reading the scale C
attached to the floating disk D. The discharging tube A ends in the overflowing tank K.

experimental control, he decided to operate below this threshold. His experimental results
are summarized by the formula

$$h = \frac{\alpha L Q + \beta Q^2}{R^4}, \tag{3.33}$$

where h is the pressure head, Q is the discharge, L is the length, α is a temperature-
dependent constant, and β is a temperature-independent constant. In true Königsberg
style, Hagen determined the coefficients and exponents by the method of least squares and
provided error estimates.[106]

Hagen correctly interpreted the quadratic term as an entrance effect, corresponding to
the live force acquired by the water when entering the tube. Assuming a conic velocity
profile ($v \propto r$), he obtained a good theoretical estimate for the β coefficient. He attributed
the linear term to friction, and justified the $1/R^4$ dependence by combining the conic
velocity profile with an internal friction proportional to the squared relative velocity of
successive fluid layers. Perhaps because this concept of friction later appeared to be
mistaken, full credit for the discovery of the QL/R^4 law has often been given to Poiseuille.
Yet Hagen's priority and the excellence of his experimental method are undeniable.[107]

3.7.2   Dr Poiseuille's capillary vessels

Jean-Louis Poiseuille, a prominent physician with a Polytechnique education, performed
his experiments on capillary-tube flow around 1840, soon after Hagen. He had no
particular interest in hydraulics, but wanted to understand 'the causes for which some
organ received more blood than another.' Having eliminated a few received explanations

[106]Hagen [1839] pp. 424, 442.

[107]*Ibid.* pp. 433, 437, 441.

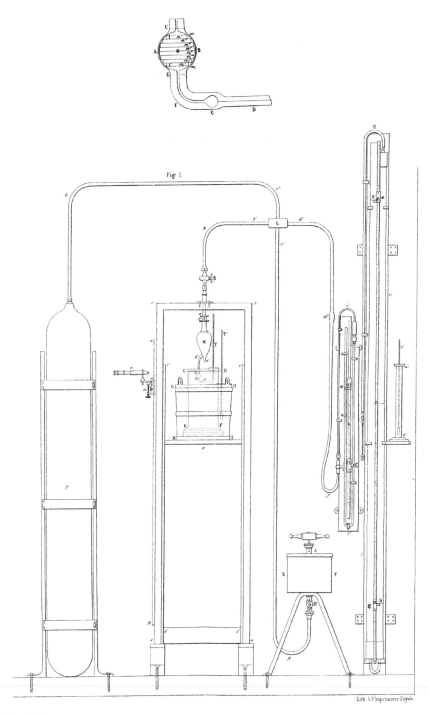

Fig. 3.9.   Poiseuille's apparatus to measure fluid discharge through capillary tubes (from Poiseuille [1844]). The reservoir P, originally filled with compressed air by the pump **AXY**, is connected to a barometric device (on the right), and to the flask M, which in turn feeds the elaborate glass part CABEFGD (enlarged above). The fluid contained in the spherical bulb **AB** is pressed to flow through the capillary tube D into a thermostatic bath.

in a previous memoir, he focused on the behavior of capillary vessels and decided to examine experimentally the effects of pressure, length, diameter, and temperature on the motion of various liquids through capillary glass tubes. He judged Girard's anterior measurements to be irrelevant, because capillary blood vessels were about one hundred times narrower than Girard's tubes.[108]

Poiseuille produced the flow-generating pressure with an airpump and reservoir, in vague analogy with the hearts of living organisms (see Fig. 3.9). He avoided the irregularities of open-air efflux and controlled temperature by immersing his capillary tubes in a thermostatic bath. He determined the discharge from the lowering of the fluid level in the feeding flask. The most delicate parts of the measurements were the optical and hydraulic control of the cylindricity of the capillary tubes, and the determination of the pressure head. Like Girard, Poiseuille overlooked the entrance effect, which is fortunately negligible for very narrow tubes. He properly took into account hydrostatic head, viscous retardation in the larger tube leading to the capillary tube, and the pressure shift in a given run. The description of his protocol was so meticulous as to include prescriptions for the filters he used to purify his liquids. His results compare excellently with modern theoretical expectations. They of course include the Poiseuille law $Q = KPR^4/L$, P being the fall of pressure and K a temperature-dependent constant.[109]

Poiseuille only mentioned Navier's theory to condemn it for leading to the wrong PR^3/L law. Unfortunately, Navier did not live long enough to know of Poiseuille's result. The Academicians who reviewed the physician's memoir (Arago, Babinet, and Piobert) did not know that Navier had already obtained the R^4 dependence in the case of a square tube of side R with zero shift at the walls. It was left to Franz Neumann, who had probably known Hagen in Königsberg, to give the first public derivation of the Hagen–Poiseuille law. Assuming zero velocity at the walls and making the internal friction proportional to the transverse velocity gradient, Neumann derived the quadratic velocity profile and integrated it to obtain the discharge. His student Heinrich Jacobson published this proof in 1860. The Basel physicist Eduard Hagenbach published a similar derivation in the same year, with an improved discussion of entrance effects and a mention of the *Erschütterungswiderstand* (agitation resistance) that occurred for larger pipes. Lastly, the French mathematician Emile Mathieu published a third similar proof in 1863.[110]

3.7.3 *A slow integration*

It would be wrong to believe that these derivations of Poiseuille's law were meant to vindicate the Navier–Stokes equation. Neumann and Mathieu did not mention Navier's theory at all. Hagenbach did, but imitated Poiseuille in globally condemning Navier's approach. Newton's old law of the proportionality between friction and transverse velocity gradient was all that these physicists needed. Hermann Helmholtz was probably the first physicist to link the Navier–Stokes equation to the Hagen–Poiseuille law.

---

[108]Poiseuille [1844]; Arago, Babinet, and Piobert [1842]. Cf. Rouse and Ince [1957] pp. 160–1, Schiller [1933] p. 89, Pedersen [1975].

[109]Poiseuille [1844] p. 519. For a modern evaluation, cf. Schiller [1933] pp. 85–9.

[110]Poiseuille [1844] p. 521; Jacobson [1860] Hagenbach [1860] Mathieu [1863].

Helmholtz's interest in fluid friction derived from his expectation that it would explain a leftover discrepancy between theoretical and measured resonance frequencies in organ pipes. In 1859, he derived a hydrodynamic equation that included internal friction, and asked his friend William Thomson whether it was the same as Stokes's, of which he had heard but had not seen. The answer was yes.[111]

In order to determine the viscosity coefficients of liquids, Helmholtz asked his student Gustav von Piotrowski to measure the damping of the oscillations of a hollow metallic sphere filled with liquid and suspended by a torsion-resisting wire. Helmholtz integrated the Navier–Stokes equation so as to extract the viscosity coefficient from these measurements and also from Poiseuille's older experiments on capillary tubes. The two values disagreed, unless a finite slip of the fluid occurred on the walls of the metallic sphere. When he learned about this analysis, Stokes told Thomson that he inclined against the slip, but did not exclude it.[112]

This episode shows that, as late as 1860, the Navier–Stokes equation did not yet belong to the physicist's standard toolbox. It could still be rediscovered. The boundary condition, which is crucial in judging consequences for fluid resistance and flow retardation, was still a matter of discussion. Nearly twenty years elapsed before Horace Lamb judged the Navier–Stokes equation and Stokes's boundary condition to be worth a chapter in a treatise on hydrodynamics. This evolution rested on the few successes met in the ideal circumstances of slow or small-scale motion, and on the confirmation of the equation by Maxwell's kinetic theory of gases in 1866. Until Reynolds's and Boussinesq's studies of turbulent flow in the 1880s, described in Chapter 7, the equation remained completely irrelevant to hydraulics.[113]

Thus, the mere introduction of viscous terms in the equations of motion did not suffice to explain the flows most commonly encountered in natural and artificial circumstances. This failure long confined the Navier–Stokes equation to the department of physico-mathematical curiosities, despite the air of necessity that its multiple molecular and non-molecular derivations gave it. As we will see in the following two chapters, a few hydrodynamicists left this equation aside and speculated that much of the true behavior of slightly-viscous fluids such as air and water could be understood without leaving the perfect-liquid context.

[111]Helmholtz to Thomson, 30 Aug. 1859, Kelvin Collection, Cambridge University Library; Thomson to Helmholtz, 6 Oct. 1859, HN. Cf. Darrigol [1998], and Chapter 4, pp. 148, 158–9.

[112]Helmholtz and Piotrowski [1860] pp. 195–214 (calculations in the spherical case), 215–17 (calculation for the Poiseuille flow); Stokes to Thomson, 22 Feb. [1862], in Wilson [1990]. Helmholtz was aware of Girard's measurements (Helmholtz and Piotrowski [1860] pp. 217–19), which he unfortunately trusted, but not of Hagen's.

[113]Lamb [1879] chap. 9. The verification of the consequences of Maxwell's kinetic theory by viscous damping experiments required new, improved solutions of the Navier–Stokes equation, cf. Hicks [1882] pp. 61–70.

4

VORTICES

> I have been able to solve a few problems of mathematical physics on which the greatest mathematicians since Euler have struggled in vain . . . But the pride I could have felt over the final results . . . was considerably diminished by the fact that I knew well how the solutions had almost always come to me: by gradual generalization of favorable examples, through a succession of felicitous ideas after many false trails. I should compare myself to a mountain climber who, without knowing the way, hikes up slowly and laboriously, often must return because he cannot go further, then, by reflection or by chance, discovers new trails that take him a little further, and who, when he finally reaches his aim, to his shame discovers a royal road on which he could have trodden up if he had been clever enough to find the right beginning. Naturally, in my publications I have not told the reader about the false trails and I have only described the smooth road by which he can now reach the summit without any effort.[1] (Hermann Helmholtz, 1891)

One way of addressing the practical failures of Euler's fluid mechanics was to introduce viscosity into the fundamental equations. This approach, described in the previous chapter, only helped in cases of laminar flow, such as the loss of head in capillary tubes or the damping of pendulum oscillations. In the 1860s, Hermann Helmholtz invented another approach based on vortex-like solutions of Euler's equations.

Helmholtz arrived at this idea while studying a specific problem of acoustics, the sounding of organ pipes. In his efforts to improve the theory of this instrument, he came to consider the internal friction of the air and its damping effect. As he was unaware of the Navier–Stokes equation, he began by analyzing the solutions of Euler's equation for which internal friction would play a role. This is the source of his famous memoir of 1858 on vortex motion.

In this study, Helmholtz included the simple case of a 'vortex sheet', that is, a continuous alignment of rectilinear vortices, and found it to be equivalent to a tangential discontinuity of the fluid velocity across the sheet. He later appealed to such discontinuous motions to explain another mystery of organ pipes, namely, the production of an *alternating* motion by a *continuous* stream of air through the mouth of the pipe. In 1868, he described the general properties of surfaces of discontinuity, the most essential one being their instability, whereby any protuberance of the surface tends to grow and to roll up spirally, as shown in Fig. 4.1.

Helmholtz reached these notions by focusing on the difficulties of a concrete application of Euler's equations to the specific system of organ pipes. By analogy, he believed that the neglection of surfaces of discontinuities or similar structures explained the failure of many

---

[1]Helmholtz [1891] p. 14.

Fig. 4.1.   Spiral rolling up of a protuberance on a surface of discontinuity. Courtesy of Greg Lawrence (in
     Fernando [1991] p. 475).

other applications of theoretical fluid mechanics. The first three sections of this chapter,
on acoustics, vortex motion, and vortex sheets, recount the emergence of the methods and
concepts that justified this conviction. Section 4.4 documents Helmholtz's interest in
meteorology and his understanding of cyclonic vortices. Section 4.5 shows how, inspired
by a singular observation in the Swiss sky, he came to apply discontinuity surfaces to the
general circulation of the atmosphere and to the theory of storms, thus foreshadowing
some central notions of modern meteorology. As is explained in the final section, Section
4.6, he predicted atmospheric waves resulting from the instability of such surfaces, and
devoted much time and effort to the analogous waves induced by wind blowing over
water.

4.1   Sound the organ

During his studies at the University of Berlin, Helmholtz read widely in physics, as is clear
from the erudition displayed in the memoir of 1847 on the conservation of force. After
obtaining the Königsberg chair of physiology, he specialized in the study of perception, at
the intersection of his interests in physics, physiology, aesthetics, and philosophy. Although his first research in this field concerned vision, in the mid-1850s he began a parallel
study of the perception of sound. Acoustics was then a developing branch of physics, and
an ideal subject for someone who loved both music and mathematics. Here is Helmholtz's
eloquent statement of his motivation:[2]

> I have always been attracted by this wonderful, highly interesting mystery: It is
> precisely in the doctrine of tones, in the physical and technical foundations of
> music, which of all arts appears to be the most immaterial, fleeting, and delicate
> source of incalculable and indescribable impressions on our mind, that the science of
> the purest and most consistent thought, mathematics, has proved so fruitful.

4.1.1   *From acoustics to mathematics*

The earliest trace of Helmholtz's interest in acoustics is a review of works 'concerning
theoretical acoustics' that Helmholtz wrote for the *Fortschritte der Physik* of 1848 and
1849. They all dealt with the physics of sound, that is, the first of the three components

---

[2]Helmholtz [1857] pp. 121–2. On Helmholtz's biography, cf. Koenigsberger [1902]. On his interest in the
perception of sound, cf. Vogel [1993], Hatfield [1993], Hiebert and Hiebert [1994].

that Helmholtz distinguished in the study of sensations, namely physical, physiological, and psychological.[3]

In particular, Helmholtz criticized Guillaume Wertheim's measurements of the velocity of sound with organ pipes. Following earlier theories by Daniel Bernoulli, Euler, and Lagrange, Wertheim assumed stationary air waves in the pipes, with a velocity node at the bottom of the pipe and an anti-node at the opening. He deduced the wavelength from the length of the tube, and multiplied it by the sound frequency to obtain the velocity. There was a difficulty in that the measured sounding frequency depended slightly on the intensity of the blowing. Also, the theoretical stationarity condition completely failed if air was replaced by water, as Wertheim demonstrated with a special water organ of his own invention. Helmholtz suggested that the boundary conditions assumed by previous theorists were oversimplified. In particular, he pointed to the need for a more realistic treatment of the motion of the fluid near the opening of the tube.[4]

Helmholtz's first publication in the field of acoustics, in 1856, concerned another problem, belonging to the physiological register. Organists had long known that when two successive harmonics of the same tone are played together loudly, the base tone is heard. Acousticians verified and generalized this result, but disagreed on the exact combination rule. According to Wilhelm Weber and Georg Simon Ohm, the combination of the frequencies mf and nf, where m and n are two integers without a common divisor, yielded the frequency f. According to Gustaf Hällström, it yielded frequencies of the form $(pn - qm)f$, where p and q are two other integers. In order to decide the issue experimentally, Helmholtz invented a clever monochromatic source by placing a tuning fork in a cavity resonator whose proper frequencies were mutually incommensurable. Playing together pure mf and nf sounds, he heard the combined frequency $(m - n)f$, and also $(m + n)f$ after he had convinced himself that it should theoretically exist. His theory was that combination tones occurred when the mechanical response of the ear was no longer linear and involved a term proportional to the square of the sound amplitude.[5]

The nature of combination tones bore on a central issue of contemporary acoustics, namely, the relevance of Fourier analysis to the perception of sounds, on which Ohm and Thomas Seebeck famously disagreed. Helmholtz's main goal was to put an end to the controversy and to base the science of acoustics on non-controversial facts. However, the lack of experimental facilities at Bonn, where he had been recently appointed, prompted him to work on the more mathematical aspects. Among the acoustic systems in urgent need of a better theory were resonant cavities, which played a central role in the production and detection of monochromatic sounds, and organ pipes, which Helmholtz used to produce strong, sustained tones in his acoustic experiments.[6]

[3]Helmholtz [1852/53]. On the tripartite structure, cf. Helmholtz [1863a] p. 7.

[4]Helmholtz [1852/53] pp. 250 (Doppler), 242–6 (Wertheim); Wertheim [1848].

[5]Helmholtz [1856] pp. 497–540. The quadratic terms include the cross-product $\cos 2\pi nft \cos (2\pi nft + \phi)$, which is the superposition of $(1/2) \cos [2\pi(m + n) ft + \phi]$ and $(1/2) \cos [2\pi(m - n) ft - \phi]$.

[6]On Bonn, cf. Helmholtz to Du Bois-Reymond, 5 Mar. 1858, in Kirsten et al. [1986]. On the aims of Helmholtz's work on combination tones, cf. Turner [1977], Vogel [1993].

4.1.2   *Organ pipes*

In 1859, Helmholtz published in Crelle's mathematical journal a major memoir 'On the motion of air in open-ended organ pipes', with an appendix on spherical resonators. In the introduction, he recalled that the 'most important mathematical physicists', including Daniel Bernoulli, Euler, Lagrange, and Poisson, had dealt with organ pipes. All of them had simplified the conditions at the opening of the pipe, generally assuming a compression node there, and complete rest for the air outside the tube. This procedure only gave a first approximation of the true motion, and neglected the damping of the vibrations through sound emission. Moreover, it disagreed with Wertheim's frequency measurements, and it contradicted recent experimental determinations of the locations of the nodes.[7]

In order to remedy these defects, Helmholtz grappled with the daunting problem of the motion of the air near the opening of the tube. He managed to determine the empirically interesting parameters of the sounding pipe—position of the nodes, frequency and intensity of the emitted sound, and phase relations—without simplifying this motion. The secret of this mathematical feast was a multiple application of Green's theorem, which was well known to the Germans since the publication of Green's *Essay* in Crelle's journal. With this theorem and a number of analytical tricks, Helmholtz not only solved a particular problem of acoustic importance, but he also inaugurated a general strategy for determining relations between controllable aspects of wave propagation when the explicit solution of the wave equation is inaccessible. Gustav Kirchhoff's diffraction theory is a direct descendent of Helmholtz's paper on organ pipes; modern scattering theory or wave-guide theory are more remote ones.[8]

Helmholtz compared his theoretical formulas for node location and sounding frequency with measurements made by Wertheim and by Friedrich Zamminer. The agreement was reasonably good for wide tubes, but poor for narrow tubes for which it should have been best (since the theory presupposed a wavelength much larger than the opening). When he published his memoir, in 1859, Helmholtz believed that the discrepancy could be explained by the known difference between the sounding frequency of blown pipes and their resonance frequency.[9]

As was well known, friction broadens the response of a resonator to periodic excitations. In an organ pipe there is friction due to the viscosity of the air. The width of the resonance should increase with this friction, and therefore with the narrowness of the pipe. Helmholtz probably had in mind this effect of viscosity when he faced the failure of his theory for narrow tubes. His improved theory of 1863, which we will consider shortly, established that viscosity implied both a broadening and a shifting of the resonance frequency of organ pipes.

4.2   Vortex motion

Helmholtz studied the general effects of internal friction on fluid motion in the same period, 1858/59, probably because he had in mind an application to organ pipes. In any

---

[7]Helmholtz [1859] pp. 303–7.   [8]Cf. Darrigol [1998] pp. 7–10.
[9]Helmholtz [1859] pp. 314–15; Wertheim [1851]; Zamminer [1856].

case, he knew that the viscosity of fluids could cause considerable deviations of experiment from theory. Being unaware of previous mathematical studies of this problem by Poisson, Navier, Saint-Venant, and Stokes, he proceeded to 'define the influence [of friction] and to find methods for its measurement'. In his opinion, the most difficult aspect of this problem was to gain an 'intuition of the forms of motion that friction brings into the fluid'. Such was the motivation of his memoir of 1858 on vortex motion.[10]

4.2.1 *Fundamental theorems*

As Helmholtz knew from Lagrange, when an incompressible, non-viscous fluid is set into motion by forces that derive from a potential or by the motion of immersed solid bodies, a potential exists for the fluid velocity. Frictional forces never derive from a potential (for they are not conservative), and are therefore able to induce states of motion for which a potential does not exist. Helmholtz's first step was to study motions of this kind, independently of the forces that caused them.

As Helmholtz explained without knowing of Stokes's earlier demonstration, the most general infinitesimal motion of the volume element of a continuous medium can be decomposed into a translation, three dilations along mutually-orthogonal axes, and a rotation around a fourth axis. In the case of a fluid, the infinitesimal rotation has the angular velocity $\omega/2$, with

$$\boldsymbol{\omega} = \nabla \times \mathbf{v}. \tag{4.1}$$

Hence the mathematical condition for the existence of a velocity potential, $\nabla \times \mathbf{v} = \mathbf{0}$, can be interpreted as the absence of local rotation in the instantaneous motion of the fluid. Conversely, the absence of a velocity potential signals the existence of vortex motion in the fluid.[11]

Helmholtz next examined how the vortices evolved in time. For this purpose he wrote Euler's equation as

$$\frac{\partial \mathbf{v}}{\partial t} + (\mathbf{v} \cdot \nabla)\mathbf{v} = -\frac{1}{\rho}\nabla P - \frac{1}{\rho}\nabla V, \tag{4.2}$$

where P is the pressure, ρ is the constant density, and V is the potential of external forces (for instance, gravitational forces). The continuity equation reads

$$\nabla \cdot \mathbf{v} = 0. \tag{4.3}$$

Applying the operation $\nabla\times$ to Euler's equation, Helmholtz obtained the further equation

$$\frac{\partial \boldsymbol{\omega}}{\partial t} + (\mathbf{v} \cdot \nabla)\boldsymbol{\omega} = (\boldsymbol{\omega} \cdot \nabla)\mathbf{v}, \tag{4.4}$$

[10]Helmholtz [1858] p. 102. That this publication antedated that on organ pipes by a few months does not exclude the reverse chronology adopted here for their genesis.

[11]Helmholtz [1858] pp. 104–8. In a letter he wrote to Moigno (quoted in *Les mondes* 17 (1868), pp. 577–8), Helmholtz named Kirchhoff [1882] (memoir on vibrating plates) as his source for the decomposition, Franz Neumann as Kirchhoff's source, and Cauchy as Neumann's probable source. Kirchhoff only used the principal dilations ([1882] pp. 246–7). Cauchy [1841] p. 321 introduced the 'rotation moyenne'. As was said in Chapter 3, Stokes [1849a] introduced the decomposition to prepare his derivation of the Navier–Stokes equation.

already known to d'Alembert and Euler. Helmholtz's brilliant innovation lay in the following kinematic interpretation.[12]

The left-hand side of eqn (4.4) represents the rate of variation of the vector $\boldsymbol{\omega}$ for a given particle of the fluid moving with velocity $\mathbf{v}$. Helmholtz first considered a particle for which the rotation $\boldsymbol{\omega}/2$ vanishes at a given instant. Then eqn (4.4) implies that the rotation of this particle remains zero at any later time. In the general case, it may be rewritten as

$$\varepsilon \, d\boldsymbol{\omega} = \mathbf{v}(\mathbf{r} + \varepsilon\boldsymbol{\omega}) \, dt - \mathbf{v}(\mathbf{r}) \, dt, \tag{4.5}$$

where $d\boldsymbol{\omega}$ is the variation of $\boldsymbol{\omega}$ on a given fluid particle during the time dt, and ε is an arbitrary infinitesimal quantity of second order. In order to interpret this relation, Helmholtz considered two fluid particles located at the points $\mathbf{r}$ and $\mathbf{r} + \varepsilon\boldsymbol{\omega}$ at time t. At time $t + dt$, the first particle has moved by $\mathbf{v}(\mathbf{r})dt$ and the second by $\mathbf{v}(\mathbf{r} + \varepsilon\boldsymbol{\omega}) \, dt$. According to eqn (4.5), the two new locations are separated by $\varepsilon\boldsymbol{\omega} + \varepsilon d\boldsymbol{\omega}$, which is parallel to the new rotation vector.[13]

For a more intuitive grasp of this result, Helmholtz defined 'vortex lines' that are everywhere tangent to the rotation axis of the fluid particles through which they pass, and 'vortex filaments' that contain all the vortex lines crossing a given surface element of the fluid. As a first consequence of eqn (4.5), two particles of the fluid that belong to the same vortex line at a given instant still do so at any later time. In other words, vortex lines follow the motion of the fluid. Equation (4.5) also implies that, during the motion of the fluid, vortex filaments stretch in the same proportion as the rotational velocity varies. Since the fluid is incompressible, this longitudinal stretching implies a sectional shrinking in inverse proportion. In other words, the product $\boldsymbol{\omega} \cdot d\mathbf{S}$ of a section of the filament by twice the amount of rotation in this section remains the same during the motion of the fluid.[14]

Lastly, this product is the same all along a given filament. In order to prove this, Helmholtz integrated the vector $\boldsymbol{\omega}$ across the closed tubular surface delimiting a piece of vortex filament. This integral is equal to the difference of the products $\boldsymbol{\omega} \cdot d\mathbf{S}$ taken at the two extremities of the piece; and it is also equal to the integral of $\nabla \cdot \boldsymbol{\omega}$ over the volume of the piece, which is zero following the definition of $\boldsymbol{\omega}$.[15]

In summary, vortex filaments are stable structures of the fluid. The product of the rotation by the section of a filament, which Helmholtz called 'intensity', does not vary in time, and is the same all along the filament. From the latter property, Helmholtz concluded that vortex filaments could only be closed on themselves or end at the limits of the liquid.[16]

A striking feature of Helmholtz's demonstration of these theorems is the intimate association of analytical relations with geometrical representations. In nineteenth-century physics, this quality seems more typically British. In fact, Stokes, Thomson, and James

---

[12]Helmholtz [1858] pp. 110–11. For d'Alembert's anticipation, see Chapter 1, p. 20.

[13]*Ibid.* pp. 111–12.

[14]*Ibid.* pp. 102–3, 112–13. 'Vortex lines' and 'vortex filaments' are Tait's translations for '*Wirbellinien*' and '*Wirbelfäden*', cf. Tait to Helmholtz, 22 Apr. 1967, HN.

[15]Helmholtz [1858] pp. 113–14.

[16]*Ibid.* p. 114. The latter conclusion is only true in topologically-simple cases (cf. Epple [1998] pp. 313–14).

Clerk Maxwell anticipated some elements of Helmholtz's reasoning. In 1845, Stokes introduced the decomposition of the instantaneous motion of an element of fluid into translation, dilations, and rotation, and gave the analytical expression for the rotation. In 1849 and for magnetism, Thomson defined 'solenoidal' distributions of the magnetic polarization $\mathbf{M}$ for which the relation $\nabla \cdot \mathbf{M} = 0$ holds; and he decomposed the corresponding magnets into elementary tubes, as Helmholtz later decomposed vortex motion into vortex filaments. In his memoir 'On Faraday's lines of force', published in 1855/56, Maxwell defined 'tubes of force' in a manner quite similar to Helmholtz's definition of vortex filaments. As a friend of Thomson's, Helmholtz may have been partly aware of this British field geometry.

4.2.2   *The electromagnetic analogy*

The British outlook on Helmholtz's paper is also evident in the next section concerned with the inverse problem of determining the velocity of the fluid when the distribution of the vorticity $\boldsymbol{\omega}$ is known. Helmholtz sought the solutions of the equations $\nabla \times \mathbf{v} = \boldsymbol{\omega}$ and $\nabla \cdot \mathbf{v} = 0$ in the form

$$\mathbf{v} = \nabla \varphi + \nabla \times \mathbf{A}. \tag{4.6}$$

The potential φ satisfies the equation $\Delta \varphi = 0$ in the fluid mass, and the vector $\mathbf{A}$ satisfies

$$\nabla(\nabla \cdot \mathbf{A}) - \Delta \mathbf{A} = \boldsymbol{\omega}. \tag{4.7}$$

Helmholtz wanted to retrieve the simpler equation $\Delta \mathbf{A} = -\boldsymbol{\omega}$, which makes the components of $\mathbf{A}$ the potentials of fictitious masses measured by the components of $\boldsymbol{\omega}/4\pi$. This is immediately possible if all the vortex filaments of the fluid are closed, since the vector potential

$$\mathbf{A}(\mathbf{r}) = \frac{1}{4\pi} \int \frac{\boldsymbol{\omega}(\mathbf{r}')}{|\mathbf{r} - \mathbf{r}'|} \, d\tau' \tag{4.8}$$

then satisfies $\nabla \cdot \mathbf{A} = 0$. In the general case, for which some vortex filaments abut on the surface of the liquid, Helmholtz prolonged the filaments beyond the real liquid so that they all became closed, which brought him back to the previous, simpler problem.[17]

Applying the operation $\nabla\times$ to the expression (4.8) for the potential $\mathbf{A}$, Helmholtz recognized the Biot–Savart formula of electromagnetism: the fluid velocity corresponding to a given distribution of vorticity $\boldsymbol{\omega}$ is exactly like the magnetic force produced by the electric-current distribution $\boldsymbol{\omega}$. Helmholtz abundantly exploited this analogy, which gave him a direct intuition for the fluid motion around vortices.[18]

Similar reasoning is easily identified in British sources. In his memoir on diffraction of 1849, Stokes introduced the decomposition (4.6) to determine a vector from its curl. In 'On Faraday's lines of force' (1855), Maxwell applied this method to the determination of the magnetic field $\mathbf{H}$ generated by the current $\mathbf{j}$. His starting-point was the equation $\nabla \times \mathbf{H} = \mathbf{j}$, which he had obtained by studying the geometry of the magnetic field around a current loop, and which corresponds to Helmholtz's $\nabla \times \mathbf{v} = \boldsymbol{\omega}$.

---

[17]Helmholtz [1858] pp. 114–117.   [18]*Ibid.* pp. 117–18.

The recourse to an analogy between electromagnetism and continuum mechanics was also a British specialty, inaugurated by Thomson. There is an interesting difference, however. Whereas Thomson and Maxwell used such analogies to shed light on electromagnetic phenomena and structure their theories, Helmholtz did the reverse. He used electromagnetic action at a distance, which was most familiar to him, for a better understanding of the motions of a mechanical continuum. This inversion explains why he did not mention that his analogy between electromagnetism and vortex motion led to the field equation $\nabla \times \mathbf{H} = \mathbf{j}$, which was unknown on the Continent.

Lastly, Helmholtz shared the British engagement in energetics, of which Thomson and himself were the main founders. For the ideal fluid on which Helmholtz reasoned, the kinetic energy

$$T = \frac{1}{2} \int \rho v^2 \mathrm{d}\tau \qquad (4.9)$$

is invariable if the walls do not move, because external forces deriving from a potential cannot perform any work on an incompressible fluid. If, in addition, the vortex motion occurs very far from the walls, recourse to eqn (4.6) and integration by parts yield

$$T = \frac{1}{2} \int \rho \boldsymbol{\omega} \cdot \mathbf{A} \, \mathrm{d}\tau. \qquad (4.10)$$

Helmholtz exploited the invariance of this integral in his subsequent discussion of the interactions between two vortices.[19]

4.2.3   Vortex sheets, lines, and rings

In the last section of his memoir, Helmholtz applied his general theorems and analogies to simple cases of vortex motion in an infinite fluid. The most trivial case is that of a uniform, plane vortex sheet. The incompressibility of the fluid implies that the normal velocity of the fluid should be the same on both sides of the sheet, while the equation $\nabla \times \mathbf{v} = \boldsymbol{\omega}$ implies a discontinuity $\varepsilon\omega$ of the tangential velocity if ω represents the average intensity of the vorticity within the sheet and ε is the infinitesimal thickness of the sheet. Within the sheet the fluid moves at a velocity intermediate between the velocities on both sides. Since vortex lines follow the motion of the fluid of which they are made, the sheet must move at a velocity which is the average of the fluid velocities on both sides.[20]

As we will see shortly, this special example of vortex motion played an essential role in Helmholtz's later hydrodynamics, at least because it showed that tangential discontinuities of the fluid motion were compatible with Euler's equations. Earlier investigators usually assumed the existence of a velocity potential, and thus excluded finite slips in the flow. Helmholtz not only demonstrated the mathematical existence of such solutions, but also indicated a way to realize them, namely, by bringing together two masses of liquid moving at different, parallel velocities.[21]

---

[19]Helmholtz [1858] pp. 123–4.   [20]*Ibid.* pp. 121–2.

[21]*Ibid.* p. 122. As will be shown in Chapter 5, pp. 185–6, Stokes repeatedly considered discontinuities of Helmholtz's type (already in Stokes [1842]), but never developed their analysis very far.

The next simple case of vortex motion is that of a single, rectilinear vortex filament. For symmetry reasons the filament must remain in a constant position. According to the electromagnetic analogy, the fluid rotates around the filament at a linear velocity inversely proportional to the distance from the filament. The next case considered by Helmholtz is that of two parallel rectilinear filaments with the intensities i_1 and i_2. The corresponding fluid velocity is the superposition of the velocities due to each filament. The velocity of the fluid in the first filament is equal to the rotation due to the second, and vice versa. As the filaments must move with the velocity of the fluid of which they are made, their mutual influence results in a uniform rotation around their barycenter for the masses i_1 and i_2.[22]

The case of a vortex ring is more complex, because the velocity imparted on the vortex by the vortex itself no longer vanishes. The rings must have a finite section for this self-interaction to remain finite. Using elliptic integrals, energy conservation, and barycentric properties, Helmholtz proved that a single circular ring must move along its axis without sensible change of size, in the direction of the flow at the center of the ring. When the section of the ring becomes infinitely thin, this velocity diverges logarithmically.[23]

When two rings are present, they are also subjected to the fluid velocity imparted by the other ring. Let us start, for instance, with two identical rings that have the same axis. Initially, they travel along this axis with the same velocity. The front ring, however, must widen under the effect of the other ring, and the back ring must shrink. Since wider rings travel slower, the back ring catches up with the front ring after a while. Then the size variations are inverted, and the relative motion slows down until the two rings are again of the same size, which brings us back to the initial configuration. In summary, the rings pass alternately through each other (see Fig. 4.2).[24]

Helmholtz indicated how to observe this dance of the vortex rings with a spoon and a calm surface of water. Immersing the spoon vertically and withdrawing it quickly creates a half vortex ring, whose two ends form small dips on the water surface. As anyone can verify, these rings do behave as Helmholtz says. Most impressive is the contrast between the simplicity of this experiment and the sophistication of the motivating physico-mathematical analysis.[25]

4.2.4 *German approbation, British enthusiasm, and French suspicion*

Helmholtz's memoir on vortex motion is now universally regarded as the historical foundation of this subject. It quickly captured the attention of eminent German mathematicians and physicists. In 1859, the Berlin mathematician Rudolph Clebsch recovered Helmholtz's theorems by variational methods. In 1860, Bernhard Riemann noted the relation between Helmholtz's theorems and his and Dirichlet's solutions for the problem of the rotating-fluid ellipsoid. The Göttingen Academy offered a prize for a Lagrangian deduction of these theorems, which Hermann Hankel won in 1861. Helmholtz's memoir provided the substance for two of Kirchhoff's famous lectures on mechanics.[26]

---

[22]Helmholtz [1858] pp. 124–7.   [23]*Ibid.* pp. 127–33.   [24]*Ibid.* p. 133.   [25]*Ibid.* p. 134.

[26]Clebsch [1859]; Riemann [1860]; Hankel [1861]; Kirchhoff [1876] lectures 15, 20. In manuscript fragments (HN, in #679 and #680), Helmholtz discussed various aspects of vortex motion, namely invariants, stability, and friction.

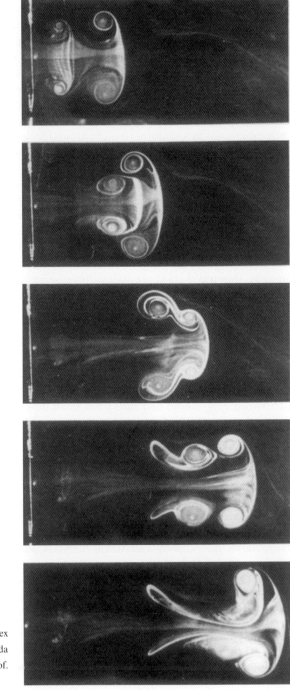

Fig. 4.2.   Leapfrogging of two vortex rings. Photograph from Yamada and Matsui [1978]. Courtesy of Prof. Yamada Hideo.

In Britain, Thomson, Maxwell, and Peter Guthrie Tait found Helmholtz's reasoning highly congenial. Tait's enthusiasm for Hamilton's quaternions arose when he realized, at the end of 1858, how well suited they were to Helmholtz's decomposition of fluid motion. In 1867, he published a translation of Helmholtz's memoir and built a 'smoke box' with an elastic membrane on one side and a hole on the other, with which he could produce spectacular shows of vortex rings and verify Helmholtz's predictions (see Fig. 4.3).[27] In a letter to Helmholtz written in January 1867, Thomson described his friend's smoke box and his own speculations on vortex-ring atoms, which were soon to be published in the *Philosophical magazine*. In July 1868, Tait announced to Helmholtz that Maxwell, 'one of the most genuinely original men I have ever met', had taken up vortex motion and proved that 'two closed vortices act on one another so that the sum of the areas of their projections on any given plane remains constant.' He also mentioned that the forthcoming *Transactions* of the Royal Society of Edinburgh would be rich in papers on this subject.[28]

The most important of these papers was a long, highly-mathematical memoir by William Thomson. There he developed Helmholtz's hints on topological aspects of fluid motion, and provided alternative proofs of his theorems based on the invariance of the integral $\oint \mathbf{v} \cdot d\mathbf{l}$ over any circuit that follows the motion of the fluid. This invariance follows from the fact that the Lagrangian time derivative of the form $\mathbf{v} \cdot d\mathbf{r}$, namely

$$\frac{d}{dt}(\mathbf{v} \cdot d\mathbf{r}) = \mathbf{v} \cdot d\mathbf{v} + \frac{d\mathbf{v}}{dt} \cdot d\mathbf{r} = d\left(\frac{1}{2}v^2\right) + \mathbf{g} \cdot d\mathbf{r} - \frac{dP}{\rho}, \qquad (4.11)$$

is an exact differential for a perfect liquid in the gravity $\mathbf{g}$. Through Stokes's theorem (of which Thomson was the true discoverer), this result contains the invariance of the product $\boldsymbol{\omega} \cdot d\mathbf{S}$ which is the essence of Helmholtz's laws of vortex motion. As Thomson knew, the reverse is not always true, because the volume occupied by the fluid may not be simply

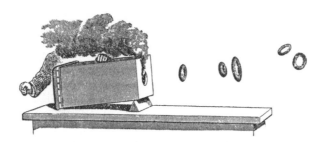

Fig. 4.3.   Tait's smoke box. From Tait [1876] p. 292.

[27]Tait to Hamilton, 7 Dec. 1858, quoted in Knott [1911] p. 127; Helmholtz [1867]; Tait [1876] p. 292 (smoke box). A similar device had already been described in Reusch [1860]. Beautiful experiments on vortices in air and in liquids are also found in Rogers [1858]. These experimenters were primarily interested in the production of vortices and were not aware of Helmholtz's predictions.

[28]Thomson to Helmholtz, 22 Jan. 1867, HN; Thomson [1867]; Tait to Helmholtz, 28 July 1868, HN. In a letter of 2 May 1859 (HN), Thomson thanked Helmholtz for his memoir, which he had read 'with very great interest' before 'falling into the vortex of [his] winter's work'. In 1866, Maxwell set Helmholtz's hydrodynamic theorems as a question to the Cambridge Mathematical Tripos (cf. Maxwell [1990] vol. 2, p. 241). On Thomson's vortex atom, cf. Silliman [1963], Smith and Wise [1989] chap. 12, Kragh [2002], and the discussion in Chapter 5, pp. 191–7.

connected. Maxwell and Tait examined the topological issues raised by Thomson, and obtained important results of the theory of knots.[29]

Thomson had long dreamt of a world made entirely of motions in a pervasive, ideal fluid. In this view, every form of energy was of kinetic origin. Every force had to be traced to dynamical effects, as fluid pressure had been reduced to molecular collisions. The rigidity of the ether with respect to light vibrations was to be explained in terms of the inertia of small-scale motions of the primitive fluid. The permanence of atoms and molecules was to be understood as the stability of special states of motion. Helmholtz's theorems, applied to vortex rings, and Thomson's further topological considerations seemed to offer a rigorous basis for developing this program.[30]

Yet, in the middle of writing his memoir on vortex motion, Thomson went through a few weeks of despair: 'It is a pity', he then wrote to Tait, 'that H^2 [Hermann Helmholtz in Maxwell's notation] is all wrong and that we all dragged so deep in the mud after him.' The cause of this lament was a criticism published by the eminent mathematician and academician Joseph Bertrand in the *Comptes rendus* of the French Academy of Sciences.[31]

Bertrand rejected Helmholtz's interpretation of $(1/2)\nabla \times \mathbf{v}$ as the rotation velocity of the elements of the fluid. Many cases of motion, Bertrand showed, could be reduced to three dilations of the fluid elements along three oblique axes. Although, intuitively, such motions involve no rotation, the corresponding $\nabla \times \mathbf{v}$ vanishes only when the three axes are orthogonal. From this remark, Bertrand concluded that the integrability of $\mathbf{v} \cdot \mathbf{dl}$ could not be identified with the absence of rotation. The consequences were devastating: 'Despite his very deep knowledge of mathematics, the author has committed a slight inadvertence at the beginning of his memoir that mars all his results by making him attach a quite excessive importance to the integrability condition $[\nabla \times \mathbf{v} = \mathbf{0}]$.'[32]

Helmholtz promptly replied that the rule according to which one decomposes a complex motion into simpler ones was to some extent arbitrary. The decomposition of the motion of a fluid element into three *orthogonal* dilations and a rotation (also a global translation) is one possibility; that into three oblique dilations (with real or imaginary axes) is another. Although the former choice seems to contradict the geometrical intuition of a rotation, it is the only one suited to fluid dynamics, because the angular momentum of the elements of fluid is determined by a rotation defined in this sense. Helmholtz added that his usage of 'rotation' was not new and could be found in Kirchhoff's memoir on vibrating plates.[33]

The source of the conflict is clear. On one side, Helmholtz and Kirchhoff (also Stokes and Thomson) adjusted their geometrical and kinematic concepts to the needs of dynamics. On the other, Bertrand refused to let physical arguments control his geometrical intuition. His first reaction to Helmholtz's rebuttal was to give a 'decisive example' of fluid motion that allegedly contradicted Helmholtz's definition of rotation: the fluid moves uniformly in planes parallel to the Oxy plane, with a velocity increasing linearly

---

[29]Thomson [1869]. On Thomson's, Tait's, and Maxwell's contributions to topology, cf. Epple [1998]. On the origin of Stokes's theorem, see *SMPP* **5**, pp. 320–1.

[30]Cf. Smith and Wise [1989] chap. 12.

[31]Thomson to Tait, 11 July 1868, quoted by Harman in Maxwell [1990] vol. 2, 399n; Bertrand [1868*a*].

[32]Bertrand [1868*a*] p. 1227.

[33]Helmholtz [1868*a*]. Helmholtz referred to Stokes's earlier analysis in his next reply to Bertrand.

with the coordinate z. According to Helmholtz's definition, this case involves a constant rotation of the fluid elements, against Bertrand's intuition of their motion.[34]

More came to irritate Bertrand. In a note to the *Comptes rendus*, Saint-Venant sided with Helmholtz and referred to Cauchy's earlier interpretation of $(1/2)\nabla \times \mathbf{v}$ as the 'average rotation' of the fluid particles. In reaction, Bertrand insisted that Helmholtz's memoir contained false theorems as well as an aberrant definition of rotation. For example, he denied that the velocity field could be determined from the vorticity field in Helmholtz's manner, because the potential φ of eqn (4.6) indirectly depended on the vorticity, through the boundary conditions (for example, the tangential component of $\nabla \varphi$ along the fixed walls of the fluid must be opposed to the tangential component of $\nabla \times \mathbf{A}$, which depends on the vorticity).[35]

This criticism was slightly more embarrassing to Helmholtz, because he had not discussed the nature of the $\nabla \varphi$ contribution to the velocity field. In his reply, he claimed that he had been aware that this contribution in general depended on the vorticity, but had nevertheless ignored it because the harmonic character of the function φ severely restricted its form. In particular, it vanished whenever the fluid mass could be regarded as infinite (and the fluid motion did not extend to infinity). All the special cases of motion treated in his paper were of that kind or could be brought back to it.[36]

In this second note, Helmholtz's tone was far less deferential than in the first; he accused Bertrand of disfiguring his theorems, and used mildly ironic phrases such as 'J'invite mon savant critique à se rappeler que...'. To make things worse, in his journal *Les mondes*, the xenophilic Abbot Moigno ridiculed Bertrand's attitude: 'While M. Bertrand persists in his inconsiderate criticism, he now wraps it in so many polite words and insistent, eloquent praises that any intelligent reader can conclude that he is certainly wrong.' In his final note, Bertrand protested his sincerity and intellectual honesty. He accused Moigno of giving a poor idea of French manners and Helmholtz of believing him. In the next issue of *Les mondes*, Moigno retorted: 'We have perfectly felt the *coups de griffe* which [M. Bertrand] gave us in his bad mood. If he maintains his jest in the *Comptes rendus*, we will have no pain to prove that of the great Academician and the humble abbot, the most serious is not the one he thinks.' More diplomatically, Helmholtz apologized for having used phrases that could lead to misinterpretations of his true intentions.[37]

Bertrand's attack was only a minor and ephemeral threat to Helmholtz's theory. Thomson quickly regained his faith in it. Between Helmholtz's first reply and Bertrand's second assault, he suggested to Tait that they should together write a letter to the French Academy to support Helmholtz. Tait preferred to let Helmholtz 'smash [Bertrand] in his own way'. Maxwell, who heard of 'Bertrand's refutation' through Thomson, declared himself completely confident in the truth of Helmholtz's theorems and continued exploring their consequences. In the following years, these theorems became standard knowledge and found applications to diverse cases of fluid motion, from Thomson's primitive world-

---

[34]Bertrand [1868*b*].   [35]Saint-Venant [1868*b*]; Bertrand [1868*c*].

[36]Helmholtz [1868*b*].   [37]*Ibid.*; Moigno [1868*b*]; Bertrand [1868*d*]. Helmholtz [1868*c*].

fluid to the atmosphere. Reviewing in 1881 the progress of hydrodynamics, the British theorist William Hicks wrote:[38]

> During the last forty years, without doubt, the most important addition to the theory of fluid motion has been in our knowledge of the properties of that kind of motion where the velocities cannot be expressed by means of a potential...Helmholtz first gave us clear conceptions in his well-known paper [on vortex motion].

4.2.5   Friction

As already mentioned, Helmholtz studied vortices in ideal fluids as a step toward including internal friction, which implies this kind of motion. We saw in the previous chapter how, in August 1859, he rediscovered the Navier–Stokes equation and used it in the analysis of the experiments he conducted with his student Piotrowski on the damping of the oscillations of a hollow metallic sphere filled with a viscous fluid.[39]

In 1869, Helmholtz returned to viscous fluids in a physiological context. Another of his students, Alexis Schklarewsky, had performed experiments on the motion of small suspended particles in capillary tubes, probably with blood circulation in mind. He found that the particles moved toward the axis of the tube, in contradiction with a theorem established by Thomson in the case of non-viscous fluids. The anomaly prompted Helmholtz to study stationary fluid motion for which the viscous term of the Navier–Stokes equation dominates the nonlinear term. Thus he derived the important theorem according to which the real motion is that for which the frictional energy loss is a minimum (for fixed boundary conditions). Although the result turned out to be irrelevant to Schklarewsky's experiments, for which the quadratic terms were non-negligible, the combination of energetic and variational principles became an important trait of Helmholtz's theoretical style, as it already was in recent British natural philosophy. The appeal of this method was that it determined general properties of physical systems without entering inaccessible or non-computable structural details.[40]

From an empirical point of view, Helmholtz's most successful discussion of the effects of viscosity concerned organ pipes. In 1863, he explained the leftover discrepancy between his earlier calculation of the resonance frequency of open pipes and Zamminer's wavelength measurements. The inclusion of the internal friction term in the equation of motion of the air diminished the propagation velocity in the required proportion. It also explained why the resonance of narrow tubes was not sharp, even though wave reflection at their open end was more perfect. Helmholtz further determined the radius of the tube for which the sum of viscous and radiation damping was minimal and the resonance was sharpest. This radius is a function of the length of the tube and the mode of resonance. As the excitation of a given mode of oscillation by the air from the bellows depends on the sharpness of the corresponding resonance, so does the timbre of the emitted sound. 'Most surprisingly', Helmholtz found that his condition of sharpest resonance justified an empirical rule discovered a century earlier by the celebrated organ-maker Andreas Silber-

---

[38]Tait to Helmholtz, 3 Sept. 1868, *HN*; Hicks [1882] 63 (quote), pp. 63–8 (detailed review of works on vortex motion until 1881). For broader reviews, cf. Truesdell [1954b], Meleshko and Aref [2007].

[39]See Chapter 3, p. 144.   [40]Helmholtz [1869].

mann, namely, that, for the timbre of a stop to be uniform, the width of the pipes has to decrease with their length, by one-half for the ninth.[41]

4.3 Vortex sheets

Helmholtz had not yet exhausted the problem of organ pipes. In a real organ pipe, such as that shown in Fig. 4.4, the sound is produced by blowing air through the slit cd against the

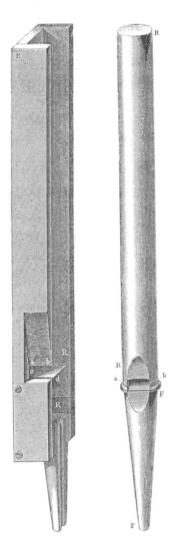

Fig. 4.4.   Organ pipes. From Helmholtz [1877] p. 149.

[41]Helmholtz [1863*b*]. Helmholtz only published the results. The full reasoning and calculations are in a long, difficult manuscript ('Die Bewegungsgleichungen der Luft', HN, #582) which Helmholtz intended for publication in *JRAM*.

'lip' ab. Helmholtz wondered by what mechanism this continuous air stream could produce an oscillatory motion in the tube. If, he reasoned, the motion of the air in the mouth of the tube admits a velocity potential, then this motion should be the same as the flow of an electric current in a uniformly-conducting medium that replaces the air. However, the latter motion is known to be smooth, steady, and progressive, without the oscillations observed in the acoustic case.[42]

Helmholtz knew other cases of fluid motions that looked very different from electric flow. As he had seen while expelling smoke with a bellow or by blowing air across a candle flame, when air flows into a wider space through a sharply-delimited opening, it forms a compact jet that gradually spreads into vortices (see Fig. 4.5), whereas electricity would flow in every direction. He did not regard internal friction as a plausible explanation for this phenomenon, for he believed it could only smooth out the flow. Instead, he turned his attention to the tangential discontinuities of motion that he knew to be possible in a non-viscous fluid.[43]

4.3.1 *Discontinuous motion*

Hydrodynamicists had generally assumed the continuity of velocity, which the differential character of Euler's equations seemed to require. The only exception was Stokes, in brief suggestions for solving paradoxes of fluid motion (see Chapter 5). Helmholtz arrived at his concept of discontinuity surfaces independently, through his study of vortex motion. As

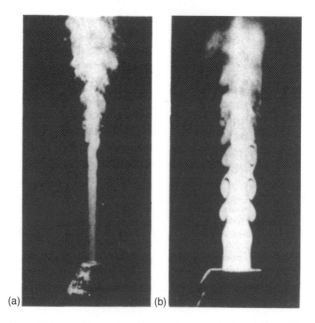

Fig. 4.5. Smoke jets, with (a) spontaneous instability, and (b) sound-triggered instability. From Becker and Massaro [1968]: plate 1. Courtesy of Henry Becker.

[42]Helmholtz [1868*d*] p. 147.   [43]*Ibid.* pp. 146–50.

we saw earlier, a thin vortex sheet implies a discontinuity of the parallel component of the velocity. Conversely, any possible discontinuity is reducible to a vortex sheet. Helmholtz used this representation to derive some basic properties of discontinuity surfaces.[44]

According to one of Helmholtz's theorems on vortex motion, the vortex filaments of the sheet must move together with the fluid particles. Let, for instance, the fluid move with the velocity v on one side of a plane and be at rest on the other side. Within the sheet the fluid moves on average with the velocity $v/2$. Therefore the sheet must move parallel to itself with the velocity $v/2$. The vortex-sheet picture also gives an immediate intuition of the effects of viscosity on a discontinuity surface. By internal friction, the rotating particles of each vortex filament gradually set into rotation the neighboring particles. Consequently, the sheet grows into a row of finite-size whirls.[45]

Most importantly, surfaces of discontinuity are highly unstable. Helmholtz referred to experiments by his friend John Tyndall that showed the astonishing sensitivity of a jet of smoky air to sound waves. 'Theory', Helmholtz went on, 'allows us to recognize that wherever an irregularity is formed on the surface of an otherwise stationary jet, this irregularity must lead to a progressive spiral rolling up of the corresponding portion of the surface, which portion, moreover, slides along the jet.' This peculiar instability of discontinuity surfaces was essential to Helmholtz's uses of them. Yet Helmholtz never explained how it resulted from theory. The behavior of discontinuity surfaces under small perturbations is a difficult problem which is still the object of mathematical research. As can be judged from manuscript sources, Helmholtz probably used qualitative reasoning of the following kind.[46]

Consider a plane surface of discontinuity, with opposite flows on each side. In this case, the velocity field is completely determined by the corresponding vortex sheet. Let a small irrotational velocity perturbation cause a protrusion of the surface. The distribution of vorticity on the deformed surface is assumed to be approximately uniform. Then the curvature of the vortex sheet implies a drift of vorticity along it, at a rate proportional to the algebraic value of the curvature. Indeed, at a given point of the vortex sheet the velocity induced by the neighboring vortex filaments is the vector sum of the velocities induced by symmetric pairs of neighboring filaments, and each pair contributes a tangential velocity as shown in Fig. 4.6(a). Consequently, vorticity grows around the inflection point on the right-hand side of the protrusion, and diminishes around the inflection point on the left-hand side of the protrusion (see Fig. 4.6(b)). The excess of vorticity on the right-hand slope induces a clockwise, rotating motion of the tip of the protrusion (see Fig. 4.6(c)). The upward component of this motion implies instability. The rightward

---

[44]Discontinuous motions are not possible in the (stationary) electric case because the current is irrotational in a uniform conductor. Helmholtz may already have been working on discontinuous fluid motion in 1862, for at that time he asked Thomson about the potential near a 'Kante', which was relevant to Helmholtz's argument for the formation of discontintuities (Thomson to Helmholtz, 29 Nov. 1862, HN).

[45]Helmholtz [1868d] pp. 151–2.

[46]Ibid. pp. 152–3. For a recent, mathematically advanced study of this problem, cf. Caflisch [1990]. Helmholtz's relevant manuscripts are 'Stabilität einer circulierenden Trennungsfläche auf der Kugel' (HN, in #681), 'Wirbelwellen' (HN, in #684), and calculations regarding a vortex sheet in the shape of a logarithmic spiral (HN, in #680).

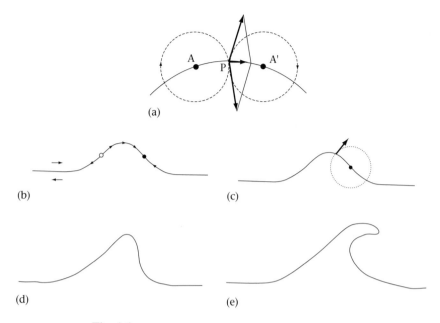

Fig. 4.6.   The growth of a protuberance on a vortex sheet.

component initiates the spiraling motion observed in actual experiments (see Fig. 4.6(d,e)).[47]

Helmholtz described physical circumstances under which the discontinuous motion necessarily occurred. According to Bernoulli's theorem, the pressure of a fluid particle diminishes when its velocity increases, by an amount proportional to the variation of its kinetic energy. Therefore, wherever the velocity of the fluid exceeds a certain upper limit, the pressure becomes negative and, according to Helmholtz, the fluid must be 'torn off'. This necessarily happens when the fluid passes a sharp edge, for the velocity of a continuous flow would be infinite at the edge. For a smoother edge, the discontinuity occurs above a certain velocity threshold.[48]

4.3.2   *Conformal mapping*

In the final section of his paper, Helmholtz managed to solve exactly a case of two-dimensional, discontinuous motion. The complex-variable method he used in this context was so influential that a digression on its origins is in order.

As we saw in Chapter 1, in his memoir of 1749 on the resistance of winds, d'Alembert ingeniously noted that for a two-dimensional flow the two conditions

$$\frac{\partial u}{\partial x} + \frac{\partial v}{\partial y} = 0 \quad \text{(incompressible flow)} \tag{4.12}$$

and

$$\frac{\partial v}{\partial x} - \frac{\partial u}{\partial y} = 0 \quad \text{(irrotational flow)} \tag{4.13}$$

were equivalent to the condition that $(u - iv)(dx + i\,dy)$ be an exact differential. In an *Opuscule* of 1761, he introduced the complex potential $\varphi + i\psi$ such that

$$(u - iv)(dx + i\,dy) = d(\varphi + i\psi). \tag{4.14}$$

The real part φ of this potential is the velocity potential introduced by Euler in 1752. As d'Alembert noted, its imaginary part ψ verifies

$$d\psi = -v\,dx + u\,dy, \tag{4.15}$$

and is therefore a constant on any line of current. This is why it is now called the 'stream function'.[49]

The first beneficiary of d'Alembert's remarks was Lagrange, in a theory of the construction of geographical maps. In order that infinitesimal figures drawn on a map (x, y) should be similar to their representation on another map (u, v), an infinitesimal displacement on one map must be related to an infinitesimal displacement on the other through relations of the form

$$du = \alpha\,dx - \beta\,dy, \quad dv = \beta\,dx + \alpha\,dy. \tag{4.16}$$

In this stipulation, Lagrange recognized d'Alembert's conditions (4.12) and (4.13) for $(u - iv)(dx + i\,dy)$ to be an exact differential, and thus reduced the problem of conformal mapping to taking the real part of any regular function of the complex variable. This is why such functions are now called 'conformal transformations'.[50]

In general, d'Alembert's remarks indicate that a function $(x, y) \rightarrow (u, -v)$ can be expressed as a regular function of the complex variable if and only if it satisfies the conditions (4.12) and (4.13). These conditions are in turn equivalent to the existence of a (holomorphic) complex potential $\varphi + i\psi$ such that $u = \partial\varphi/\partial x = \partial\psi/\partial y$ and $v = \partial\varphi/\partial y = -\partial\psi/\partial x$. They are now called the Cauchy–Riemann conditions, because Augustin Cauchy and Bernhard Riemann exploited them in their beautiful theories of functions in the complex plane. Physicists were initially less receptive. In a mathematical study of incompressible fluids of 1838, Samuel Earnshaw gave the general integral of $\Delta\varphi = 0$ as $\varphi = f(\alpha x + \beta y) + g(\alpha x - \beta y)$ with $\alpha^2 + \beta^2 = 0$, and $\varphi = \ln r$ and $\varphi = \theta$ as particular solutions; but he did not introduce the stream function or the complex potential. In 1842, Stokes introduced the stream function as a way of solving the incompressibility condition (4.12), but he did not appeal to complex functions.[51]

Helmholtz was the first hydrodynamicist to take full advantage of d'Alembert's marvelous discovery. His reading of Riemann's dissertation of 1861 may have alerted him to the tremendous power of the theory of complex functions. In his lectures on deformable

[47]A more precise argument of the same kind is given in Batchelor [1967] pp. 511–17.

[48]Helmholtz [1868*d*] pp. 149–50. Negative pressure, or tension, is in fact possible as a metastable condition of an adequately prepared fluid: cf. Reynolds [1878] and earlier references therein.

[49]D'Alembert [1752] pp. 60–2, [1761] p. 139. See Chapter 1, pp. 21–2.

[50]Lagrange [1779].

[51]Earnshaw [1838] pp. 207–12; Stokes [1842] p. 4. Lagrange ([1781] p. 720) also introduced ψ, but without the geometrical interpretation.

media, Arnold Sommerfeld reports that, during a vacation in the Swiss Alps, the Berlin mathematician Karl Weierstrass asked Helmholtz to take a look at Riemann's dissertation, for he could not make sense of it. Helmholtz found it very congenial, presumably because Riemann's considerations had their roots in the study of physics problems. Also, Helmholtz's idea of characterizing potential flow through singular surfaces of discontinuity has some similarity with Riemann's program of characterizing analytic functions through their singularities in the complex plane.[52]

As suggested by d'Alembert, Helmholtz reduced the solution of his problem of two-dimensional flow to the search for a complex potential that was a holomorphic function of the variable $x + iy$. This potential (unlike its derivative $u - iv$) satisfies a simple boundary condition, namely that its imaginary part, the stream function, must be a constant along the frontier of an immersed solid and along a line of discontinuity. It must, of course, be singular on the lines of discontinuity, the form of which is not a priori given. Helmholtz astutely started with a simple analytic form of the inverse function $\varphi + i\psi \rightarrow x + iy$, so that geometrically simple boundary conditions could be imposed on the flow.[53]

Helmholtz first tried the simple form

$$z = A(w + e^w),\tag{4.17}$$

with $z = x + iy$ and $w = \varphi + i\psi$. This gives $x = A(\varphi - e^\varphi)$ and $y = \pm A\pi$ for $\psi = \pm \pi$, so that the two parallel, interrupted straight lines defined by $x \leq -A$ and $y = \pm A\pi$ can be regarded as walls along which the fluid is constrained to run. Consequently, eqn (4.17) expresses the motion of a liquid flowing from an open space into a canal bounded by two thin parallel walls. At the extremities of these walls, for which $\varphi = 0$, the fluid velocity is easily seen to diverge. Helmholtz modified the expression (4.17) so that the lines of current $\psi = \pm \pi$ run along the outer walls of the canal (from the left) and become discontinuity surfaces after passing the extremity of the walls. The condition of constant pressure on these surfaces led him to the not-so-simple expression

$$z = A(w + e^w) + A \int \sqrt{2e^w + e^{2w}}\, dw,\tag{4.18}$$

which gives the flow shown in Fig. 4.7. As Kirchhoff later remarked, the dead water may be replaced by air without altering the boundary solutions. With this modification, Helmholtz's flow represents the jet formed by water issuing from a large container through a so-called Borda mouthpiece. The contraction of the fluid vein is exactly one-half, as Charles de Borda had proved a century earlier by balancing the momentum flux of the jet with the resultant of the pressures on the walls of the container.[54]

Helmholtz's amazing exploitation of complex numbers quickly attracted the attention of contemporary physicists. Gustav Kirchhoff and Lord Rayleigh soon derived other cases of discontinuous, two-dimensional motion by Helmholtz's method. Kirchhoff dealt with free fluid jets, such as those emerging from a water nozzle. He also solved the

[52]Sommerfeld [1949] p. 135 (Sommerfeld got the anecdote from Adolf Wüllner).

[53]Helmholtz [1868d] pp. 153–7.

[54]Ibid.; Kirchhoff [1869]; Borda [1766]. On Borda's reasoning, cf. Truesdell [1955] pp. LXXIII–LXXV.

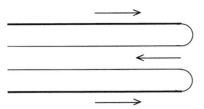

Fig. 4.7.   Helmholtz's discontinuity surface (thin line) for the flow (arrows) from an infinite container into a
      pipe (thick lines). From Kirchhoff [1869] p. 423.

problem of a plane blade immersed obliquely in a uniform flow. Rayleigh did the same and
calculated the dragging force of the fluid on the blade. Between the lines of discontinuity
issuing from the edges of the blade, the fluid remains at rest, so that the pressure on the
rear side of the plate is smaller than that on the front. Rayleigh thus offered a new escape
from d'Alembert's paradox.[55]

Helmholtz believed his short paper on discontinuous fluid motion to be 'of great
importance,' no doubt because it filled some of the gap between the fundamental equations of hydrodynamics and the fluid motions observed in nature. He regarded the
surfaces of discontinuity and their spiral rolling up as basic features of flow around
solid obstacles. His contemporaries were divided on this issue. As we will see in the next
chapter, Stokes shared Helmholtz's view that discontinuous flows correctly schematized
the behavior of nearly-inviscid fluids. William Thomson, despite his friendship with
Helmholtz, believed that they contradicted basic dynamical principles.[56]

4.3.3   *Pipe blowing*

Helmholtz's most immediate concern was the blowing of organ pipes (see Fig. 4.4).
According to the three first editions of the *Tonempfindungen*, the air stream from the
mouth cd of the pipe produces a hissing noise while breaking on its lip ab, and the Fourier
components of this noise near the resonance frequency of the tube excite the vibrations of
the air column. In his hydrodynamic paper of 1868, Helmholtz mentioned that the true
explanation of the blowing of organ pipes should instead be based on discontinuous air
motion. The details are found in the fourth edition (1877) of the *Tonempfindungen*.[57]

The mouth of the pipe, Helmholtz explained, produces an air blade that would hit the
lip ab if no additional motion intervened. Now suppose that the air in the tube is already
oscillating, with alternating compression and expansion. Owing to this motion, air streams
back and forth perpendicular to the blade and forces it alternately in and out of the tube

---

[55]Kirchhoff [1869]; Rayleigh [1876*a*] (resistance), [1876*b*] (*vena contracta*). Works on discontinuous fluid
motion are reviewed in Hicks [1882] pp. 68–71 and Lamb [1895] pp. 100–11. For a recent assessment of Rayleigh's
solution to the blade problem, cf. Anderson [1997] pp. 100–6.

[56]Helmholtz to Du Bois-Reymond, 20 Apr. 1868 ('Grundgedanken ... von grosser Tragweite'), in Kirsten
et al. [1986]. See Chapter 5, pp. 197–207.

[57]Helmholtz [1863*a*] p. 150; Helmholtz [1877] pp. 154–7, 629–31.

(since the corresponding vortex sheets must follow the motion of the air). Owing to the instability of vortex sheets, the air of the blade 'mixes with the oscillating air of the pipe'. The air from the bellows is thus fed into the tube during the compression phase, whereas this air avoids the tube during the expansion phase. For a strong supply of air from the bellows, the blade moves suddenly from one side of the lip to the other because of the high intensity of the deflecting stream. Consequently, the driving force is a crenellated function of time. If the tube is narrow enough not to suppress higher harmonics, the resulting vibrations of the air column are a saw-shaped function of time, as is the case in bowed-string instruments. This is how Helmholtz justified the names *Geigenprincipal*, *Viola di Gamba*, *Violoncell*, and *Violonbass*, that German organ-builders had given to the stops made of strongly-blown, narrow tubes.[58]

We have now gone full circle through Helmholtz's hydrodynamic studies. Schematically, the insufficiencies of his theory of organ pipes led him to investigate internal friction and vortex motion, the latter study being a preliminary of the former. One simple case of vortex motion brought him to discontinuous motions, which turned out to be relevant to the blowing of organ pipes. Here we find the thematic interconnectivity, and the oscillation between the theoretical and the practical that characterized much of Helmholtz's work.

4.4   Foehn, cyclones, and storms

4.4.1   *Outdoor thermodynamics*

Organ pipes are a simple, small-scale, and man-made device that can be manipulated in the laboratory and subjected to physico-mathematical analysis. The emergence of modern physics largely depended on the focus on systems of this kind; so did the later progress of hydrodynamics. Yet the study of complex, large-scale, natural systems never came to a halt. It constituted what Wolfgang Goethe called the 'morphological sciences', including botany, comparative anatomy, geology, and meteorology. In these sciences, the application of general physical laws was rarely attempted and the approach was mostly descriptive and non-mathematical.[59]

Toward the mid-nineteenth century, the methodological gap between small- and large-scale physics became smaller, mainly thanks to the efforts of British natural philosophers. One important novelty was the introduction of mimetic experimentation, the imitation in the laboratory of natural phenomena such as clouds, rain, and thunder. Another was the new thermodynamics, which controlled the global evolution of complex macro-systems independently of their detailed constitution. Meteorology nevertheless retained its descriptive, empirical character. Attempts to subject it to general physical principles were rare.

---

[58]Helmholtz [1877] pp. 155–6, 629–30. On p. 630n Helmholtz referred to several anticipations of his notion of *blattförmiger Luftstrom*—by Heinrich Schneebeli (*Luftlammelle*), by Hermann Smith (air reed), and by W. Sonrek (*Anblasestrom*). Schneebeli [1874] demonstrated the air blade experimentally, by means of movable lips, smoked air, and silk paper, and theoretically justified it through Helmholtz's discontinuous surfaces. Helmholtz did not explain how the motion in the tube was started. Schneebeli and Smith did so in two different manners (cf. Ellis [1885] pp. 396–7). Ellis mentions that the famous French organ-builder Aristide Cavaillé Coll had presented the notion of *anche libre aérienne* to the French Academy of Science in 1840.

[59]Cf. Mertz [1965] pp. 200–26.

As late as 1890, the American meteorologist Cleveland Abbe still deplored this state of affairs: 'Hitherto, the professional meteorologist has too frequently been only an observer, a statistician, an empiricist—rather than a mechanician, mathematician and physicist.'[60]

Helmholtz loved mountaineering and boating, and was a keen observer of natural phenomena. From Thomson's yacht, he watched and measured the waves on the sea. In the Alps, he scrutinized cloud and storm formations and admired the *Mer de glace*: 'A truly magnificent show, this motion, so slow, so constant, and so powerful and irresistible.' Like his British friends Thomson and Tyndall, in these beautiful, sometimes strange phenomena he saw an opportunity to demonstrate the powerful generality of physical laws. The Italian physicist Pietro Blaserna, who often accompanied Helmholtz on his hikes, recalled:[61]

> He loved to climb mountains and glaciers and to enjoy the wonderful views that nature generously offers from their heights. He was a strong and confident climber, for whom four to six hours climbing was nothing... It was very interesting to walk a glacier with him. His eyes were everywhere, and he immediately turned any remarkable phenomenon or formation that ice could offer into an object of investigation.

In 1865, Helmholtz lectured on 'Ice and glaciers', elaborating on James Forbes's and John Tyndall's studies. At the beginning of his talk, he discussed the temperature gradient of the atmosphere, which determines the existence and height of the snow line. This led him to a brief thermodynamic explanation of a peculiar meteorological phenomenon, the foehn. In the first step of this explanation, warm, humid air from the Mediterranean Sea expands adiabatically while rising over the Alps and thus cools down. Owing to the precipitations occurring around the summits, the air becomes warmer and drier. Then it flows down the northern side of the Alps, and the resulting adiabatic compression makes it even warmer and drier. At that stage it is experienced as the foehn wind.[62]

This explanation of the foehn was original at the time Helmholtz proposed it. Since about 1850, Swiss meteorologists believed in a Saharan origin of the foehn. More recently, the leading German meteorologist, Heinrich Dove, had proposed an equatorial origin of this wind, against the Swiss evidence for its dryness. The controversy between Dove and the Swiss lasted even after the publication of Helmholtz's theory, which was apparently ignored. Only after Julius Hann independently proposed and powerfully defended the same theory did meteorologists change their mind.[63]

Even though the foehn is a minor meteorological phenomenon, the Helmholtz–Hann explanation of it has broader historical significance, as one of the first successful applications of thermodynamics to the atmosphere. Although the meteorological importance of

---

[60]Abbe [1890] p. 77. On mimetic experimentation, cf. Galison [1997] pp. 80–1, Schaffer [1995]. On the state of meteorology, cf. Garber pp. [1976] pp. 52–3, Kutzbach [1979] pp. 1–3, a book on which I heavily rely. Other useful sources are Khrgian [1970], Schneider-Carius [1955], Brush and Landsberg [1985].

[61]Helmholtz [1865] p. 111; Blaserna, quoted in Koenigsberger [1902] vol. 2, p. 66.

[62]Helmholtz [1865] p. 97. On Helmholtz's glacier theory, cf. Darrigol [1998] pp. 31–3.

[63]Cf. Kutzbach [1979] pp. 58–62. James Espy had proposed a theory of the foehn similar to Helmholtz's around 1840, without success, cf. Khrgian [1970] pp. 165–6.

adiabatic convection had been known since the 1840s, the process was then analyzed in terms of the caloric theory of heat. The first applications of the new thermodynamics to adiabatic convection in the atmosphere occurred in the mid-1860s.[64]

In 1862, William Thomson first applied thermodynamics to the rate of temperature decline in the atmosphere, and showed that the adiabatic convection of saturated air would result in a smaller rate than for dry air, in conformance with observations. In 1864, the German mathematician Theodor Reye independently published a thorough analysis of the role of the adiabatic expansion of saturated air in the formation of ascending currents, to which we will return shortly. Originally, these considerations attracted even less attention than the Helmholtz–Hann theory of the foehn. They became standard meteorological knowledge after Reye included them in his influential *Die Wirbelstürme, Tornados und Wettersäulen*, published in 1872.[65]

4.4.2 *General circulation*

Helmholtz returned to meteorology in 1875, in a popular lecture on 'cyclones and storms'. The incentive was probably Reye's book, which he admired for its insights into the role of adiabatic processes. He may also have been struck by the magnificent pictures of atmospheric vortices that Reye provided (see Fig. 4.8).[66]

At the beginning of his lecture, Helmholtz briefly discussed the difficulty of applying the general laws of physics to atmospheric phenomena. The beholder of a cloudy sky, he noted, could not help feeling that 'the rebellious and absolutely unscientific demon of chance' was at work. Yet physicists did not doubt that meteorological phenomena obeyed the laws of hydrodynamics and thermodynamics. The true difficulty, Helmholtz explained, was that the very nature of the system forbade detailed predictions:

> The only natural phenomena that we can pre-calculate and understand in all their observable details, are those for which small errors in the input of the calculation bring only small errors in the final result. As soon as unstable equilibrium interferes, this condition is no longer met. Hence chance still exists in our [predictive] horizon; but in reality chance only is a way of expressing the defective character of our knowledge and the 'roughness of our combining power.'

As an example of an unstable system with high sensitivity to the initial conditions (as today's physicists would put it), Helmholtz gave a vertical, rigid bar that can freely rotate around its lower, fixed extremity. Clearly, the smallest departure from the vertical position has dramatic consequences on the future of this system. Helmholtz judged that many meteorological situations led to such instabilities.[67]

There already existed, however, successful applications of mechanical and thermal principles to the atmosphere. The most trivial that Helmholtz mentioned was the seasonal

---

[64]Cf. Garber [1976] pp. 53–7, Kutzbach [1979] pp. 22–7 (on Espy in the 1840s), 45–58 (on Thomson, Reihe, and Peslin in the 1860s).

[65]Thomson [1865]; Reye [1864], [1872]. Cf. McDonald [1963a], Garber [1976] pp. 56–7, Kutzbach [1979] pp. 46–58.

[66]Helmholtz [1876].

[67]*Ibid.* pp. 140, 162, 151. Helmholtz borrowed the bar on a fingertip from Reye [1872] p. 40.

Fig. 4.8.   Whirling winds over burning reed bushes. From Reye [1872].

variation of average temperature, obviously a consequence of the average height of the Sun over the horizon. More generally, meteorologists could hope to explain regularities in spatial and temporal averages. For instance, it was known that trade winds arise from the combined effect of the heating of equatorial air and the rotation of the Earth. According to George Hadley's theory of 1735, the hot air around the equator rises into the upper atmosphere and the resulting rarefaction of the lower atmosphere induces low-altitude winds converging toward the equator. During the latter motion, the air moves toward lower latitudes for which the linear velocity of the surface of the Earth is larger. Consequently, the air flow relative to the Earth is shifted toward the west. The resulting winds are NE above, and SE below the equator. Hadley further noted that the effect of the rotation of the Earth would be much too large were it not diminished by friction on the ground. Lastly, he described the inverse effects in the upper atmosphere, resulting in SW winds in the northern hemisphere and NW winds in the southern hemisphere. Beyond the trade-wind belt, these upper trade winds cool down and fall back on the Earth, which explains the dominance of western winds in mid-latitudes.[68]

Helmholtz improved this theory by taking into account a mechanical effect overlooked by Hadley, namely that, when air moves toward the equator, its angular momentum is conserved, and therefore its absolute linear velocity diminishes as the inverse of its distance from the axis of the Earth. This effect adds to Hadley's relative velocity. Helmholtz also offered an explanation of equatorial calms: the lower trade-wind air can only rise after its relative westward motion has been halted by friction, so that its centrifugal force becomes

[68]Helmholtz [1876] p. 142; Hadley [1735]. For a critical assessment of this theory, cf. Lorenz [1967] pp. 1–3.

as high as possible (for a relative westward motion, the absolute rotation of the air is smaller than that of the Earth).[69]

Helmholtz described similar effects around the poles. The descent of cold air at the poles implies a diverging flow of air in the lower atmosphere and a converging one in the upper atmosphere. Owing to the rotation of the Earth, the resulting winds are NE around the North Pole and SE around the South Pole. Helmholtz thereby explained Dove's observation that Germany was alternately exposed to these dry cold winds and to warm, humid SW winds from the equator.[70]

4.4.3 *Tropical storms*

Having described this global circulation system, Helmholtz turned to its perturbations. Those were easiest to explain, he judged, where they were the rarest, that is, in the tropical zone. He borrowed from Reye the basic scenario for the formation of these storms.[71]

Reye himself owed the basic idea of a thermal origin of storms to an American meteorologist, James Espy, who developed his views in the 1830s. According to Espy, hot saturated air at the ground level is in an unstable equilibrium. Any local perturbation induces an ascending motion of this air in a column. In this process the air expands and cools down, so that water is condensed. The formation of clouds is thus explained. Moreover, the condensation frees the caloric of the vapor and thus prevents the rising air from cooling too fast. As Espy could determine from his own experiments on the adiabatic expansion of saturated air, this heating of the air column by condensation was sufficient to keep the density of the rising air smaller than that of the surrounding air and thus sufficient to sustain the rising motion.[72]

This ascending convection of warm, saturated air not only accounted for the formation of clouds, but it also provided an explanation of storms. The warm column of ascending air, Espy noted, needed to be fed at its base by converging air. The corresponding horizontal winds were those observed in storms. Their radial pattern was still acceptable in Espy's times, owing to the imprecision and confusion of data. In around 1860, another American meteorologist, William Ferrel, recognized the importance of the effects of the Earth's rotation on atmospheric motion, gave its precise mathematical expression, and modified Espy's theory accordingly. Espy's converging winds had to be deflected, in a counterclockwise manner in the northern hemisphere. The resulting motion was a whirl with growing rotation toward the center of the storm.[73]

---

[69]Helmholtz [1876] pp. 142–5. In the late 1850s, the American meteorologist William Ferrel had already given a correct mathematical formulation of the effect of the Earth's rotation on the motion of the atmosphere (Kutzbach [1979] pp. 35–41). As his works were largely unknown in Europe, Helmholtz's use of the conservation of angular momentum in this context may have been original.

[70]Helmholtz [1876] p. 146. In the 1850s, Ferrel had already described the 'cold polar vortex' (Kutzbach [1979] pp. 39–40; Khrgian [1970] pp. 239–40). James Thomson's general-circulation system of 1857 was similar to Helmholtz's (Khrgian [1970] pp. 239, 242). For a critical history of general-circulation systems, cf. Lorenz [1967] pp. 59–78.

[71]Helmholtz [1876] p. 148.

[72]Espy [1841]. Cf. McDonald [1963b], Garber [1976] pp. 53–5, Kutzbach [1979] pp. 22–7.

[73]Ferrel [1860]. Cf. Kutzbach [1979] pp. 35–41, and p. 37n for literature on the introduction of the Coriolis force.

Reye's theory of whirling storms was a modernized version of Espy's theory, based on a thermodynamic analysis of the saturated adiabatic process and including some insight by the Australian-based English naturalist Thomas Belt. The first step of Reye's theory was the formation of an inferior layer of hot, humid air, owing to solar heating. In conformity with some of Belt's theory, Reye proved the instability of this layer for a sufficient temperature gradient, even if its air was not saturated. The rest of Reye's theory resembled Espy's. When the equilibrium is broken at some place, the hot humid air begins to rise. At some elevation, it becomes saturated and it keeps rising through the surrounding dry air because saturated air is more expandable than dry air (its density diminishes faster for the same decrease of pressure). This rising column of air induces a radial motion of the lower air toward the foot of the column. Due to the rotation of the earth, this air whirls around the column.[74]

Helmholtz reproduced this theory with a superior understanding of the relevant dynamics. For example, he explained the whirling motion by the conservation of angular momentum. A ring of air entering the depression without initial rotational velocity with respect to the Earth has an absolute rotational velocity (increasing with the latitude) and a corresponding angular momentum. When the ring converges, its angular momentum is a constant, so that its absolute angular velocity must grow. Consequently, it gains a rotational velocity with respect to the Earth. In the northern hemisphere, the rotation seen from the sky is counterclockwise.[75]

Most originally, Helmholtz deduced the direction of motion of a tornado. In his paper on vortex motion, he had shown that a linear vortex parallel to a plane wall moved in a parallel plane in the direction of the fluid flow between the vortex and the wall. The reason is that the velocity field is the same as it would be if the fluid were unlimited and had another vortex, the mirror image of the real vortex with respect to the wall (the flow in the median plane of the fictitious two-vortex system is parallel to this plane, in conformance with the boundary condition of the real system). According to one of Helmholtz's theorems, the real vortex must move with the velocity induced by its mirror image. A similar reasoning applies to the case when the vortex is no longer parallel to the wall. Then the vortex moves in the direction of the fluid motion in the sharp angle made by its axis and the wall (of course, there is no motion when the vortex is perpendicular to the wall).[76]

For a tornado in the northern hemisphere, the lower and upper trade winds tilt the whirling axis: the upper part of the tornado is dragged NE and the lower part SW. Consequently, the flow in the sharp angle made by the vortex and the surface of the Earth is directed NW. In the absence of external winds, the tornado must move in the same direction.

Lastly, Helmholtz offered some vague considerations on the more complex storms in the temperate zone. Here he departed from Reye, who believed in a subtropical origin of all storms, independent of the general circulation system. Instead, Helmholtz shared Dove's opinion that mid-latitude perturbations were due to the encounter of polar air

---

[74]Reihe [1872] pp. 40–6 (instability and weather columns), 137–8 (rotation of the Earth); Belt [1859]. Cf. Kutzbach [1979] pp. 88–96. Reihe was unaware of Ferrel's works.

[75]Helmholtz [1876] pp. 151–7. Cf. Kutzbach [1979] pp. 96–9.

[76]Helmholtz [1858] p. 127, [1876] p. 159.

and equatorial air. This is not to say that he followed Dove in every respect. The patriarch of German meteorology still believed that the mechanical conflict between the two air currents was the cause of mid-latitude storms, whereas Helmholtz knew well that only the thermal contrast between the two air masses could account for the energy of the storms.[77]

Helmholtz proposed that the equatorial air rose over the polar air, cooled down adiabatically, and thus precipitated rain. Wherever air rose, a depression occurred and induced whirling winds. The modern reader may recognize here some elements of the polar-front theory of mid-latitude storms.[78]

4.5   Trade winds

4.5.1   *A revelation*

Ten years elapsed before Helmholtz returned to meteorology. The incentive was an observation he made during a vacation in the Swiss Alps. His short report for the Berlin Akademie began as follows:

> Clouds and thunderstorm formation.

> On one day of the first half of September this year Mount Rigi offered a clear perspective toward Jura. At a height somewhat lower than the viewing point—the Rigi's Känzli—there was the quite regular, upper horizontal limit of a heavier and more turbid air layer; this limit was indicated by a thin layer of small clouds which went from North to South in narrow stripes, and which revealed the whirls formed by perturbation and rolling up of the limiting surface.

Where the average mountaineer would only have seen one more pretty scene of nature, Helmholtz registered a concrete realization of a central hydrodynamic concept, namely, a vortex sheet in the sky! (compare with Fig. 4.9).[79]

From this observation, Helmholtz conceived that discontinuous motion played an important role in atmospheric phenomena. He could thus explain a paradox that must have been on his mind for some time. Hadley's theory of trade winds made friction responsible for the moderate value of the eastern or western component of these winds in comparison to the exceedingly high values that the Earth's rotation would by itself imply. This explanation seemed plausible for the lower trade winds, which experience friction on the surface of the Earth. However, it failed for the upper trade winds, which are exposed only to internal friction. As Helmholtz confirmed by an application of the principle of mechanical similarity, the effects of the viscosity of the air on atmospheric

[77]Helmholtz [1876] p. 160. On Dove's views, cf. Khrgian [1970] pp. 168–71, Kutzbach [1979] pp. 11–16, 81–2.

[78]Helmholtz [1876] p. 160. In 1841, the American meteorologist Elias Loomis had proposed a similar mechanism, without the whirling and without the connection between the two currents and the general circulation. In the 1870s, the Norwegian meteorologist Henrik Mohn also used a two-current thermal mechanism, but without the cold front; Mohn assumed that the opposite cold and warm air currents mixed before the latter could rise. Cf. Kutzbach [1979] pp. 27–35 (on Loomis), 76–80 (on Mohn).

[79]Helmholtz [1886].

Fig. 4.9.   A rare cloud formation at Denver, Colorado, on 14 February 1953. Photograph by P.E. Branstine, in
Colson [1954] p. 34.

motion are extremely small or confined to the surface of the Earth. Yet, undamped upper
trade winds would give western winds as impossibly high as 130 m / s at a latitude of $30°$.[80]
   The key to the paradox, Helmholtz realized after contemplating a vortex sheet in the Swiss
sky, was discontinuous motion. The boundary between the upper and lower tradewinds had
to be a surface of discontinuity, analogous to those occurring in the mouth of an organ pipe.
This unstable surface waved and eventually rolled up into a series of vortices. The two
different air strata thus became thoroughly mixed. In the case of organ pipes, the mixing
allowed the transfer of momentum from the blown air to the oscillating air column. In the
atmospheric case, it produced the required damping of the upper trade winds, as well as heat
exchange between the upper and the lower air. In Helmholtz's words:[81]

> The principal obstacle to the circulation of our atmosphere, which prevents the
> development of far more violent winds than are actually experienced, is to be found
> not so much in the friction on the Earth's surface as in the mixing of differently
> moving strata of air by means of whirls that originate in the rolling up of surfaces of
> discontinuity. In the interior of such whirls the originally separate strata of air are
> wound in continually more numerous and therefore thinner layers spiraling about
> each other; the enormously extended surfaces of contact allow a more rapid exchange
> of temperature and the equalization of their movement by friction.

[80]Helmholtz [1888] pp. 290–3. If the spatial variations of velocity are smooth, the large scale of atmospheric
motions implies a negligible effect of viscosity. However, when there are abrupt variations of velocity (on the
surface of the Earth, or between two strata of air), the surface friction does not depend on the scale and can have
a non-negligible damping effect.

[81]Helmholtz [1888] p. 308. In modern terms, the discontinuity surface evolves into a fractal object.

4.5.2   *The thermal genesis of discontinuity surfaces*

In order to establish this conclusion, Helmholtz needed 'to show how by means of continually effective forces [owing to gravitation, solar heat, rotation of the Earth, and friction on the Earth's surface] surfaces of discontinuity were formed in the atmosphere.' For this purpose, he introduced two idealizations. Firstly, he ignored the longitudinal variations of the state of the atmosphere. Secondly, he assumed that, within sufficiently thin layers of the atmosphere, the air was in (dry) adiabatic equilibrium and had a uniform motion. Accordingly, the basic concept of his analysis was that of ring-shaped air layers rotating around the axis of the Earth with a uniform angular momentum σ per unit mass and a 'heat content' θ.[82]

By the latter quantity, Helmholtz meant the temperature that a sample of the air would take when brought adiabatically to the normal pressure. At least since his theory of the foehn, and even more since his reading of Reihe, Helmholtz was aware of the importance of adiabatic processes in atmospheric phenomena. Perhaps he also knew of Thomson's paper on adiabatic convection and rate of temperature decrease with elevation. By an argument similar to the one used for viscosity, he showed that ordinary thermal conduction was negligible in smooth, continuous atmospheric motions, so that only Thomson's convective equilibrium, and not ordinary thermal equilibrium, applied to homogenous air layers. For the former kind of equilibrium, the 'heat content', and not the ordinary temperature, is uniform. This concept, under a different name, quickly became central in meteorological thermodynamics. Dove's successor at the head of the Prussian Meteorological Institute, Wilhelm von Bezold, called it 'potential temperature', which had no overtones of the old caloric theory.[83]

From Euler's equation of motion and the law of adiabatic compression applied within a given layer, Helmholtz derived the relation

$$\alpha\theta P^{(\gamma-1)/\gamma} = \frac{1}{2}\frac{\sigma^2}{d^2} + \frac{G}{r} - \frac{1}{2}\Omega^2 d^2 + \beta \qquad (4.19)$$

between the pressure P at a point of the layer, the distance r of this point from the center of the Earth, and its distance d from the axis of the Earth (α and β are two constants, γ is the ratio of the specific heats at constant pressure and constant volume, G is the gravitational constant, and Ω is the rotational velocity of the Earth). For two such layers to be in equilibrium the pressures at their contact surface must be the same. This condition yields a relation between r and d which defines the trace of this surface on a meridian plane.[84]

Helmholtz required this equilibrium to be stable in the sense that the imbalance of pressure caused by a protrusion of the surface must counteract this protrusion. This condition implies that the potential temperature must be higher in the layer that is closer to the celestial pole. When the rate σ^2/θ is an increasing function of the distance ρ from

[82]Helmholtz [1888] pp. 308 (quote), 298 (rings).

[83]*Ibid.* p. 293; Bezold [1888] p. 1189. Cf. Garber [1976] pp. 59–62, Kutzbach, [1979] pp. 143–4. Helmholtz corresponded with Bezold and supported him for the Buys–Ballot medal of meteological merits in 1893 (Julius Hann won), cf. Hörz [1997] pp. 201–24.

[84]Helmholtz [1888] p. 298.

the axis, which is true in the normal state of the atmosphere, stability also implies that the separating surface should depart from the ground at an angle between the horizon and the polar direction (see Fig. 4.10). Hence, thin layers, or strata of homogenous air, lean to such an extent that their lower part is further from the Earth's axis than their higher part. From this simple rule, Helmholtz drew essential conclusions.[85]

Suppose that the stratum appears on Earth as an easterly wind. Then its rotation is slower than that of the Earth. Friction on the Earth's surface therefore increases the absolute rotation of the lower part of the stratum. This part of the stratum being also the furthest from the axis, the effect of friction will remain confined to it, because the excess of centrifugal force pulls the air away from the axis (and because internal friction is negligible).[86] Consequently, the stratum slides toward the equator and its velocity becomes more and more heterogeneous: whereas the momentum of the upper part remains unchanged and corresponds to higher and higher easterlies, that of the lower part, near the ground, increases until the wind vanishes. According to Helmholtz, this process feeds high-momentum air into the zone of calms, which consequently grows to touch the upper part of the converging easterly strata (see Fig. 4.11). Since momentum is conserved for the two kinds of air thus coming into contact (internal friction again being negligible), a surface of discontinuity is born.[87]

4.5.3 Corrections and additions

Helmholtz expressed himself in so condensed a manner that his arguments are sometimes difficult to follow. For example, he claims to be explaining the formation of a surface of discontinuity from an initial, continuous state of motion, whereas in fact he starts

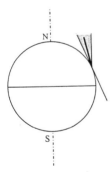

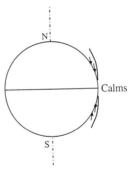

Fig. 4.10. Permitted inclination of an air stratum (thick line) according to Helmholtz's atmospheric circulation theory.

Fig. 4.11. Formation of a discontinuity surface (thick line) bordering the zone of calms.

---

[85]*Ibid.* pp. 299–301.

[86]Helmholtz compared this process with the heating of a volume of air from its top, which affects only the upper air layer because of the lack of convection.

[87]Helmholtz [1888] pp. 304–5.

with a state of contiguous air layers with different angular momenta. At first glance, it would seem that he has only proved that existing discontinuities can be amplified. In reality, his stronger claim for the genesis of discontinuities holds. As he made clear at some point, he included the case of infinitely-thin strata in his analysis. The above reasoning extends to this case, and still leads to the discontinuity at the upper limit of the calm zone.

Another difficulty of Helmholtz's memoir is that it blurs the distinction between intuitive arguments, empirical data, and strict dynamical deduction. For example, should the existence of the zone of calms be regarded as empirically given, or does it result from dynamical reasoning? In this paper, with its emphasis on the formation of discontinuities, Helmholtz seems to be taking the first option, whereas in his earlier discussion of trade winds in 1875 he explained the calm zone as an indirect consequence of the centrifugal force.

At least in one case, Helmholtz dangerously confused intuition and deduction. He argued that the mixed air produced by instability at the border between the lower and upper trade winds had to move toward the equator, because its intermediate temperature and velocity belonged to lower altitudes and latitudes. As he admitted the following year, this conclusion was wrong. A rigorous treatment of the conditions of equilibrium of the mixed air with the two mixing layers implies an ascending motion of this air.[88]

During this upwards expulsion of the mixed air, originally-remote parts of the two mixing layers come into contact. Owing to the conservation of momentum, the shifted parts of the polar-side layer lose velocity, whereas those on the equatorial side gain velocity. Hence the discontinuity surface is renewed, even if the remote parts of the layers were originally at rest with respect to the Earth. As Helmholtz explained, in his first paper he had shown how and where discontinuities were formed in an originally continuously-moving atmosphere. He could now show that the mixing process at a surface of discontinuity renewed this surface instead of destroying it.[89]

A short section of the 1888 memoir sketched the production of surfaces of discontinuity around the poles. Due to the cooling of the Earth near the poles, cold air strata diverge from the pole at low altitude. Owing to the rotation of the Earth, these strata appear as north-easterlies. As was shown for the lower trade winds, their inferior parts experience friction on the Earth's surface and the resulting increase of the centrifugal force drags them further south. Owing to the inclination of the strata, this cold air remains close to the surface of the Earth, in conformance with the fact that in northern Germany the north-east winter winds do not reach the summits of mountains. 'At the front border of these easterlies advancing into warmer zones,' Helmholtz went on, 'the same circumstances that produce discontinuities of motion between upper and lower currents in the advancing trade winds are effective, bringing about a new cause for the formation of vortices.'[90]

[88]Helmholtz [1888] p. 306; Helmholtz [1889] pp. 312–15.

[89]Helmholtz [1889] p. 315n.   [90]Helmholtz [1888] pp. 307–8.

4.5.4 *Anticipations*

Reading these lines, modern meteorologists could speculate that Helmholtz introduced the now fundamental notion of the polar front. The extreme concision of his statement does not allow any such judgment. The main purpose of his paper was to find a mechanism for damping the winds induced by the rotation of the Earth. Unlike some of his followers, he did not have in mind a theory of storms based on surface discontinuities. Most likely, he still believed that mid-latitude storms were too complex to be subjected to dynamical analysis.

Yet there is no doubt that Helmholtz was the first to realize the essential importance of surface discontinuities in meteorology, before horizontal and vertical field measurements made them clearly visible toward the end of the century. With some delay, his 1888 paper was a major source of inspiration for the meteorologists who applied the concept to the theory of storms. Early in this century, the Austrian Max Margules integrated Helmholtz's discontinuity surfaces in his atmospheric energetics and generalized Helmholtz's formula for the slope of surfaces of discontinuity. Subsequently, Felix Exner and his Viennese school of meteorological dynamics heavily relied on Margules's extensions of Helmholtz's meteorological concepts.[91]

The Norwegian meteorologist Vilhelm Bjerknes also owed much to Helmholtz. This is true for two of his breakthroughs in dynamical meteorology and weather forecasting. His circulation theorem, giving the rate of variation of the vorticity for a compressible fluid, was a simple extension of Helmholtz's theorem on the conservation of the vorticity in incompressible fluids. His atmospheric kinematics emphasized the singularities of the velocity field that Helmholtz first discovered. The concept of a 'cold front', which is so central to modern meteorological forecasting, occurred to him while studying Helmholtz's 1888 paper in his Leipzig seminar.[92]

Helmholtz's works were not the only resources exploited by Margules, Bjerknes, and other founders of modern meteorology. As Kutzbach has shown, the various thermal theories of cyclones, their late-nineteenth-century difficulties, the enormous improvement of the quality and quantity of weather data, and observations of the higher atmosphere were all important factors of progress. Against this view, later meteorologists have usually regarded the polar-front theory as a sharp break from the past and ignored the many continuities with the past, including its Helmholtzian roots.[93] They may have been blinded by the spectacular progress in weather forecasting that this theory brought about. Or

[91]Margules [1906]; Exner [1925] has many references to Helmholtz's works, on pp. 92 (mechanical similarity), 203–10 (air rings), 214n (general circulation), 234–5 (empirical verification of Helmholtz's stability conditions), 334 (Helmholtzian origin of Bjerknes's polar front). On the observation of surface discontinuities, cf. Kutzbach [1979] pp. 175 (Bigelow), 181–3 (Shaw), 194–7 (Margules). On Margules's extensions of Helmholtz's results, cf. *ibid*. pp. 197–9.

[92]Bjerknes [1898] for the circulation theorem; Bjerknes *et al.* [1910] for the kinematics; Bjerknes *et al.* [1933] pp. 784 (reading Helmholtz and the polar front), 785 (reading Helmholtz and the wave theory of cyclones). The first frontal cyclone model was published by Bjerknes's son Jacob, who used Margules's generalization of Helmholtz's slope formula for the surface of discontinuity. Cf. Kutzbach [1979] pp. 158–71 (circulation theorem), 206–18 (J. Bjerknes's model), Khrgian [1970] pp. 215–16, Friedman [1989].

[93]Cf. Kutzbach [1979] pp. 218–20. Two notable exceptions are Wenger [1922] and Bernhardt [1973], who gave competent reviews of Helmholtz's meteorological works.

perhaps they wished to glorify the founders of the newer schools of meteorology, especially the Bergen school, at the expense of earlier investigators. Yet there is no doubt that Helmholtz anticipated some central concepts of modern meteorology.[94]

4.6 Wave formation

4.6.1 *From atmospheric waves to water waves*

Both for organ pipes and for the general circulation of the atmosphere, an essential property of Helmholtz's surfaces of discontinuity is their instability, which allows mixing of the layers in contact. In the atmospheric case, the two layers usually have different temperatures, and therefore different densities. Consequently, Helmholtz noted in his memoir of 1888, the instability is similar to the one induced by wind blowing on a quiet sea. For moderate winds, the surface of the sea oscillates periodically. For larger velocities, whirls are formed and the tips of the waves break into foam and droplets. Helmholtz imagined similar turbulence to occur at the contact surface between two atmospheric layers and to permit intimate mixing of their contents.[95]

From Mount Rigi, Helmholtz had seen stratified and whirling clouds that directly suggested the analogy between atmospheric and sea waves. He justified this analogy in 1889, in a sequel to his paper on atmospheric motion. By a similarity argument, he showed that the scale of the waves varied as the square of the wind velocity, and that similar waveforms occurred when the ratio of the kinetic-energy densities of the two media was the same in the reference system for which the waves are stationary. These rules imply that typical waves in the atmosphere are much larger than waves on the ocean, since the density ratios are much smaller in the atmospheric case. For example, waves of one meter in length on the ocean correspond to waves of about two kilometers in length between two layers of the atmosphere under the same wind (relative velocity) and with a temperature difference of 10° Celsius.[96]

Helmholtz thus related waves in the sky to the better-known waves on the sea. The theory of the latter kind of waves was fairly developed, thanks to the efforts of British physicists. For example, in 1871 William Thomson had given a theory of small waves on a calm sea, including the influence of capillarity. Helmholtz must have been aware of part of this work, since he helped Thomson measure the minimum velocity of such waves during a yacht trip. Thomson's calculations included the effect of a horizontal wind, and showed that in the linear approximation initially-small waves grew indefinitely when the wind velocity exceeded a certain, small limit. In other words, a plane water surface became unstable under a sufficiently strong wind.[97]

---

[94]In Khrgian's words, 'Helmholtz's works … are referred to only quite rarely now, but it should be remembered that these studies helped lay the basis for present day synoptic meteorology' (Khrgian [1970] p. 208).

[95]Helmholtz [1889] pp. 305–6.

[96]*Ibid.* pp. 316–22. As Helmholtz noted on p. 310n, Luvini [1888] pp. 370–1 independently introduced atmospheric waves and billows.

[97]See above, pp. 87–8, for the fishing line; see below, pp. 188–90, for the wind wave instability.

Helmholtz did not know the latter aspect of Thomson's ripple studies.[98] He was, nevertheless, correct in regarding the production of *finite* waves by wind as an open mathematical question.[99] With his usual analytical power, he attacked this formidable nonlinear problem. Confining himself to two-dimensional, periodic waves of steady form with irrotational flow, he applied the conformal method of his memoir on discontinuous fluid motion. He thus determined the profile and the velocity of the waves to third order in their relative height (height over wavelength), with the following results.[100]

Under a given wind, the wavelength can vary within certain limits that grow with the wind strength. For a given wavelength, the remaining characteristics of the wave are completely determined. The longest possible waves are slowly-propagating, low-amplitude sine waves. Shorter waves are higher, faster, and more abrupt.[101] At the lowest wavelength, the profile of the waves becomes discontinuous. Around the resulting ridge, the infinite velocity results in violent projections of water. This is the frothing of waves.

Under any given wind, an obvious solution of the equations of motion corresponds to flat, undisturbed water. If some of the steady waves under the same wind have a smaller energy, then this solution is unstable. Helmholtz proved that the instability occurred for arbitrarily-small winds and for all permitted wavelengths except those closest to the frothing point.[102] Hence the faintest wind can produce waves on calm water. If the wind grows, the height of these waves increases. The shortest ones break into foam, because they were already on the verge of instability. Another cause of breaking is the superposition of waves of different length and velocity. Remembering his old acoustic works, Helmholtz did not fail to notice that nonlinear superposition generated waves of longer wavelength, just as the ear generates combination tones.[103]

4.6.2   *A minimum principle*

In order to reach his two main conclusions, the instability of a plane water surface under a constant wind and the breaking for high waves, Helmholtz used truncated power series of

---

[98]Cf. Rayleigh to Helmholtz, 29 Oct. 1889, HN.

[99]As was discussed in Chapter 2, pp. 70–2, 83, Stokes and Rayleigh studied water waves of finite height, but only in the absence of wind. Stokes [1880c] used an analytical method somewhat intermediate between those of Helmholtz [1868d] and [1889]. Helmholtz was probably aware of this paper, since he knew a result of Stokes [1880c] published next to [1880c] in the same volume, namely, the 120 degrees of the highest possible wave (Helmholtz [1889] p. 328, where Stokes, however, is not named).

[100]Helmholtz [1889] pp. 323–8. There is some confusion in Helmholtz's notation (besides numerous typographical errors). For example, in some formulas the letter *b* stands for the velocity potential, and in others for the velocity. Helmholtz published only an outline of his calculations. Details are found in manuscripts, as well as calculations for other types of wave (HN, #682, #884).

[101]By an unfortunate slip, Helmholtz stated the opposite on p. 328 of his paper [1889]; however, he gave the right variations on p. 331, in conformance with his equations.

[102]Helmholtz [1889] pp. 329–32. This does not contradict Thomson's earlier result, because Thomson's wind threshold vanishes if capillarity is neglected. The existence of a lower limit for the wavelength seems to contradict the calculation in Thomson [1871a], according to which (capillarity being neglected) the plane water surface is unstable under any perturbation of small wavelength. In fact, it does not, because the growth of these perturbations does not necessarily lead to stable finite-height waves of the same wavelength.

[103]Helmholtz [1889] p. 332.

the relative height of the waves. Since breaking occurs precisely when these series diverge, the second conclusion was rather fragile. Moreover, Helmholtz had only exhibited one class of steady solutions to the hydrodynamic equations. His conclusions did not necessarily apply to more general solutions.[104] The following year, he offered a more rigorous approach based on a new variational principle, akin to the principle of least action on which he was trying to base all physics in these years.[105]

In the wind-over-water problem, the water surface is steady if and only if the pressure is the same on both sides of the surface. For irrotational flow, Helmholtz found this condition to be equivalent to the stationarity of the difference $V - T$ between the potential and the kinetic energy of the motion under infinitesimal deformations of the surface, the total air and water fluxes being kept constant. This condition is similar to the condition of static equilibrium, which requires the stationarity of the potential energy. Helmholtz extended this analogy to the discussion of stability: while in the static case the equilibrium is stable if and only if V is a minimum, in the steady-wave case the motion is stable if and only if $V - T$ is a minimum.[106]

By ingenious qualitative reasoning, Helmholtz determined how the shape of the surface representing the variations of $V - T$ (with respect to the parameters of the waves) changed with the wind velocity. He found that, for a given wavelength, no minimum could occur if the wind velocity was too high. In other words, stable, steady waves of a given length are only possible if the wind velocity does not exceed a certain limit. Nor can they occur if the wave velocity and the wind velocity with respect to the waves are both below certain limits. These theorems agreed with Helmholtz's earlier, less rigorous result about the finite range of wavelengths that corresponds to a given wind strength.[107]

Lastly, Helmholtz discussed the energy and momentum conditions for the initial formation and the growth of waves. He found that the initial small waves could only have a very short wavelength (about 10 cm for a wind of 10 m / s). In order that higher and longer waves could be formed, the wind had to keep blowing in the same direction and to communicate (by some unspecified mechanism) some of its energy and momentum to the waves. Another cause of growth was the nonlinear superposition of different waves, as Helmholtz had already suggested in his previous paper.[108]

---

[104]The waves discussed in Helmholtz's paper are not the only possible steady waves, since in the no-wind case they do not include the Rayleigh–Stokes solution (which can have any wavelength). Helmholtz probably only meant to give the form of *forced* waves, although nowhere did he explain how free and forced solutions of the equations of motion should be discriminated. He soon became aware of the necessity of a broader class of waves, as is attested by a footnote in his collected paper (*HWA* 3, p. 325n); there he suggested to require the balance of pressure only to second order, so that the ratio between wind velocity and wave velocity could be freely chosen. Wilhelm Wien realized this program in Wien [1900] pp. 169–99.

[105]Helmholtz [1890]. *Ibid.* p. 334, Helmholtz noted a connection between the steady-wave problem and the theory of polycyclic systems that he was developing to give a Lagrangian form to thermodynamics and electrodynamics.

[106]*Ibid.* pp. 335–40.   [107]*Ibid.* pp. 340–4.

[108]*Ibid.* pp. 349–55. Helmholtz's zero-momentum condition for the initial formation of waves excludes the waves discussed in his previous paper, which never have zero momentum for a finite height. This contradiction disappears for the more general waves considered later by Helmholtz (see footnote 104).

4.6.3 *Dubious idealizations*

In March 1890, Helmholtz applied to the Staatsminister for a one-month journey to the French Riviera. He not only wished to bring back his family who had spent the winter there, but also 'to perform a few scientific observations on the behavior of sea waves ... in order to test the truth of a few new theoretical propositions on the interaction between wind and waves.' He spent a whole week of April at the tip of the Cap d'Antibes, measuring the wind with a portable anemometer and counting the number of approaching billows. His main purpose was to verify that these two quantities varied in inverse proportion, as resulted from the similitude argument of his first paper on wind and waves. The results were embarrassingly unconvincing. Helmholtz had to admit that the wave count on the shore mainly depended on the off-shore winds, with a delay corresponding to the propagation time. He nevertheless reported his measurements to the Berlin Academy in July 1890, perhaps to justify the official character of his time on the Riviera.[109]

Unlike his other hydrodynamic works, Helmholtz's papers on wind and waves had little follow-up. Their historical impact was limited to the concept of atmospheric waves, which soon became a basic meteorological reality. Experts in hydrodynamics paid little attention to Helmholtz's theory of wave formation, as can be judged from the very brief and fragmentary mention in Lamb's otherwise thorough treatise. Wilhelm Wien seems to have been the only important physicist to pursue Helmholtz's line of thought, with no significant progress. Oceanographers only had a passing interest in it. The 1911 edition of Otto Krümmel's *Handbuch der Ozeanographie* included a praiseful summary of Helmholtz's findings. Later treatises on water waves systematically ignored them.[110]

The reasons for this neglect are not too difficult to guess. One may be that Helmholtz wrote his two papers on wind and waves in a hurry, neglected to provide intelligible summaries, and did not carefully read the proofs. In the long run, a more fundamental reason to ignore Helmholtz's conclusions was that they depended on a number of arbitrary idealizations. He neglected capillarity forces, although they affect the energy and stability of short waves. He only considered steady waves, whereas more general waves could have a different range of stability. He took the air and water flows to be irrotational, whereas the actual air flow is always turbulent. Until the mid-twentieth century, oceanographers could only complain that no theory properly took into account this complexity of wind waves. Reasonable models later became available for the interaction between the turbulent air flow and the oscillatory water surface. Essential to their success was the consideration of the random nature of ocean waves, which nineteenth-century theorists completely ignored. All of this explains why Helmholtz's ingenious memoirs on waves and wind have fallen into oblivion.[111]

---

[109]Helmholtz to Bötticher, 9 Mar. 1890, in Koenigsberger [1902] vol. 3, p. 27; Helmholtz [1890] pp. 353–5. Helmholtz planned a third paper on this topic, see the manuscript fragment 'Forme Sationärer Wogen' (HN, #684).

[110]Lamb [1895] 409n, pp. 421–3; Wien [1900], and previous papers listed therein; Krümmel [1911] vol. 2, pp. 61–4. Cf. also Forchheimer [1905] pp. 429–32 for a summary of Helmholtz's results, and Baschin [1899] p. 410 for a generous assessment of Helmholtz's contribution: 'a theory ... which in a single blow explains all the circumstances of wave formation that are observed in nature.'

[111]On older failures, cf. Russell and MacMillan [1952] pp. 61–2; on modern successes, cf. Kinsman [1965].

Following Helmholtz's strange itinerary across worlds of fluid motion, we began with the pitch of organ pipes, spent a while on atmospheric motion, and ended with water waves. Helmholtz jumped from one domain to the next through an amazing series of conceptual innovations and analogies. Most importantly, he identified the vortex filament as a fundamental, invariant structure of inviscid incompressible flow, and inaugurated a powerful approach to hydrodynamics in which vortices and discontinuity surfaces controlled the flow. With this new perspective, he elucidated basic processes of jet formation, shear instability, and mixing.

Although nineteenth-century physicists and mathematicians recognized the depth of Helmholtz's contributions to hydrodynamics, their practical importance only became apparent in the twentieth century. In Helmholtz's times, the vortex theorems offered more to British theorists of ether and matter than they did to hydraulic engineers. As we will see in the next chapter, Rayleigh's solution to d'Alembert's paradox in terms of Helmholtz's discontinuity surfaces turned out to be quantitatively inadequate; Rayleigh himself did not believe in it, and Kelvin completely dismissed it. As we will see in the last chapter, its connection to a practically useful treatment of fluid resistance was only understood in the early twentieth century. Although Helmholtz offered many new insights into the motion of perfect liquids, he did not know precisely how to relate such motions with those occurring in the slightly-viscous fluids of nature.

5

INSTABILITY

> There is scarcely any question in dynamics more important for Natural Philosophy than the stability of motion.[1] (William Thomson and Peter Guthrie Tait, 1867)

In the previous chapter, we encountered a special kind of instability, now called the Kelvin–Helmholtz instability, which occurs when two fluid masses slide on each other, for instance along smoke jets or on a plane water surface under wind. Helmholtz arrived at this instability by reasoning on vortex sheets and used it to explain phenomena that seemed to elude Euler's equations. He was neither the first nor the last theorist to emphasize the role of instabilities in fluid mechanics. The nineteenth-century interest in this question was for two reasons. Firstly, the discrepancy between actual fluid behavior and known solutions of the hydrodynamic equations suggested the instability of these solutions. Secondly, the British endeavor to reduce all physics to the motion of a perfect liquid presupposed the stability of the forms of motion used to describe matter and ether. Instability in the former case and stability in the latter case needed to be proved.

In nineteenth-century parlance, kinetic instability broadly meant a departure from an expected regularity of motion. In hydrodynamics alone, this notion included unsteadiness of motion, non-uniqueness of the solutions of the fundamental equations under given boundary conditions, sensibility of these solutions to infinitesimal local perturbation, sensibility to infinitesimal harmonic perturbations, sensibility to finite perturbations, and sensibility to infinitely-small viscosity. Although this spectrum of meanings is much wider than a modern treatise on hydrodynamic stability would tolerate, it must be respected in a historical study that does not artificially separate issues that nineteenth-century writers conceived as a whole.

The first section of this chapter is devoted to Stokes's pioneering emphasis on hydrodynamic instability as the probable cause of the failure of Eulerian flows to reproduce the essential characteristics of the observed motions of slightly-viscous fluids (air and water). Stokes believed instability to occur whenever the lines of flow diverged too strongly, as happens in a suddenly-enlarged conduit or past a solid obstacle. Section 5.2 recounts how William Thomson, in 1871, discussed the instability of a water surface under wind, independently of Helmholtz and with a different method.

In this and Helmholtz's case, instability was derived from the hydrodynamic equations. In Stokes's case, it was only a conjecture. Yet the purpose was the same, namely, to explain observed departures from exact solutions of Euler's equations. In contrast, Thomson's vortex theory of matter required stability for the motions he imagined in the primitive

---

[1] Thomson and Tait [1867] par. 346.

perfect liquid of the world. These considerations, which began in 1867, are discussed in Section 5.3. As Thomson could only prove the stability of motions simpler than those he needed, for many years he contented himself with an analogy with the observed stability of smoke rings. At last, in the late 1880s, he became convinced that vortex rings were unstable.

Owing to their different interests, Stokes and Thomson had opposite biases about hydrodynamic (in)stability. This is illustrated in Section 5.4 through an account of their long, witty exchange on the possibility of discontinuity surfaces in a perfect liquid. From his first paper (1842) to his last letter to Thomson (1901), Stokes argued that the formation of surfaces of discontinuity provided a basic mechanism of instability for the flow of a perfect liquid past a solid obstacle. Thomson repeatedly countered that such a process would violate fundamental hydrodynamic theorems and that viscosity played an essential role in Stokes's alleged instabilities. The two protagonists never came to an agreement, even though they shared many cultural values within and outside physics.

Section 5.5 deals with the (in)stability of parallel flow. The most definite nineteenth-century result on this topic was Lord Rayleigh's criterion of 1880 for the stability of two-dimensional parallel motion in a perfect liquid. The context was John Tyndall's amusing experiments on the sound-triggered instability of smoke jets. In 1883, Osborne Reynolds's precise experimental account of the transition between laminar and turbulent flow in circular pipes motivated further theoretical inquiries into parallel-flow stability. Cambridge authorities, including Stokes and Rayleigh, selected this question for the Adams prize of 1889. This prompted Thomson to publish proofs of instability for two cases of parallel, two-dimensional viscous flow. Rayleigh soon challenged these proofs. William Orr proved their incompleteness in 1907, thus showing the daunting difficulty of the simplest questions of hydrodynamic stability.

5.1. Divergent flows

5.1.1 *Fluid jets*

Pioneering considerations of hydrodynamic stability are found in Stokes's first paper, published in 1842 and devoted to two-dimensional and cylindrically-symmetric steady motions of a perfect liquid obeying Euler's equation. From an analytical point of view, most of Stokes's results could already be found in Lagrange's or J. M. C. Duhamel's writings. Stokes's discussion of their physical significance was, nonetheless, penetrating and innovative. Struck by the difference between computed and real flows, he suggested that the possibility of a given motion did not imply its necessity; there could be other motions compatible with the same boundary conditions, some of which could be stable and some others unstable. 'There may even be no stable steady mode of motion possible, in which case the fluid would continue perpetually eddying.'[2]

As a first example of instability, Stokes cited the two-dimensional flow between two similar hyperbolas. An experiment of his own showed that the theoretical hyperbolic flow only held in the narrowing case. He compared this result with the fact that a fluid passing through a hole from a higher pressure vessel to a lower pressure vessel forms a jet, instead

---

[2]Stokes [1842] pp. 10–11.

of streaming along the walls as the most obvious analytical solution would have it (see Fig. 5.1). Although Mariotte, Bernoulli, and Borda already knew of such effects, Stokes was the first to relate them to a fundamental instability of fluid motion and to enunciate a general tendency of a fluid 'to keep a canal of its own instead of spreading out'.[3]

In the case represented in Fig. 5.1, Stokes argued that, according to Bernoulli's theorem, the velocity of the fluid coming from the first vessel was completely determined by the pressure difference between the two vessels. This velocity was therefore homogeneous, and the moving fluid had to form a cylindrical jet in order to comply with flux conservation. Dubious though it may be (for it presupposes a uniform pressure in the second vessel), this reasoning documents Stokes's early conviction that nature sometimes preferred solutions of Euler's equation that involved surfaces of discontinuity for the tangential component of the velocity.

This conviction reappears in a mathematical paper that Stokes published four years later. There he considered the motion of an incompressible fluid enclosed in a rotating cylindrical container, a sector of which has been removed (see Fig. 5.2). For an acute sector, the computed velocity is infinite on the axis of the cylinder. Stokes judged that in this case the fluid particles running toward the axis along one side of the sector would 'take

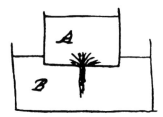

Fig. 5.1. The formation of a jet as a liquid is forced through a hole in a vessel A into another vessel B. From Stokes to Kelvin, 13 Feb. 1858, *ST*.

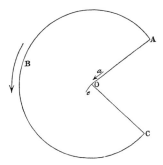

Fig. 5.2. The formation of a surface of discontinuity (O*e*) during the rotation of a cylindrical container (section OABC). After sliding along OA, the fluid particle *a* shoots off at the edge O. From Stokes [1847*b*] p. 310.

[3]Stokes [1842] p. 11.

off' to form a surface of discontinuity. For the rest of his life, Stokes remained convinced of the importance of such surfaces for perfect-fluid motion. Yet he never offered a mathematical theory of their development.[4]

5.1.2 *The pendulum*

As we saw in Chapter 3, much of Stokes's early work was motivated by a more concrete problem of fluid motion, namely, the effect of the ambient air on the oscillations of a pendulum. When applied to the spherical bulb of the pendulum, Euler's hydrodynamics gave no damping at all. In 1843, Stokes considered two kinds of instabilities that could explain the observed resistance. Firstly, he imagined that the fluid particles along the surface of the sphere would come off tangentially at some point, forming a surface of discontinuity. Secondly, he evoked his earlier conviction that divergent flow was unstable:

> It appears to me very probable that the *spreading out* motion of the fluid, which is supposed to take place behind the middle of the sphere or cylinder, though dynamically possible, nay, the *only* motion dynamically possible when the conditions which have been supposed are accurately satisfied, is unstable; so that the slightest cause produces a disturbance in the fluid, which accumulates as the solid moves on, till the motion is quite changed. Common observation seems to show that, when a solid moves rapidly through a fluid at some distance below the surface, it leaves behind it a succession of eddies in the fluid.

Stokes went on to ascribe fluid resistance to the *vis viva* of the tail of eddies, as Poncelet and Saint-Venant had already done in France. To make this more concrete, he recalled that a ship had the least resistance when it left the least wake.[5]

In the following years, Stokes realized that these instabilities did not occur in the pendulum case. The true cause of damping was the air's internal friction. In 1845, Stokes solved the linearized Navier–Stokes equation for an oscillating sphere and cylinder, representing the bulb and thread, respectively, of a pendulum. The excellent agreement with experiments left no doubt about the correctness and stability of his solutions.[6]

For the sake of completeness, Stokes also examined the case of a uniform translation of the sphere and cylinder, which corresponds to the zero-frequency limit of the pendulum problem. In the case of the sphere, he derived the resistance law that bears his name. In the case of the cylinder, he encountered the paradox that the resulting equation does not have a steady solution (in a reference system bound to the cylinder) that satisfies the boundary conditions. Stokes explained:

> The pressure of the cylinder on the fluid continually tends to increase the quantity of fluid which it carries with it, while the friction of the fluid at a distance from the sphere continually tends to diminish it. In the case of the sphere, these two causes eventually counteract each other, and the motion becomes uniform. But in the case of a cylinder, the increase in the quantity of fluid carried continually gains on the decrease due to the friction of the surrounding fluid, and the quantity carried increases indefinitely as the cylinder moves on.

---

[4]Stokes [1847*b*] pp. 305–13.   [5]Stokes [1843] pp. 53–4. See Chapter 3, pp. 136–7.
[6]See Chapter 3, pp. 139–40.

Therein Stokes saw a symptom of instability:

> When the quantity of fluid carried with the cylinder becomes considerable compared
> with the quantity displaced, it would seem that the motion must become unstable, in
> the sense in which the motion of a sphere rolling down the highest generating line of
> an inclined cylinder may be said to be unstable.

If the cylinder moved long enough in the same direction (as would be the case for the suspending wire of a very slow pendulum) then 'the quantity of fluid carried by the wire would be diminished, portions being continually left behind and forming eddies.' Stokes also mentioned that in such an extreme case the quadratic term of the Navier–Stokes equation might no longer be negligible. According to a much later study by Carl Wilhelm Oseen, this is the true key to the cylinder paradox.[7]

5.1.3 *Ether drag*

Air and water were not the only imperfect fluid that Stokes had in mind. In 1846 and 1848, he discussed the motion of the ether in reference to the aberration of stars. In his view the ether behaved as a fluid for sufficiently slow motions, since the Earth and celestial bodies were able to move through it without appreciable resistance. However, its fluidity could only be imperfect, since it behaved as a solid for the very rapid vibrations implied in the propagation of light. Stokes explained the aberration of stars by combining these two properties in the following manner.[8]

He first showed that the propagation of light remained rectilinear in a moving medium, the velocity of which derived from a potential. Hence any motion of the ether that met this condition would be compatible with the observed aberration. Stokes then invoked Lagrange's theorem, according to which the motion of a perfect liquid always meets this condition when it results from the motion of immersed solid bodies (starting from rest). For a nearly-spherical body like the Earth, Stokes believed the Lagrangian motion to be unstable (for it implies a diverging flow at the rear of the body). However, his ether was an imperfect fluid, with tangential stresses that quickly dissipated any departure from gradient flow: 'Any nascent irregularity of motion, any nascent deviation from the motion for which [$\mathbf{v} \cdot d\mathbf{r}$] is an exact differential, is carried off into space, with the velocity of light, by transversal vibrations.'[9]

In the course of this discussion, Stokes noted that his solution of the (linearized) Navier–Stokes equation in the case of the uniformly-moving sphere did not depend on the value of the viscosity parameter and yet did not meet the gradient condition. Hence an arbitrarily-small viscous stress seemed sufficient to invalidate the gradient solution. Stokes regarded this peculiar behavior as a further symptom of the instability of the gradient flow.

In summary, in the 1840s Stokes evoked instability as a way to reconcile the solutions of Euler's equations with observed or desired properties of real fluids, including the ether. He regarded a divergence of the lines of flow (in the jet and sphere cases) and fluid inertia (in the cylinder case) as a destabilizing factor, and imperfect fluidity (viscosity or jelly-like behavior) as a stabilizing factor (explicitly in the ether case, and implicitly in the pendulum

---

[7]Stokes [1850*b*] pp. 65–7. Cf. Lamb [1932] pp. 609–17.

[8]Stokes [1846*c*], [1848*b*]. Cf., e.g., Wilson [1987] pp. 132–45.

[9]Stokes [1848*b*] p. 9.

bulb case). His intuition of unstable behavior derived from common observations of real flows and from the implicit assumption that ideal flow behavior should be the limit of real fluid behavior for vanishing viscosity.

Stokes did not attempt a mathematical investigation of the stability of flow. He did offer a few formal arguments, which today's physicist would judge fallacious. His deduction of jet formation was based on an unwarranted assumption of uniform pressure in the receiving vessel. The steady flow around a cylinder, which he believed to be impossible, is in fact possible when the quadratic terms in the Navier–Stokes equation are no longer neglected. The argument based on the zero-viscosity limit of the flow around a sphere fails for a similar reason. Stokes's contemporaries did not formulate such criticisms. Rather, they noted his less speculative achievements, namely, new solutions of the hydrodynamic equations that bore on the pendulum problem, and rigorous, elegant proofs of important hydrodynamic theorems.

5.2   Discontinuous flow

In Chapter 4, we saw how Helmholtz made discontinuity surfaces a basic element of perfect-liquid dynamics and derived the spiral growth of any bump on such a surface in 1868. This instability, to which Helmholtz attributed important physical consequences including fluid mixing, wave formation, and meteorological perturbations, is now called the 'Kelvin–Helmholtz' instability, owing to its similarity with another instability studied by William Thomson in 1871.

Thomson's consideration is related to the strange episode recounted in Chapter 2, that while slowly cruising on his personal yacht and fishing with a line, he observed a beautiful wave pattern and explained it by the combined action of gravity and capillarity. In a natural extension of this theory, he took into account the effect of wind over the water surface, and showed that the waves grew indefinitely when the wind velocity exceeded a certain, small limit that vanished with the surface tension. In other words, the plane water surface is unstable for such velocities. The calculation proceeds as follows.[10]

A solution of Euler's equation is sought for which the separating surface takes the plane monochromatic waveform

$$y = \eta(x,t) = a e^{i(kx-\omega t)}, \tag{5.1}$$

the x-axis being in the plane of the undisturbed water surface, and the y-axis being normal to this plane and directed upwards. Neglecting the compressibility of the two fluids, and assuming irrotationality, their motions have harmonic velocity potentials φ and φ'. By analogy with Poisson's wave problem, Thomson guessed the form

$$\varphi = C e^{ky+i(kx-\omega t)} \tag{5.2}$$

for the water, and

---

[10]Thomson [1871a], [1871b], [1871c]. See Chapter 2, pp. 87–8. Some commentators, including Lamb ([1932] p. 449), have Thomson say that the plane surface is stable for lower velocities, which leads to an absurdly high threshold for the production of waves (about twelve nautical miles per hour). Thomson did not and could not state so much, since he only considered irrotational perturbations of perfect fluids.

$$\varphi' = vx + C'e^{-ky+i(kx-\omega t)} \tag{5.3}$$

for the air, where v is the wind velocity.

A first boundary condition at the separating surface is that a particle of water originally belonging to this surface must retain this property. Denoting by $x(t)$ and $y(t)$ the coordinates of this particle at time t, this gives

$$y(t) = \eta(x(t),t) \tag{5.4}$$

at any t, or, differentiating with respect to time,

$$\frac{\partial \varphi}{\partial y} = \frac{\partial \varphi}{\partial x}\frac{\partial \eta}{\partial x} + \frac{\partial \eta}{\partial t} \quad \text{when } y = \eta(x,t). \tag{5.5}$$

A similar condition must hold for the air. The third and last boundary condition is the relation between pressure difference, surface tension, and curvature. For simplicity, capillarity is neglected in the following so that the pressure difference vanishes. The water pressure P is related to the velocity potential φ by the equation

$$P + \frac{1}{2}\rho(\nabla\varphi)^2 + \rho gy + \rho\frac{\partial\varphi}{\partial t} = \text{constant}, \tag{5.6}$$

obtained by the spatial integration of Euler's equation. A similar relation holds for the air.

Substituting the harmonic expressions for φ, φ', and η into the boundary conditions and retaining only first-order terms (with respect to a, C, and C') leads to the relations

$$Ck = -ia\omega, \quad C'k = ia(\omega - kv) \tag{5.7}$$

and

$$\rho(ga - iC\omega) = \rho'[ga - iC'(\omega - kv)]. \tag{5.8}$$

Eliminating a, C, and C' gives

$$\rho\omega^2 + \rho'(\omega - kv)^2 = gk(\rho - \rho'). \tag{5.9}$$

The discriminant of this quadratic equation in ω is negative if

$$v^2 > \frac{g}{k}\frac{\rho^2 - \rho'^2}{\rho\rho'}. \tag{5.10}$$

Hence there are exponentially-diverging perturbations of the separation surface for any value of the velocity v; and the water surface is unstable under any wind, no matter how small.[11]

---

[11]To every growing mode there corresponds a decaying mode by taking the complex-conjugate solution of eqn (5.9). This seems incompatible with the growth derived in the vortex-sheet consideration of Chapter 4, pp. 161–2. In fact, it is not, because Thomson's harmonic perturbations imply an initially heterogeneous distribution of vorticity on the separating surface, whereas the vortex-sheet argument assumes an initially homogenous distribution (to first order). For Thomson's decaying modes, the initial distribution has an excess of vorticity on the left-hand side of every positive arch of the sine-shaped surface, and a defect on the right-hand side.

This conclusion only holds when capillarity is neglected. As Thomson showed, the surface tension implies a wind-velocity threshold for the exponential growth of short-wave, irrotational perturbations. Thomson did not discuss the limiting case of equal densities for the two fluids. This limit could not mean much to him: as will be seen shortly, he did not believe in the possibility of discontinuity surfaces in homogeneous fluids.[12] In 1879, Rayleigh examined this very limit and derived the existence of exponentially-growing perturbations at any wavelength. Thus, he showed the similarity of the instabilities discovered by Helmholtz and Thomson. The modern phrase 'Kelvin–Helmholtz instability' captures the same connection, with an unfortunate permutation of the names of the two founders.[13]

5.3 Vortex atoms

5.3.1 *Hydrodynamic analogies*

Even though Thomson observed and measured waves while sailing and fishing, his main interest in hydrodynamics derived from his belief that the ultimate substance of the world was a perfect liquid. His earliest use of hydrodynamics, in the 1840s, was merely analogical: he developed formal analogies between electrostatics, magnetostatics, and the steady motion of a perfect liquid, mainly for the purpose of transferring theorems from one field to another. His correspondence of this period contains letters to Stokes in which he enquired about the hydrodynamic results he needed. In exchange, he offered new hydrodynamic theorems that his development of the energetic aspects of electricity suggested.[14]

One of these theorems is worth mentioning, for it played an important role in Thomson's later discussions of kinetic stability. Consider a perfect liquid limited by a closed surface that moves from rest in a prescribed manner. If the equation of this surface is $F(\mathbf{r}, t) = 0$, then the condition that a fluid particle initially on this surface should remain on it reads

$$\mathbf{v} \cdot \nabla F + \frac{\partial F}{\partial t} = 0. \tag{5.11}$$

According to a theorem by Lagrange, the motion $\mathbf{v}$ taken by the fluid derives from a potential φ. Now consider any other motion $\mathbf{v}'$ that satisfies the boundary condition at a given instant. The kinetic energy for the latter motion differs from the former by

$$T' - T = \int \frac{1}{2}\rho(\mathbf{v} - \mathbf{v}')^2 \mathrm{d}\tau + \int \rho\mathbf{v} \cdot (\mathbf{v}' - \mathbf{v})\mathrm{d}\tau. \tag{5.12}$$

Partial integration of the second term gives

$$\int \rho\nabla\varphi \cdot (\mathbf{v}' - \mathbf{v}) \, \mathrm{d}\tau = \int \rho\varphi\nabla \cdot (\mathbf{v} - \mathbf{v}') \, \mathrm{d}\tau - \int \rho\varphi(\mathbf{v}' - \mathbf{v}) \cdot \mathrm{d}\mathbf{S}. \tag{5.13}$$

[12]See Helmholtz to Thomson, 3 Sept. 1868, quoted in Thompson [1910] p. 527.

[13]Thomson [1871a] p. 79; Rayleigh [1879] pp. 365–71.

[14]See the letters of the period March–October 1847, *ST*. Cf. Smith and Wise [1989] pp. 219–27, 263–75, Darrigol [2000] chap. 3.

The volume integral vanishes because the fluid is incompressible. The surface integral also vanishes because the surface element dS is parallel to ∇F and both motions satisfy condition (5.11). Consequently, $T' - T$ is always positive. *The energy of the motion that the fluid takes at a given time owing to the motion impressed on its boundary is less than the energy of any motion that satisfies the boundary condition at the same time.*[15]

Even though in these early years Thomson constantly transposed such theorems to electricity and magnetism, he did not yet assume a hydrodynamic nature of electricity or magnetism. His attitude changed around 1850, after he adopted Joule's conception of heat as a kind of motion. In this view, the elasticity of a gas results from hidden internal motion, so that an apparently potential form of energy turns out to be kinetic. Thomson and a few other British physicists speculated, for the rest of the century, that every energy might be of kinetic origin. The mechanical world view would thus take a seductively simple form.[16]

The kind of molecular motion that William Rankine and Thomson then contemplated was a whirling, fluid motion around contiguous molecules. Gas pressure resulted from the centrifugal force of molecular vortices. Thomson elaborated this picture to account for the rotation of the polarization of light when traveling through magnetized matter, for electromagnetic induction, and even for the rigidity of the optical ether. In private considerations, he imagined an ether made of 'rotating motes' in a perfect liquid. The gyrostatic inertia of the whirls induced by these motes provided the needed rigidity. In 1857, Thomson confided these thoughts to his friend Stokes, with an enthusiastic plea for a hydrodynamic view of nature:[17]

> I have changed my mind greatly since my freshman's years when I thought it so much more satisfying to have to do with electricity than with hydrodynamics, which only first seemed at all attractive when I learned how you had fulfilled such solutions as Fourier's by your boxes of water.[18] Now I think hydrodynamics is to be the root of all physical science, and is at present second to none in the beauty of its mathematics.

5.3.2   *A new theory of matter*

A year after this pronouncement, Helmholtz published his memoir on vortex motion. In early 1867, Thomson saw the 'magnificent way' in which his friend Peter Guthrie Tait produced and manipulated smoke rings.[19] He gathered that Helmholtz's theorems offered a fantastic opportunity for a theory of matter based on the perfect liquid. Instead of rotating motes, he now considered vortex rings, and assimilated the molecules of matter with combinations of such rings. The permanence of matter then resulted from the

---

[15]Thomson [1849]. Thomson stated two corollaries (already known to Cauchy): (i) the existence of a potential and the boundary condition completely determine the flow at a given instant; (ii) the motion at any given time is independent of the motion at earlier times. See also Thomson and Tait [1867] pp. 312, 317–19.

[16]Cf. Smith and Wise [1989] chap. 12, Stein [1981].

[17]Thomson to Stokes, 20 Dec. 1857, *ST*. Cf. Smith and Wise [1989] pp. 402–12, Knudsen [1971]. As noted in Yamalidou [1998], the hydrodynamic view of nature implied a non-molecular idealization for the primitive fluid of the world.

[18]This is an allusion to Stokes's calculation ([1843] pp. 60–8) of the inertial moments of boxes filled with perfect liquid and his subsequent experimental verification of the results by measuring the torsional oscillations of suspended boxes of this kind. Cf. Chapter 3, p. 136.

[19]Thomson to Helmholtz, 22 Jan. 1867, quoted in Thompson [1910] p. 513.

conservation of vorticity. The chemical identity of atoms became a topology of mutually-embracing or self-knotted rings. Molecular collisions appeared to be a purely kinetic effect resulting from the mutual convection of two vortices by their velocity fields. In a long, highly mathematical memoir, Thomson developed the energy and momentum aspects of the vortex motions required by this new theory of matter.[20]

The most basic property of matter being stability, Thomson faced the question of the stability of vortex rings. Helmholtz's theorems only implied the permanence of the individual vortex filaments of which the rings were made. They did not exclude significant changes in the shape and arrangement of these filaments when subjected to external velocity perturbations. Thomson had no proof of stability, except in the case of a columnar vortex, that is, a circular-cylindric vortex of uniform vorticity. He showed that a periodic deformation of the surface of the column propagated itself along and around the vortex with a constant amplitude. An extrapolation of this behavior to thin vortex rings did not seem too adventurous to him. Moreover, Tait's smoke-ring experiments indicated stability as long as viscous diffusion did not hide the ideal behavior.[21]

During the next ten years, Thomson had no decisive progress to report on his vortex theory of matter. The simplest, non-trivial problem he could imagine, that of a cylindric-ally-symmetric distribution of vorticity within a cylindrical container, proved to be quite difficult. In 1872/73, he exchanged long letters with Stokes on this question, with no definite conclusion.[22] Thomson's arguments were complex, elliptic, and non-rigorous. As he admitted to Stokes, 'This is an extremely difficult subject to write upon.' A benevolent and perspicacious Stokes had trouble guessing what his friend was hinting at. I have fared no better.[23]

A stimulus came in 1878 from Alfred Mayer's experiments on floating magnets. The American professor had shown that certain symmetric arrangements of the magnets were mechanically stable. Realizing that the theoretical stability criterion was similar to that of a system of vortex columns, Thomson exulted: 'Mr Mayer's beautiful experiments bring us very near an experimental solution of a problem which has for years been before me unsolved—of vital importance in the theory of vortex atoms: to find the greatest number of bars which a vortex mouse-mill can have.' Thomson claimed to be able to prove the steadiness and stability of simple regular configurations, mathematically in the triangle and square cases, and experimentally in the pentagon case.[24]

These considerations only shed light on the stability of a mutual arrangement of vortices with respect to a disturbance of this arrangement, and not on their individual stability. They may have prompted Thomson's decision to complete his earlier, mostly unpublished

[20]Thomson [1867], [1869]. Cf. Silliman [1963], Smith and Wise [1989] pp. 417–25, Kragh [2002]. See also Chapter 4, pp. 154–5.

[21]Thomson [1867] p. 4, [1880a]. As John Hinch told me, the relevance of the latter observation is questionable; the smoke rings may not indicate the actual distribution of vorticity, because the diffusivity of vorticity is much more efficient than that of smoke particles.

[22]Thomson to Stokes, 19 Dec. 1872, 1–2, 8, 11, 21–22 Jan. 1873; Stokes to Thomson, 6, 18, 20 Jan. 1873, ST. Cf. Smith and Wise [1989] pp. 431–8.

[23]Thomson to Stokes, 19 Dec. 1872, ST.

[24]Thomson [1878] p. 135. The subject was further discussed by Alfred Greenhill in 1878, J. J. Thomson in 1883, and William Hicks in 1882. Cf. Love [1901] pp. 122–5, Kragh [2002].

considerations on the stability of cylindrical vortices. In harmony with the energy-based program developed in his and Tait's *Treatise on natural philosophy*, Thomson formulated an energetic criterion of stability. In problems of statics, stable equilibrium corresponds to a minimum of the potential energy. In any theory that reduces statics to kinetics, there should be a similar criterion for the stability of motion. For the motion of a perfect liquid of unlimited extension, Thomson stated the following theorem. *If, with the vorticity and impulse given, the kinetic energy is stationary, then the motion is steady. If it is a (local) minimum or maximum, then the motion is not only steady but stable.*[25]

Some thinking is necessary to understand what Thomson had in mind, since he did not care to provide a proof. For simplicity, I only consider the case of a fluid confined in a rigid container with no particular symmetry. Then the condition of a given impulse must be dropped, and 'steadiness' has the ordinary meaning of constancy of the velocity field. The condition of a given vorticity, Thomson tells us, is the fixity of the number and intensity of the vortex filaments (it is *not* the steadiness of the vorticity field). In more rigorous terms, this means that the distribution of vorticity at any time can be obtained from the original distribution by pure convection.

The variation $\delta\mathbf{v} = \boldsymbol{\omega} \times \delta\mathbf{r}$, with $\nabla \cdot \delta\mathbf{r} = 0$, of the fluid velocity meets this condition, since it has the same effect on the vorticity distribution $\boldsymbol{\omega}$ as a displacement $\delta\mathbf{r}$ of the fluid particles.[26] Therefore, the integral

$$\delta T = \int \rho\mathbf{v} \cdot (\boldsymbol{\omega} \times \delta\mathbf{r}) \, \mathrm{d}\tau = \int \rho\delta\mathbf{r} \cdot (\mathbf{v} \times \boldsymbol{\omega}) \, \mathrm{d}\tau \qquad (5.14)$$

must vanish for any $\delta\mathbf{r}$ such that $\nabla \cdot \delta\mathbf{r} = 0$. This implies that

$$\nabla \times (\mathbf{v} \times \boldsymbol{\omega}) = \mathbf{0}. \qquad (5.15)$$

Combined with the vorticity equation (the curl of Euler's equation)

$$\frac{\partial\boldsymbol{\omega}}{\partial t} - \nabla \times (\mathbf{v} \times \boldsymbol{\omega}) = \mathbf{0}, \qquad (5.16)$$

this gives the steadiness of the vorticity distribution. The fluid being incompressible, this steadiness implies the permanence of the velocity field, as was to be proved.

Thomson declared the other part of his theorem, the stability of the steady motion when the kinetic energy is a maximum or a minimum, to be 'obvious'. Any motion that differs little from an energy extremum at a given time, Thomson presumably reasoned, should retain this property in the course of time, for its energy, being a constant, should remain close to the extremum value. Metaphorically speaking, a hike at a constant elevation slightly below that of a summit cannot lead very far from the summit. Thomson did not worry that the proximity of two fluid motions was not as clearly defined as the proximity of two points of a mountain range.[27]

---

[25]Thomson, letters to Stokes (1872–73), *ST*; Thomson [1876], [1880*b*] (energetic criterion); Thomson and Tait [1867] (cf. Smith and Wise [1989] chap. 11). For a modern interpretation of Thomson's criterion, cf. Arnol'd [1966], Drazin and Reid [1981] pp. 432–5.

[26]Rigorously, a gradient term must be added to $\boldsymbol{\omega} \times \delta\mathbf{r}$ in order that $\delta\mathbf{v}$ be parallel to the walls of the container. However, this gradient term does not contribute to the variation δT of the kinetic energy.

[27]Thomson [1876] p. 116.

At any rate, Thomson's energetic criterion helped little in determining the stability of vortex atoms. The energy of a vortex ring turned out to be a 'minimax' (saddle point), in which case the energy consideration does not suffice to decide stability.[28] Presumably to prepare another attack on this difficult problem, he dwelt on the simpler problem of cylindrically-symmetric motions within a tubular container. In this case, a simple consideration of symmetry shows that a uniform distribution of vorticity within a cylinder coaxial to the container corresponds to a maximum energy in the above sense. Similarly, a uniform distribution of vorticity in the space comprised between the walls and a coaxial cylinder has minimum energy. These two distributions are therefore steady and stable.[29]

5.3.3 *Labyrinthine degradation*

Thomson had already studied the perturbations of the former distribution, the columnar vortex, in the absence of walls. He now included a reciprocal action between the vortex vibration and a 'visco-elastic' wall. He thus seems to have temporarily left the ideal world of his earlier reasoning to consider what would happen to a vortex in concrete hydrodynamic experiments for which the walls of the container necessarily dissipate part of the energy of the fluid motion.[30]

Thomson described how, owing to the interaction with the visco-elastic walls, 'the waves [of deformation of the surface of the vortex] of shorter length are indefinitely multiplied and exalted till their crests run out into fine laminas of liquid, and those of greater length are abated.' The container thus becomes filled with a very fine, but heterogeneous mixture of rotational fluid with irrotational fluid, which Thomson called a 'vortex sponge'.[31] At a later stage, the compression of the sponge leads to the minimum energy distribution for which the irrotational fluid is confined in an annular space next to the wall. A few years later, George Francis FitzGerald and Thomson himself based a reputed theory of the ether on the intermediate vortex-sponge state.[32]

Some aspects of the dissipative evolution of a columnar vortex are relatively easy to understand. According to Helmholtz's vortex theorems, the rotational and irrotational parts of the fluid (which have, respectively, the vorticity ω of the original vortex column and zero vorticity) behave like two incompressible, immiscible fluids. Since the original configuration is that of maximum energy, the dissipative interaction with the visco-elastic wall leads to a lesser-energy configuration for which portions of the rotational fluid are closer to the walls. As the ω-fluid is incompressible, this evolution implies a corrugation of the vortex surface. As Thomson proved in his study of columnar vortex vibrations, the corrugation rotates at a frequency that grows linearly with its inverse wavelength (and linearly with the

---

[28]Thomson [1876] p. 124. For a given vorticity and a given impulse, the energy of a thin vortex ring (with quasi-circular cross-section) is decreased by making its cross-section oval; it is increased by making the ring thicker in one place than in another.

[29]Thomson [1880*b*] p. 173.

[30]Thomson to Stokes, 19 Dec. 1872, *ST*; Thomson [1880*b*] pp. 176–80.

[31]Thomson [1880*b*] p. 177. In his correspondence of 1872, Thomson imagined a different process of 'labyrinthine' and 'spiraling' penetration of the rotational fluid into the irrotational fluid.

[32]Cf. Hunt [1991] pp. 96–104. FitzGerald first wrote on the vortex-sponge ether in FitzGerald [1885]. Thomson first wrote on this topic in Thomson [1887*e*]. See Chapter 6, pp. 242–3.

vorticity ω). Since the energy-damping effect of the walls is proportional to the frequency of their perturbation, the energy of the smaller corrugation waves diminishes faster. As for these special waves (unlike sea waves) a smaller energy corresponds to a higher amplitude, the shorter waves must grow until they reach the angular shape that implies frothing and mixing with the irrotational fluid.[33] On the latter point, Thomson probably reasoned by analogy with the finite-height sea-wave problem, which he had been discussing with Stokes.

Thomson expected a similar degradation to occur for any vortex in the presence of visco-elastic matter. 'An imperfectly elastic solid', he noted in 1872, 'is slow but sure poison to a vortex. The minutest portion of such matter, would destroy all the atoms of any finite universe.' Yet Thomson did not regard this peculiar instability as a threat to his vortex theory of matter. Visco-elastic walls did not exist at the scale of his ideal world fluid: all matter, including container walls, was made of vortices in the same fluid. In Thomson's imagination, the interactions of a dense crowd of vortices only resembled visco-elastic degradation to the extent needed to explain the condensation of a gas on the walls of its container.[34]

5.3.4 *Delusion*

For a few more years, Thomson contented himself with the observed stability of smoke rings and with the demonstrated stability of the columnar vortex. By 1889, however, he encountered difficulties that ruined his hope of a vortex theory of matter. This is attested by a letter he wrote to the vortex-sponge enthusiast FitzGerald: 'I have quite confirmed one thing I was going to write to you (in continuation with my letter of October 26), viz. that rotational vortex cores must be absolutely discarded, and we must have nothing but irrotational revolution around vacuous cores.' He adduced the following reason: 'Steady motion, with crossing lines of vortex columns, is impossible with rotational cores, but is possible with vacuous cores and purely irrotational circulations around them.'[35]

Crossing lines of vortex columns occurred in FitzGerald's and Thomson's vortex ether. They were also a limiting case of the mutually-embracing vortex rings that Thomson contemplated in his theory of matter. Their unsteadiness was therefore doubly problematic. Thomson was pessimistic: 'I do not see much hope for chemistry and electromagnetism.' Although vacuous-core vortices with zero vorticity still remained possible, Thomson was much less eager to speculate on vortex atoms than he had been earlier. In subsequent letters, he tried to persuade FitzGerald to abandon the vortex ether.[36]

Considerations of stability also played a role in Thomson's renunciation. Since 1867, his friend Stokes had been warning him about possible instabilities: 'I confess', Stokes wrote in January 1873, 'I am skeptical about the stability of many of the motions which you appear to contemplate.' In a letter to Stokes of December 1898, Thomson described the frittering and diffusion of an annular vortex, with the comment:[37]

---

[33]According to Thomson [1880*b*] pp. 176–7, this process only occurs if the canister offers no resistance to rotation (so that the angular momentum of the fluid is constant).

[34]Thomson to Stokes, 1872, *ST*, pp. 378–9.

[35]Thomson [1889] p. 202.

[36]*Ibid.* p. 204. Cf. Hunt [1991] p. 102.

[37]Stokes to Thomson, 8 Jan. 1873; Thomson to Stokes, 27 Dec. 1898, *ST*.

> I now believe that this is the fate of vortex rings, and of every kind of irrotational [rotational?] motion (with or without finite slips anywhere) in a limited portion of an inviscid mass of fluid, which is at rest at great distances from the moving parts. This puts me in mind of a thirty-year-old letter of yours with a drawing in black and red ink suggesting instability of the motion of a columnar vortex, which I did not then believe. I must see if I can find the letter.

According to Thomson's later recollections, he became aware of the instability of vortex rings in unpublished work of 1887:[38]

> It now seems to me certain that if any motion be given within a finite portion of an infinite incompressible liquid originally at rest, its fate is necessarily dissipation to infinite distances with infinitely small velocities everywhere; while the total kinetic energy remains constant. After many years of failure to prove that the motion in the ordinary Helmholtz circular ring is stable, I came to the conclusion that it is essentially unstable, and that its fate must be to become dissipated as now described. I came to this conclusion by extensions not hitherto published of the considerations described in a short paper entitled: 'On the stability of steady and periodic fluid motion', in the *Phil. Mag.* for May 1887.

In this short paper, Thomson proved that the energy of any vortex motion of a fluid confined within deformable walls could be increased indefinitely by doing work on the walls in a systematic manner. More relevantly, he announced that the energy of the motion would gradually vanish if the walls were viscously elastic. It is not clear, however, why this result would have been more threatening to vortex atoms than the degradation of a vortex column surrounded by viscously-elastic walls already was.[39]

Another paper of the same year seems more relevant. Therein Thomson considered the symmetric arrangement of vortex rings represented in Fig. 5.3 as a possible model of a rigid ether. He worried:

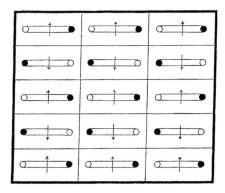

Fig. 5.3.   William Thomson's arrangement of vortex rings as a tentative model of the optical ether. The arrows represent the axes of the rings, and the black and white dots their intersections with the plane of the figure. From Thomson [1887*e*] p. 317.

---

[38]Thomson [1905] pp. 370n–371n.   [39]Thomson [1887*b*].

It is exceedingly doubtful, so far as I can judge after much anxious consideration from time to time during these last twenty years, whether the configuration represented [in Fig. 5.3] or any other symmetrical arrangement, is stable when the rigidity of the ideal partitions enclosing each ring separately is annulled through space … The symmetric motion is unstable, and the rings shuffle themselves into perpetually varying relative positions, with *average homogeneousness*, like the ultimate molecules of a homogeneous liquid.

This instability threatened not only the vortex theory of ether—on which Thomson pronounced 'the Scottish verdict of *not proven*'—but also any attempt at explaining chemical valence by symmetric arrangements of vortex rings. After twenty years of brooding, Thomson's hope for a grand theory of ether and matter was turning into disbelief.[40]

5.4 The Thomson–Stokes debate

5.4.1 *Conflicting ideals*

When, in 1857, Thomson was contemplating an ether made of a perfect liquid and rotating motes, his friend Stokes warned him about the instability of the motion of a perfect liquid around a solid body.[41] Thomson confidently replied: 'Instability, or a tendency to run to eddies, or any kind of dissipation of energy, is impossible in a perfect fluid.' As he had learned from Stokes ten years earlier and as Cauchy had proved in 1827, the motion of solids through a perfect liquid completely determines the fluid motion if the solids and fluid are originally at rest. Following Lagrange's theorem, the latter motion is irrotational and devoid of eddying. Following Thomson's theorem of 1849, it is the motion that has at every instant the minimum energy compatible with the boundary conditions. Thomson believed these two results to imply stability.[42]

Stokes disagreed. He insisted: 'I have always inclined to the belief that the motion of a perfect incompressible liquid, primitively at rest, about a solid which continually progressed, was unstable.' The theorems of Lagrange, Cauchy, and Poisson, he argued, only hold 'on the *assumption of continuity*, and I have always been rather inclined to believe that surfaces of discontinuity would be formed in the fluid.' The formation of such surfaces would imply a loss of *vis viva* in the wake of the solid and thus induce a finite resistance to its motion. A surface of discontinuity, he told Thomson, is surely formed when fluid passes from one vessel to another through a small opening (see Fig. 5.1), which implies the instability of the irrotational, spreading-out motion. Similarly, Stokes went on, the spreading-out motion behind a moving sphere (see Fig. 5.4) should be unstable. Stokes was only repeating the considerations he had used in 1842/43 to reconcile perfect- and real-fluid behaviors.[43]

---

[40]Thomson [1887e] pp. 318, 320.

[41]This is inferred from the letter from Thomson to Stokes of 17 June 1857, *ST*: 'I think the instability you speak of cannot exist in a perfect … liquid.'

[42]Thomson to Stokes, 23 Dec. 1857. Presumably, Thomson believed that a slightly-perturbed motion would remain close to the original motion because its energy would remain close to that of the minimum-energy solution. However, this is only true in a closed system for which there is no external energy input. As Stokes later argued, such an input may feed the perturbation.

[43]Stokes to Thomson, 12–13 Feb. 1858, *ST*.

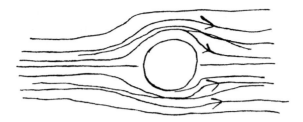

Fig. 5.4.   The spreading motion of a fluid behind a sphere. From Stokes to Thomson, 13 Feb. 1858, *ST*.

In general, Stokes drew his ideas on the stability of perfect-liquid motion from the behavior of real fluids with small viscosity, typically water. In 1880, while preparing the first volume of his collected papers, he reflected on the nature of the zero-viscosity limit. His remark of 1849 on the discontinuity surface from an edge, he then noted, depended on the double idealization of a strictly inviscid fluid and an infinitely-sharp edge:[44]

> A perfect fluid is an ideal abstraction, representing something that does not exist in nature. All actual fluids are more or less viscous, and we arrive at the conception of a perfect fluid by starting with fluids such as we find them, and then in imagination making abstraction of the viscosity. Similarly, any edge we can mechanically form is more or less rounded off, but we have no difficulty in conceiving of an edge perfectly sharp.

Stokes then considered the flow for a finite viscosity μ and a finite curvature radius a of the edge, and argued that the limit of this flow when a and μ reached zero depended on the order in which the two limits were taken. If the limit $\mu \to 0$ is taken first, then the resulting flow is continuous and irrotational, and it obviously remains so in the limit $a \to 0$. If the limit $a \to 0$ is taken first, then the resulting flow is that of a viscous fluid passing an infinitely-sharp edge. The viscous stress is easily seen to imply the formation of a trail of vorticity from the edge. In the limit $\mu \to 0$ this trail becomes infinitely narrow, and a vortex sheet or discontinuity surface is formed. Stokes believed the latter double limit to be the only one of physical interest, because the result of the former was unstable in the sense that an infinitely-small viscous stress was sufficient to turn it into a widely different motion.[45]

Stokes returned to his idea of the double limit in several letters.[46] In 1894, it led him to an instructive comment on the nature of his disagreement with Thomson: 'Your speculations

---

[44]*SMPP* **1**, pp. 311–12.

[45]*Ibid.* As Thomson later pointed out, in this alleged instability there is an apparent contradiction between the vanishing work of the viscous stress and the finite energy difference between the two compared motions. Stokes replied with a metaphor (27 Oct. 1894, *ST*): 'Suppose there is a railway AB which at B branches off towards C and towards D. Suppose a train travels without stopping along AB and onwards. Will you admit that the muscular exertion of the pointsman at B is the merest trifle of the work required to propel the train along BC or CD? Now I look on viscosity in the neighborhood of a sharp, though not absolutely sharp, edge as performing the part of the pointsman at B.'

[46]Stokes to Thomson, 1 Nov. 1894, 22–23 Nov. 1898, 14 Feb. 1899, *ST*. In this last letter, Stokes considers the state of things at time t from the commencement of motion and at distance r from an edge, and argues that the limit $t \to 0$ gives the 'mike' (minimum kinetic energy) solution, whereas the limit $r \to 0$ gives a discontinuity surface.

about vortex atoms led you to approach the limit in the first way [$\mu \to 0$ first]; my ideas, derived from what one sees in an actual fluid, led me to approach it in the other way [$a \to 0$ first].' Indeed, Thomson's reflections on stability mostly occurred in the context of his theory of ether and matter. He was therefore prejudiced in favor of stability, and generally expected important qualitative differences between real- and perfect-fluid behavior.

5.4.2 Coreless vortices, goring, and dead water

In 1887, Thomson publicly rejected the possibility of surfaces of discontinuity, arguing that they could never be formed by any natural action. In his opinion, continuity of velocity was always obtained when two portions of fluid where brought into contact. He now agreed with Stokes and Helmholtz that the flow around a solid obstacle was unstable when the velocity exceeded a certain value, but denied that this instability had anything to do with surfaces of discontinuity. For a perfect liquid, the determining effect was the separation of the fluid from the solid surface.[47]

In the case of flow around a sphere, Thomson described the instability as follows. The fluid separates at the equator when the asymptotic velocity V of the fluid exceeds the value for which the pressure at the equator becomes negative ($\frac{5}{8}\rho V^2$ according to Bernoulli's law applied to the irrotational solution of Euler's equation).[48] A coreless vortex is formed, as indicated in Thomson's drawing (see Fig. 5.5). This vortex grows until it separates from the sphere and follows the flow. The whole process repeats itself indefinitely and results in a 'violently disturbed motion'.[49]

Stokes did not comment on this cavitational instability, which was known to occur on the edges of swiftly-moving immersed solids, for instance ship propellers. He did, however,

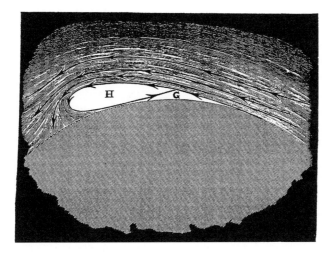

Fig. 5.5.   The formation of a coreless vortex H near the equator G of a sphere immersed in a moving liquid. From Thomson [1887a] p. 151.

[47]Thomson [1887a].   [48]On negative pressure, see Chapter 4, footnote 48, p. 163.
[49]Ibid. p. 149.

contest Thomson's assertion that discontinuity surfaces could not be formed by any
natural process. In Stokes's view, a drop of perfect liquid falling on a calm surface of
the same liquid led to discontinuous motions. So did the 'goring' of fluid on itself, as
drawn in Fig. 5.6.[50]

Thomson rejected these suggestions, as well as Helmholtz's idea of bringing into contact
two parallel plane surfaces bounding two portions of liquid moving with different velocities. In every case, he argued, the contact between the two different fluid portions always
begins at an isolated point, and the boundary of the fluid evolves so that no finite slip ever
occurs. The drawings of Fig. 5.7 illustrate his understanding of the goring and raindrop
cases. In Helmholtz's plane-contact process, the imperfect flatness of the surfaces achieves
the desired result.[51]

Seven years later, Thomson published another provocative article in *Nature* against the
'doctrine of discontinuity'. This time his target was the alleged formation of a surface of
discontinuity past a sharp edge. The relief from infinitely-negative pressure at the sharp
edge, Thomson declared, never was the formation of a surface of discontinuity, which
contradicted his minimum-kinetic-energy theorem. The true compensatory factors were
finite viscosity, finite compressibility, or the yielding boundary of the fluid. Thomson
illustrated the compensations using the example of a thin moving disc. When the first
factor dominates, a layer of abrupt velocity change, or, equivalently, a vortex sheet with
small thickness, is formed behind the moving solid. When the third factor dominates, a
succession of thin hollow rings is created behind the disk in a manner similar to that which

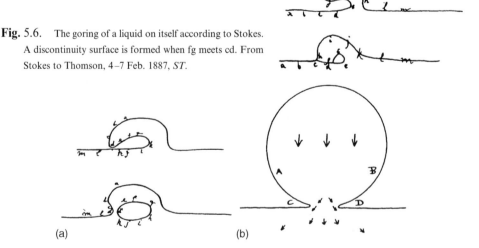

Fig. 5.6.   The goring of a liquid on itself according to Stokes.
A discontinuity surface is formed when fg meets cd. From
Stokes to Thomson, 4–7 Feb. 1887, *ST*.

(a)   (b)

Fig. 5.7.   (a) The goring of a liquid on itself and (b) the fall of a drop on a plane water surface, according to
Thomson. From Thomson to Stokes, 6–9 Feb. 1887, *ST*.

---

[50]Stokes to Thomson, 4, 7 Feb. 1887, *ST*. For cavitation around ship propellers, as discussed in Reynolds
[1873], see later on p. 246 (in real fluids vapor fills the cavities).

[51]Thomson to Stokes, 6, 9 Feb. 1887, *ST*.

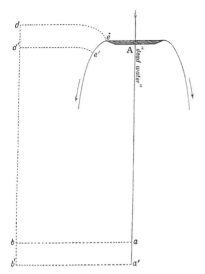

Fig. 5.8.   Thomson's drawing for Rayleigh's 'dead water' theory of fluid resistance. A discontinuity surface *ee'* is formed at the edge of the disc A, the axis of which is in the plane of the figure. From Thomson [1894] p. 220.

Thomson described in 1887 for the moving sphere. Both processes imitate a surface of discontinuity when the fluid is nearly perfect. However, the imitation is always imperfect.[52]

The strict doctrine of discontinuity, Thomson went on, leads to an absurd theory of resistance. His target was Rayleigh's 'dead-water' theory of resistance of 1876, according to which the fluid remains at rest (with respect to the solid) in the space limited by a tubular surface of discontinuity extending from the edges to infinity (see Fig. 5.8). The pressure in the dead water immediately behind the solid is inferior to the pressure on the front of the body, so that a finite resistance results. Whereas Rayleigh offered this picture as a solution to d'Alembert's old paradox, Thomson denounced its gross incompatibility with experiment. The dead water, if any, could not realistically extend indefinitely rearwards. Moreover, the resistance measured by William Dines for a rectangular blade under normal incidence was three times larger than that indicated by Rayleigh's calculation in this case. Truncation of the discontinuity surface, Thomson showed, did not remove this discrepancy. As a last blow to the dead-water theory, he conceived a special case in which it gave zero resistance (see Fig. 5.9).[53]

5.4.3   *Birth of discontinuity surfaces*

Stokes's reaction was strong and immediate. He had never supported the dead-water theory, and believed instead that the main cause of resistance was the formation of eddies.

---

[52]Thomson [1894]. Thomson had already expressed this opinion in a letter to Helmholtz of 3 Sept. 1868, quoted in Thompson [1910] p. 527: 'Is it not possible that the real cause of the formation of a vortex-sheet may be viscosity which exists in every real liquid, and that the ideal case of a perfect liquid, perfect edge, and *infinitely thin* vortex sheet, may be looked upon as a limiting case of more and more perfect fluid, finer and finer edge of solid, and consequently thinner and thinner vortex-sheet?'

[53]Thomson [1894]; Rayleigh [1876*b*]. See Chapter 4, p. 165.

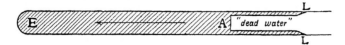

Fig. 5.9.   Case of motion for which the dead-water theory gives zero resistance. The hatched tube EA moves to
the left through a perfect liquid, leaving a dead-water wake in its rear cavity and within the cylindrical surface
of discontinuity which begins at LL. The longitudinal resultant of pressure on the front part E is very nearly
equal to the pressure at infinity times the transverse section of the tube, because the cylindrical part of the
tube is much larger than its curved front part. The same equality holds exactly at the rear of the tube, because
the pressure is continuous across the discontinuity surface and constant within the dead water. Therefore, the
net longitudinal pressure force on the tube vanishes. From Thomson [1894] p. 228.

He nonetheless maintained that the continuous, irrotational, and steady motion of a
perfect liquid around a solid body with sharp edges was unstable. After conceding to
Thomson that this motion was that of minimum energy under the given boundary
conditions, he interjected: 'But what follows from that? There is the rub.' Instability, he
explained, was still possible:[54]

> What is meant by the motion being unstable? I should say, the motion is said to be
> stable when whatever small deviation from the phi motion [the minimum-energy
> motion, for which there exists a velocity potential φ] is supposed to be produced, and
> the fluid thenceforth not interfered with, the subsequent motion differs only by small
> quantities from the phi motion, and unstable when the small initial deviation goes on
> accumulating, so that presently it is no longer small.—I have a right to take for my
> small initial deviation one in which the fluid close to the edge shoots past the edge,
> forming a very minute surface of discontinuity. The question is, Will this always
> remain correspondingly minute, or will the deviation accumulate so that ultimately it
> is no longer small? I have practically satisfied myself that it will so accumulate, and
> the mode of subsequent motion presents interesting features.

Thomson replied that the would-be surface of discontinuity would 'become instantly
ruffled, and rolled up into an 'ανηριθμον γελασμα' (by the last word I mean laughing at
the doctrine of finite slip)'[55] and would be washed away and left in the wake. Stokes
declared himself undisturbed by this objection. He knew well the instability of discontinuity surfaces, but their spiral unrolling was not a priori incompatible with their continual
formation at the edge of a body. 'The rub' was still Thomson's pretense to derive stability
from his minimum-energy theorem. The theorem, Stokes explained, did not require that
the actual motion should be that of minimum energy, because the additional energy
needed to create the discontinuity surface could result from work done by the external
pressures that sustained the flow.[56]

[54]Stokes to Thomson, 11 Oct. 1894, *ST*.

[55]Cf. Aeschylus, *The Prometheus bound*, verses 89–90, 'κυμάτων ἀνήριθμον γέλασμα', which literally means
'a countless smile of waves'. The whole strophe reads (in George Thomson's translation, Cambridge, 1932, p. 55):
'O divine Sky, and swiftly-winging Breezes,/O River-springs, and multitudinous gleam/Of smiling Ocean—to thee,
All-Mother Earth,/And to the Sun's all-seeing orb I cry:/See what I suffer from the gods, a god!'

[56]Thomson to Stokes, 14 Oct. 1894; Stokes to Thomson, 27 Oct. 1894, *ST*. See also Stokes to Thomson, 22–23
Nov. 1898, *ST*.

Perhaps, Stokes wondered, there was another 'Kelvinian theorem' that truly excluded the discontinuity. The only one that came out in later letters was the theorem that the angular momentum of every spherical portion of a liquid mass in motion, relative to the center of the sphere, is always zero, if it is so at any one instant for every spherical portion of the same mass. The theorem, Stokes judged, no more excluded the formation of a surface of discontinuity than Lagrange's and Cauchy's theorems (regarding fluid motion produced by moving immersed solids) already did, for its proof required the continuity of the fluid motion near the walls.[57]

After a pause of four years, Stokes resumed the discussion with some considerations on the growth of a 'baby surface of discontinuity' at a sharp edge. Presumably, Thomson had objected that the continuity of pressure across the baby surface was incompatible with the discontinuity of velocity. Stokes explained that the growth of the surface and the resulting unsteadiness of the flow implied an additional term $\partial\varphi/\partial t$ in the pressure equation (Bernoulli's law) that counterbalanced the discontinuity of $\frac{1}{2}\rho v^2$. He also repeated his conviction that Thomson's minimum-energy theorem was not incompatible with the formation of discontinuity surfaces.[58]

Thomson replied with a thought experiment (see Fig. 5.10):

> To keep as closely as possible to the point (edge!) of your letter of the 22nd, let E be an edge fixed to the interior of a cylinder, with two pistons clamped together by a connecting-rod as shewn in the diagram, and the space between them filled with incompressible inviscid liquid. Let the radius of curvature of the edge be 10^{-12} of a centimeter.

The curvature still being finite, Thomson thought that Stokes would agree about the perfectly-determinate and continuous character of the fluid motion induced by pushing the double piston. A moderate velocity of the piston would then imply an enormous pressure, tending to break the connecting rod. Although Thomson did not say why, he probably reasoned by combining Bernoulli's law and the impossibility of negative pressure at the edge, as he had done earlier for the flow around a globe. In the real world, Thomson went on, the connecting rod would either break, or yield slightly, thus allowing the liquid to leave the solid wall before it comes to the edge. In neither case would there be a slip of liquid over liquid.[59]

The argument backfired. In his response (20–21, 26 Dec. 1898, *ST*), Stokes placed the cylinder and pistons vertically, and counterpoised the double piston and liquid by means of a string, pulley, and weight (see Fig. 5.11). Then a housefly perching on the upper piston would suffice to break a connecting rod of large, but finite, resistance to traction. Stokes's solution to this paradox was the formation of a surface of discontinuity past the edge, despite the lack of a strict angular point.[60]

[57]Stokes to Thomson, 27 Oct. 1894, 26 Dec. 1898; Thomson to Stokes, 23 Dec. 1898, *ST*. See also their letters of 13, 18–20 Dec. 1900, and 4 Jan. 1901, *ST*.

[58]Stoke to Thomson, 22–23 Oct. 1898, *ST*.   [59]Thomson to Stokes, 25 Nov. 1898, *ST*.

[60]Stokes to Thomson, 20–21, 26 Dec. 1898 (paradox), 14 Feb. 1899 (solution). Another escape from the paradox would be to note that the fly cannot communicate a finite velocity to the piston, and therefore cannot induce an infinite pressure of the fluid if the 'mike' solution still applies.

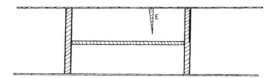

Fig. 5.10.   Thomson's diagram for a thought experiment regarding flow around a sharp edge. From Thomson to Stokes, 25 Nov. 1898, *ST*.

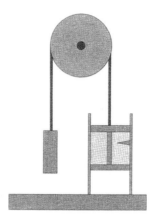

Fig. 5.11.   Stokes's device for his housefly paradox, from the description in Stokes to Thomson, 20–21, 26 Dec. 1898, *ST*.

5.4.4   *Separation and boundary layer*

From the beginning, Stokes believed that surfaces of discontinuity were formed even behind smoothly-shaped obstacles. In previous letters, he had only focused on the infinitely-sharp edge because the instability of the 'mike' (minimum kinetic energy) solution was the easiest to understand in this case. Two days after he enunciated the housefly paradox, he re-expressed his conviction that the 'mike' solution for a uniform flow around a cylinder was unstable at the rear of the cylinder and challenged Thomson for a proof of stability in this case. He referred to the turbulent flow behind the pillars of a bridge as an instance of this instability. 'It is hard to imagine', he reflected, 'that the instability which the commonest observation shows to exist is wholly due to viscosity, especially as an increase of viscosity seems to tend to increased stability, not the reverse.'[61]

A week later, Stokes described how surfaces of discontinuity could be generated even without a sharp edge:[62]

> I can see in a general way how it is that it is towards the rear of a solid moving through a fluid that a surface of discontinuity is formed. I find that at the point of a solid which is the birthplace of such a surface ... the flowing fluid must go off at a

[61]Stokes to Thomson, 22 Dec. 1898, *ST*.   [62]Stokes to Thomson, 27 Dec. 1898, *ST*.

tangent, and the fluid at the other side of the surface of discontinuity must just at the birthplace be at rest.

In a crossing letter, Thomson denied instability in the perfect-liquid case, and proceeded to explain the practical instability for a real fluid of small viscosity and negligible compressibility, such as water. He first considered the fluid motion induced by a sudden acceleration (from rest) of an immersed solid body:[63]

> The initial motion of the water will be exceedingly nearly that of an incompressible inviscid liquid (the motion of minimum kinetic energy). There will be an exceedingly thin stratum of fluid round the solid through which the velocity of the water varies continuously from the velocity of the solid to the velocity in the solution for inviscid fluid. It is in this layer that there is instability. The less the viscosity, the thinner is this layer for a given value of the initial acceleration; but the surer the instability. Not very logical this.

Thomson did not say why he thought the thin layer of vorticity to be unstable. He only alluded to his earlier argument about the practical instability of the plane Poiseuille flow (parallel flow between two fixed parallel plates), to be discussed shortly.[64] He moved on to consider what would happen to the fluid if the acceleration ceased and the body (now a globe) was kept moving uniformly:

> If the velocity is sufficiently great, the motion of the fluid at small distances from its surface all round will always be very nearly the same as if the fluid were inviscid, and the difference will be smaller near the front part than near the rear of the globe.

Here we have a description of what Ludwig Prandtl later called the boundary layer. The rest is more personal to Thomson:

> If now the whole fluid suddenly becomes inviscid and the globe be kept moving uniformly, the rotationally moving fluid will be washed off from it, and left moving turbulently in the wake, and mixing up irrotationally moving fluid among it.

Thus, Thomson made viscosity responsible for the formation of an unstable state of motion, but regarded the instability of this state as unrelated to viscosity and therefore felt free to 'turn off' viscosity to discuss it. Although, for a given state of motion at a given instant, viscosity could only have a stabilizing effect, it could make a stable state evolve into an unstable one.[65]

In his reply to this letter, Stokes expressed his agreement with everything Thomson had said, except for what would happen if the viscosity were suddenly brought to zero. In his opinion,

> the streams of right-handedly revolving and left-handedly revolving fluid at the two sides would have the rotationally moving fluid washed away, at least in the side trails, and the streams would give place to streams bounded by surfaces of finite slip, commencing at the solid, and then being paid out from thence. The subsequent

[63]Thomson to Stokes, 27 Dec. 1898, *ST*.   [64]Thomson [1887c]. See later on pp. 211–3.

[65]Thomson to Stokes, 27 Dec. 1898, *ST*.

motion would doubtless be of a very complicated character [owing to the Helmholtz–Kelvin instability].

Again, Stokes wanted the inviscid behavior to be a limit of the low-viscosity behavior. If a discontinuity surface was formed in the ideal inviscid fluid case, then it had to play a role in the practical case of a slightly-viscous fluid.[66]

5.4.5 *Epilogue*

The debate continued until Stokes's last letter to Thomson, dated 23 October 1901. In this late period the two old friends stuck at their positions. They could not even agree on the (in)compatibility of Lagrange's theorem with the formation of discontinuity surfaces. Stokes refined his picture of the formation of a discontinuity surface behind a moving solid sphere, so as to reach 'continuity in the setting of discontinuity'. In the new picture the contact line of the solid and surface began as a tiny circle around the rearward pole of the sphere, and then widened out until the surface took its final, steady shape. Stokes also made the spiral unrolling of the discontinuity surface the true cause of eddying behind a solid obstacle:[67]

> It seems evident that the mere viscosity of water would be utterly insufficient to account for [the eddies] when they are formed on a large scale, as in a mill pool or whirlpool … Of course eddies are *modified* by viscosity, but except on quite a small scale I hold that viscosity is subordinate. Of course, it prevents a finite slip, which it converts into a rapid shear, but viscosity tends to stability, not to instability.

Throughout their long, playful disagreement, Stokes and Thomson were driven by different interests. Whereas Stokes wanted to understand the behavior of real liquids, Thomson primarily reasoned on the ultimate perfect liquid of the world. Thus, they had opposite prejudices on the stability properties of the flow of a perfect liquid past a solid obstacle. As the intrinsic mathematical difficulty of the subject prevented a settling of issues by a rigorous argument, they relied on intuition and past experience. Stokes appealed to the natural world and conjectured that the behavior of perfect liquids should reflect that of real liquids with small viscosity and compressibility. Thomson instead appealed to the energy-based dynamics that founded his natural philosophy. Hence he promoted the minimum-energy flow and an energy-based criterion of stability.

The Thomson–Stokes debate is not only instructive for the kind of theoretical prejudices it reveals, but also as an indication of the powers and limits of intuitive discussions of hydrodynamic instability. The modern reader may wonder which of the protagonists was right, and whether they anticipated later insights into low-viscosity fluid behavior. Here is a brief answer to these ahistorical questions.

Consider first the formation of discontinuity surfaces. As Stokes correctly argued, none of the theorems invoked by Thomson prohibits the formation of such surfaces, even in the

---

[66]Stokes to Thomson, 30 Dec. 1898, *ST*. The modern reader may recognize Prandtl's separation process for the boundary layer.

[67]Stokes to Thomson, 5 Jan. 1899, 19–20 Dec. 1900 (quote), *ST*.

absence of a sharp edge. These theorems presuppose the continuity of the motion. For example, the demonstration of Lagrange's theorem requires the finiteness of the term $(\boldsymbol{\omega} \cdot \nabla)\mathbf{v}$ in the vorticity equation and therefore the continuity of the velocity.[68] If the flow is continuous at a given time, then it remains so at subsequent times. If, however, a tiny surface of discontinuity is grafted onto the wall, then Helmholtz's theorems and the electromagnetic analogy imply that it should grow at a rate given by the velocity discontinuity at its origin, with a spiral unrolling of its extremity.[69]

In order that the discontinuity be finite, the fluid should be stagnant at one side of the origin of the discontinuity surface, and move continuously on the other side. Consequently, the surface must depart tangentially from the wall (in the case of an edge, it is, at any time, tangent to one side of the edge). As far as Marcel Brillouin and Felix Klein could see, there is nothing in Euler's equations that contradicts this growth process. Neither is there anything in this equation that restricts the points from which an embryonic surface would grow (at least in the two-dimensional case). In summary, in an Eulerian fluid surfaces of discontinuity can be formed as Stokes wished, but their departure point is more arbitrary than experiments on real fluids would suggest.[70]

Another important issue of the Stokes–Thomson debate is the connection between inviscid and viscous behavior. According to Ludwig Prandtl's later views, at high Reynolds numbers the flow of a real fluid along a solid obstacle is irrotational beyond a thin boundary layer of intense shear. Unless the solid is specially streamlined, this layer separates from the body at some point (line) of its profile. The resulting flow resembles the surfaces of discontinuity imagined by Stokes for the Eulerian fluid. However, the separation point can only be determined through the Navier–Stokes equation (even though it does not depend on the value of the viscosity parameter!). Hence Stokes was right to expect a resemblance between the low-viscosity limit of real flows and discontinuous Eulerian flow; but Thomson was also right to lend viscosity a decisive role in forming the thin vortex layers that imitate discontinuity surfaces.[71]

---

[68]This is emphasized in Stokes [1849a] pp. 106–13.

[69]Jacques Hadamard ([1903] pp. 355–61) gave a proof that surfaces of discontinuity could not be formed in a perfect fluid as long as cavitation is excluded. This proof, however, does not exclude the growth of a pre-existing, tiny surface of discontinuity. Marcel Brillouin [1911] made this point, described the growth process, and extended the conformal methods of Helmholtz, Kirchhoff, and Levi-Civita to curved obstacles devoid of angular points. Felix Klein [1910] described the evolution of a surface of discontinuity formed by immersing an infinitely-thin blade (concretely, a rudder) perpendicularly to the liquid surface, pulling it at uniform speed in the direction of its normal, and suddenly withdrawing it. He resolved the apparent contradiction between Helmholtz's vorticity theorems and the formation of discontinuity processes as follows: 'Clearly the source of [the contradiction] is that we have now admitted the confluence of two originally separated fluid masses, whereas the usual foundation of the theorem presupposes that fluid particles that once belonged to the surface of the fluid must indefinitely belong to this surface.'

[70]Brillouin [1911]; Klein [1910]. According to Brillouin, in the two-dimensional case the departure point of a *steady* surface of discontinuity must be beyond a certain point of the surface of the body.

[71]Prandtl [1905]. For a viscous fluid, separation is not an instability issue. However, it is so in the ideal fluid case according to Stokes.

5.5   Parallel flow

5.5.1   *From dancing flames to the inflection theorem*

In the course of his acoustic studies, the London professor John Tyndall heard about the sensitivity of flames to sound that his American colleague John Le Conte had observed at a gas-lit musical party. The flames from 'fish-tail' gas burners danced gracefully as the musicians played a Beethoven trio, so that 'a deaf man might have seen the harmony'. In 1867, Tyndall displayed this strange phenomenon at the Royal Institution, as well as a similar effect with smoke jets, and published an account in the *Philosophical magazine*. When subjected to various sounds, the jet shortened to form a stem with a thick bushy head (see Fig. 5.12). The length of the stem depended on the pitch, and high-pitch notes were ineffective. Tyndall made this instability the true cause of the dancing of flames, but he did not propose any theoretical explanation.[72]

Tyndall's work attracted Lord Rayleigh's attention. This country gentleman had an uncommon disposition for physics, both mathematical and experimental. Coached by Edward Routh and inspired by Stokes's lectures at Cambridge, he emerged as senior wrangler and Smith's Prizeman in 1866. Until his appointment as Cavendish Professor on Maxwell's death (1879), his main research interests were in optics and acoustics. His elegant and masterful *Theory of sound*, first published in 1877, became one of the fundamental treatises of British physics, and remains an important reference to this day.[73]

Rayleigh, the theorist of sound, was naturally interested in Tyndall's observations as well as in Félix Savart's and Joseph Plateau's earlier experiments on the sound-triggered instability of water jets. In the latter case, the determining factor is the capillarity of the water surface, which favors a varicose shape of the jet and its subsequent disintegration into detached masses whose aggregate surface is less than that of the original cylinder. In 1879, Rayleigh determined the condition for the growth of an infinitesimal sinusoidal perturbation of the jet surface, as Thomson had done in the case of wind over water. He also gave a theory of smoke-jet instability, in even closer analogy to Thomson's wave theory. The relevant instability is that of a cylindrical surface of discontinuity for the air's

Fig. 5.12.   Smoke jets subjected to sounds of various pitch. From Tyndall [1867] p. 385.

[72]Le Conte [1858] p. 235; Tyndall [1867].   [73]Cf. Lindsay [1976].

motion. Neglecting capillarity, Rayleigh showed that, on a jet of velocity V, a sinusoidal perturbation with the spatial period λ grew as $e^{Vt/\lambda}$.[74]

This result contradicted Tyndall's observation that short sound waves were ineffective. Rayleigh traced the discrepancy to the viscosity of the air. In the case of two-dimensional parallel motion, the Navier–Stokes equation implies that the vorticity ω evolves according to the equation

$$\frac{\partial \omega}{\partial t} = \frac{\mu}{\rho} \Delta \omega \tag{5.17}$$

(the convective terms vanish), so that vorticity is 'conducted' through the fluid according to the same laws as heat. Consequently, any vortex sheet or discontinuity surface evolves into a layer of vorticity of finite thickness. Rayleigh then examined the stability of a finite layer of uniform vorticity. Switching off viscosity, he found that the layer became stable when its thickness somewhat exceeded the wavelength of the perturbation. This result made it likely that viscosity, by smoothing out the velocity discontinuity, should stabilize a jet for high-pitched sounds.[75]

After thus resolving the discrepancy between fluid mechanics and Tyndall's experiments, Rayleigh proceeded to the theoretically similar problem of two-dimensional parallel flow between fixed walls. He first studied the stability of successive finite layers of uniform vorticity with perturbed separating surfaces, using Helmholtz's analogy between vorticity and electric current. The result suggested that, for a continuous variation of the vorticity ω, stability would depend on the constancy of the sign of the variation $d\omega/dy$ between the two walls. In other words, the curvature d^2U/dy^2 of the velocity profile could not change sign.[76]

Rayleigh then offered the more direct approach to the stability problem that has now become standard. Let Ox denote an axis parallel to the flow, Oy the perpendicular axis, $U(y)$ the original velocity, and $u(x, y)$ and $v(x, y)$ the components of a small velocity perturbation. The vorticity equation gives

$$\frac{\partial \omega}{\partial t} + (U + u)\frac{\partial \omega}{\partial x} + v\frac{\partial \omega}{\partial y} = 0, \tag{5.18}$$

with

$$\omega = \frac{\partial v}{\partial x} - \frac{\partial u}{\partial y} - \frac{dU}{dy}. \tag{5.19}$$

Retaining only first-order terms in u and v, assuming that u and v vary as $e^{i(kx-\sigma t)}$, and eliminating u by means of the continuity equation $\partial u/\partial x + \partial v/\partial y = 0$, Rayleigh reached the stability equation

$$\left(U - \frac{\sigma}{k}\right)\left(\frac{\partial^2 v}{\partial y^2} - k^2 v\right) - \frac{d^2 U}{dy^2}v = 0. \tag{5.20}$$

He derived his stability criterion in the following ingenious manner.[77]

---

[74]Rayleigh [1879]. See also Rayleigh [1896] pp. 362–5.

[75]Rayleigh [1880] pp. 474–83.   [76]*Ibid.* pp. 483–4.   [77]*Ibid.* pp. 484–7.

The stability equation has the form $v'' + \alpha v = 0$, with

$$\alpha = -k^2 - \frac{U''}{U - \sigma/k}. \tag{5.21}$$

Multiplying by the complex conjugate v^* of v, and integrating from wall to wall gives

$$\int |v'|^2 \, dy + \int \alpha |v|^2 \, dy = 0. \tag{5.22}$$

Hence, the imaginary part of the function α must satisfy the condition

$$\int \mathrm{Im}(\alpha)|v|^2 \, dy = 0, \tag{5.23}$$

or

$$\mathrm{Im}(\sigma) \int \frac{|v|^2}{|U - \sigma/k|^2} U'' \, dy = 0. \tag{5.24}$$

If the sign of U'' is constant (and if the perturbation v does not uniformly vanish), then the integral is nonzero, so that the imaginary part of σ must vanish and the perturbation cannot grow exponentially. Rayleigh concluded that parallel flow without inflexion of the velocity profile was stable. As he noted, the criterion is of no help in the jet case for which U'' changes sign.[78]

5.5.2   Reynolds's instabilities

In this discussion of parallel flow between fixed walls, Rayleigh probably had in mind a two-dimensional approach to the stability of pipe flow.[79] Yet he did not discuss this application, presumably because of the lack of relevant experiments. As we will see in the next chapter, Osborne Reynolds filled this gap in 1883 with a thorough study of the transition between 'direct' and 'sinuous' flow in straight circular pipes. Reynolds had the turbulent eddying in his pipes depend on an excess of the inertial term of the Navier–Stokes equation over the viscous term. When the flow depends on only one characteristic length L (the pipe diameter) and on the average velocity V, the ratio between the two terms is governed by the ratio LV / ν, where ν is the kinematic viscosity μ/ρ. This ratio is now called the Reynolds number.[80]

Through color-band experiments, Reynolds verified that the critical transition depended on this number. He thereby noticed the surprisingly sudden character of this transition: violent eddying occurred as soon as the critical Reynolds number was reached. Moreover, the flow appeared to be unstable with respect to finite perturbations well before the critical number was reached:[81]

---

[78]Rayleigh [1880] p. 487. Rayleigh also gave (without proof) the criterion in the cylindrical case that 'the rotation either continually increases or continually decreases in passing outwards from the axis.'

[79]Rayleigh states this in *RSP* **3**, p. 576.

[80]Reynolds [1883] pp. 54–5. A more detailed account will be given in Chapter 6, pp. 249–52.

[81]*Ibid.* p. 61. Also, *ibid* pp. 75–6: 'The fact that the steady motion breaks down suddenly, shows that the fluid is in a state of instability for disturbances of the magnitude which cause it to break down. But the fact that in some

The critical velocity was very sensitive to disturbance in the water before entering the tubes ... This showed that the steady motion was unstable for large disturbances long before the critical velocity was reached, a fact which agreed with the full-blown manner in which the eddies appeared.

From casual observations of conflicting streams of water, Reynolds was aware of the existence of another kind of instability for which the transition from direct to sinuous motion was gradual and independent of the size of the disturbances. In his memoir of 1883, he recounted an elegant experiment in which he had a lighter fluid slide over a heavier one with a variable velocity difference. For a certain critical velocity, the separating surface began to oscillate. The waves then grew with the sliding velocity, until they curled and broke.[82]

Reynolds was unaware of relevant theoretical considerations by Helmholtz, Kelvin, and Rayleigh. He was therefore 'anxious' to find a theoretical explanation of the two kinds of instabilities he had encountered. He first studied the stability of the solutions of Euler's equation, with the result that 'flow in one direction was stable, flow in opposite directions unstable.' As he could only imagine a stabilizing effect of viscosity, the instability of pipe flow puzzled him for a long time. At last, he attempted a similar study in the more difficult case of the Navier–Stokes equation. He then found that the boundary condition for viscous fluids (vanishing velocity at the walls) implied instability for sufficiently-small values of the viscosity: 'Although the tendency of internal viscosity of the fluid is to render direct or steady motion stable, yet owing to the boundary condition resulting from the friction at the solid surface, the motion of the fluid, irrespective of viscosity, would be unstable.'[83]

Reynolds never published his stability calculations. He could conceivably have handled the inviscid case in a manner similar to Rayleigh's, although the roughness of his statement of the criterion suggests some erring. That he could derive a boundary-layer instability in the viscous case seems highly implausible, considering the subtlety of the later considerations of that sort by Prandtl, Heisenberg, and Tollmien.[84]

5.5.3 *Thomson's proofs of stability in viscous cases*

In his presidential address to the British Association meeting of 1884, Rayleigh praised Reynolds's contribution to the study of the transition between laminar and turbulent flow. His view of the future of the subject was singularly optimistic: 'In spite of the difficulties which beset both the theoretical and the experimental treatment, we may hope to attain before long to a better understanding of a subject which is certainly second to none in scientific as well as practical interest.' It is likely that he and Stokes were responsible for the subject of the Adams prize for 1889: 'On the criterion of the stability and instability of

---

condition it will break down for a large disturbance, while it is stable for a smaller disturbance, shows that there is a certain residual stability, so long as the disturbances do not exceed a given amount ... It was a matter of surprise to me to see the sudden force with which the eddies sprang into existence, showing a highly unstable condition to have existed at the time the steady motion broke down.—This at once suggested the idea that the condition might be one of instability for disturbances of a certain magnitude, and stable for small disturbances.'

[82]*Ibid.* pp. 61–2.   [83]*Ibid.* pp. 62–3.

[84]On the later considerations, cf. Drazin and Reid [1981] chap. 4, and Chapter 7, pp. 294–6.

the motion of a viscous fluid'. After a reference to Reynolds's work, the announcement of the prize read:[85]

> It is required either to determine generally the mathematical criterion of stability, or to find from theory the value [of the critical Reynolds number] in some simple case or cases. For instance, the case might be taken of steady motion in two dimensions between two fixed planes, or that of a simple shear between two planes, one at rest and one in motion.

The only theorist to claim success in solving these two cases was no beginner in need of the £170 prize; it was Sir William Thomson.[86] In the second case (plane Couette flow),[87] the simpler one because of its constant vorticity, Thomson provided a fairly explicit procedure for deriving the evolution of an arbitrary small perturbation of the flow. From the Navier–Stokes equation and the incompressibility condition, he first obtained the linearized equation

$$\left(\frac{\partial}{\partial t} + U\frac{\partial}{\partial x} - \nu\Delta\right)\Delta v = 0, \tag{5.25}$$

which only contains the second component, v, of the velocity perturbation of the basic flow $U = \beta y$ (the y-axis being perpendicular to the plates, and the x-axis being parallel to the motion of the moving plate). As he astutely noted, this equation and those for the other components u and w can be solved explicitly for any initial value of the perturbed velocity which is compatible with the incompressibility condition, if only the real boundary condition (vanishing relative velocity on the plates) is replaced with the sole condition of vanishing *normal* velocity at the plates for $t = 0$. This may be called the relaxed solution.

Thomson next used Fourier analysis to find the 'forced solution' of the linearized equations for which the velocity perturbation on the plates was a given function of time, vanishing for negative time and opposite to the velocity of the relaxed solution on the plates for positive time. As Thomson believed the latter solution to vanish identically for negative time, he regarded the sum of the relaxed and the forced solutions to be the requested solution of the real initial-value problem. The relaxed solution is easily seen to decrease exponentially in time. This implies the same behavior for the forced and the complete solutions. Thomson concluded that the simple shear flow of the prize question was stable.

In the other case of the Adams prize (plane Poiseuille flow), Thomson could no longer obtain the relaxed solution. Instead, he directly applied Fourier analysis to the real initial-

---

[85]Rayleigh [1884*b*] p. 344; G. Taylor, G. H. Darwin, G. G. Stokes, and Lord Rayleigh (examiners), 'The Adams Prize, Cambridge University', *PM* **24** (1887), pp. 142–3.

[86]Thomson [1887*c*]. According to *The Cambridge review* **9** (1889), p. 156, the prize was not adjudged in default of candidates.

[87]The Couette flow is the steady viscous flow between two concentric parallel cylinders, one of which is rotating at a constant speed. Following a suggestion by Max Margules, in 1890 Maurice Couette measured the viscosity of various fluids from the torque exerted on a cylinder immersed in the fluid contained in a rotating, coaxial cylinder (Couette [1890*a*]). This method permitted a better control of pressure (in the gas case), better precision, and a wider range of velocity gradients than Coulomb's and Maxwell's earlier methods (thus permitting a more extensive confirmation of the Navier–Stokes equation). Couette [1890*b*] described the instability of this flow beyond a critical velocity of the rotating cylinder.

value problem. He seems to have believed that both the boundary condition and the initial condition could be satisfied by superposing Fourier components of the form $f(y)e^{i(\sigma t+kx+mz)}$, where the frequencies σ, k, and m are real numbers. Accordingly, he contented himself with proving that, for any nonzero value of the viscosity parameter and for any values of σ, k, and m, convergent power-series expansions could be found for the y dependence of the Fourier components. From this result and from the real character of the frequencies σ, he concluded that the plane Poiseuille flow was also stable.[88]

Lastly, Thomson dealt with the practical instability of pipe flow. In conformance with Reynolds's observation that the growth of perturbations in this case depended on their size, he proposed that the flow was probably stable for infinitesimal perturbations (as he thought it was in two dimensions) but unstable for finite ones. It would be so, he argued, if the inviscid flow with Poiseuille velocity profile was unstable, and if viscosity could only damp sufficiently-small perturbations. The margin of stability would then increase for higher viscosity, as Reynolds had observed.[89]

The instability of inviscid flow with a parabolic velocity profile clearly contradicted Rayleigh's inflection theorem. Thomson believed, however, that a 'disturbing infinity vitiate[d] [Rayleigh's] seeming proof of stability.' As Rayleigh himself noted, the stability equation

$$\left(U-\frac{\sigma}{k}\right)\left(\frac{\partial^2 v}{\partial y^2}-k^2 v\right)-\frac{d^2 U}{dy^2}v=0 \tag{5.20}$$

becomes singular for values of the coordinate y for which the velocity σ/k of the plane-wave perturbation is identical to the velocity $U(y)$ of the unperturbed flow (and U'' does not simultaneously vanish). At such a point, the flow is obtained by superposing a sine-wave velocity pattern with a shearing motion. For an observer moving along the fluid, the flow has the 'cat's eye' outlook of Fig. 5.13, which Thomson published in 1880.[90]

From then on, Thomson attached great importance to the disturbing infinity: 'The "awkward infinity" ', he wrote to George Darwin in August 1880, 'threatens quite a revolution in vortex motion (in fact a *revolution* where nothing of the kind, nothing but the laminar rotational movement, was even suspected before), and has been very bewildering.' Thomson believed the elliptic whirls of this flow to be the source of the turbulence observed by Reynolds. Any simple perturbation of the fluid boundary necessarily contained Fourier components for which elliptic whirling would disturb the laminar flow.[91]

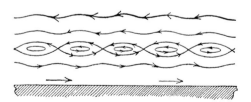

Fig. 5.13.   Thomson's 'cat's eye' flow pattern. From Thomson [1880c] p. 187.

[88]Thomson [1887d].   [89]*Ibid.* p. 335.

[90]Thomson [1887d] p. 334; Rayleigh [1880] p. 486; Thomson [1880c].

[91]Thomson to Darwin, 22 Aug. 1880, in Thompson [1910] p. 760. Thomson does not address the question of the growth of the whirls.

5.5.4   *Challenging Thomson*

Rayleigh valiantly defended his stability criterion against Thomson's 'disturbing infinity':

> Perhaps I went too far in asserting that the motion was thoroughly stable; but it is to be observed that if [the frequency σ] be complex, there is no 'disturbing infinity'. The argument, therefore, does not fail regarded as one for excluding complex values of [σ]. What happens when [σ] has a real value such that [$\sigma - kU$] vanishes at an interior point, is a subject for further examination.

Equation (5.20) is indeed non-singular for a complex value of σ, so that an exponential increase of infinitesimal perturbations and a constant sign of U'' are truly incompatible. Rayleigh conceded, however, that the impossibility of an exponential increase did not rigorously establish stability. Perhaps a less rapid increase of perturbations was still possible owing to the 'disturbing infinity'. Perhaps higher-order terms in the stability equation implied a departure from the first-order behavior. In sequels to his 1880 study, Rayleigh provided arguments that made these escapes implausible. Modern writers on hydrodynamic stability no longer question the validity of his stability criterion.[92]

In return to Thomson's criticism of his criterion, Rayleigh politely questioned Thomson's proofs of stability of plane viscous flow:

> Naturally, it is with diffidence that I hesitate to follow so great an authority, but I must confess that the argument does not appear to me demonstrative. No attempt is made to determine whether in free 'disturbances of the type [$e^{i\sigma t}$] the imaginary part of [σ] is finite, and if so whether it is positive or negative.' If I rightly understand it, the process consists in an investigation of forced vibrations of arbitrary (real) frequency, and the conclusion depends on the tacit assumption that if these forced vibrations can be 'expressed in periodic form, the steady motion from which they are deviations cannot be unstable.'

Rayleigh went on to show that the tacit assumption was wrong in the case of a (rigid) pendulum situated near the highest point of its orbit. Whether he correctly interpreted Thomson's intentions is questionable. He was right, however, to judge Thomson's reasoning incomplete.[93]

The Irish mathematician William Orr clearly identified the gaps in 1907. Consider first Thomson's proof of stability of plane Poiseuille flow. This proof assumes that a superposition of harmonic solutions (with respect to t, x, and z) that satisfies the boundary conditions is sufficient to reproduce any initial value of the velocity perturbation. This does not need to be true, because the boundary conditions might restrict the harmonic solutions too much. Thomson's proof also fails in the case of plane Couette flow. The forced solution in this proof does not need to vanish for $t = 0$, even though it is forced to vanish on the boundaries of the fluid for any negative time. Indeed, the boundary conditions completely determine the Fourier-type solution, thus leaving no room for a further restriction of the initial motion. Consequently, the complete solution may not have the requested initial value.[94]

---

[92]Rayleigh [1892] p. 380, [1887], [1895]. Cf. Drazin and Reid [1981] pp. 126–47.

[93]Rayleigh [1892] p. 582. Yet, in 1895 Rayleigh (unwisely) endorsed Thomson's 'special solution' for disturbances of the plane Couette flow.

[94]Orr [1907]. This paper also contains an unconvincing interpretation of Rayleigh's criticism of 1892.

5.5.5 *Rayleigh's paradox*

Thomson himself had become aware of the weakness of his reasoning, as appears in a letter he wrote to Stokes in December 1898: 'Several papers of mine in *Phil. Mag.* about 1887 touch inconclusively on this question [of the stabilizing effect of viscosity].' Yet he still believed that the instability observed by Reynolds depended on the instability of the parabolic velocity profile at zero viscosity. In contrast, Rayleigh never really doubted the truth of his inflection theorem, which forbade this sort of instability. This led him to enunciate the basic paradox of pipe flow:

> If [my criterion] is applied to a fluid of infinitely small viscosity, how are we to explain the observed instability which occurs with moderate viscosities? It seems very unlikely that the first effect of increasing viscosity should be to introduce an instability not previously existent, while, as observation shows, a large viscosity makes for stability.

Rayleigh offered a few suggestions to explain this discrepancy. Firstly, irregularities of the wall surface could play a role. Secondly, instability could occur for *finite* disturbances even when the Rayleigh criterion gave stability. Thirdly, the three-dimensional case of Reynolds's experiments could qualitatively differ from the two-dimensional case studied by Rayleigh and Thomson. Fourthly, Rayleigh wrote, 'it is possible that, after all, the investigation in which viscosity is altogether ignored is inapplicable to the limiting case of a viscous fluid when the viscosity is supposed infinitely small.'[95]

The main purpose of Rayleigh's paper was to exclude the third possibility by extending his stability criterion to cylindrically-symmetric flow. In retrospect, his short comments on the fourth conjecture are most interesting:[96]

> There is more to be said in favour of this view than would at first be supposed. In the calculated motion there is a finite slip at the walls [when viscosity is ignored], and this is inconsistent with even the smallest viscosity. And further, there are kindred problems relating to the behaviour of a viscous fluid in contact with fluid walls for which it can actually be proved that certain features of the motion which could not enter into the solution, were the viscosity ignored from the first, are nevertheless independent of the magnitude of viscosity, and therefore not to be eliminated by supposing the viscosity to be infinitely small.

Rayleigh had in mind the explanation he had given in 1883 of an acoustic anomaly discovered by Savart in 1820 and studied by Faraday in 1831, namely that, when a plate sprayed with light powder is set into vibration, the powder gathers at the antinodes of the motion, whereas Chladni's earlier experiments with sand gave the expected nodal figures. Faraday traced this anomaly to the action of currents of air, rising from the plate at the antinodes, and falling back at the nodes.[97]

In his confirming calculation, Rayleigh assumed a plane monochromatic standing wave for the motion of the plate and solved the Navier–Stokes equation for the fluid motion above the plate perturbatively, taking the nonlinear $(\mathbf{v} \cdot \nabla)\mathbf{v}$ term as the perturbation. The

---

[95]Thomson to Stokes, 27 Dec. 1898; Rayleigh [1892] pp. 576–7.

[96]Rayleigh [1892] p. 577.

[97]Cf. Rayleigh [1883*a*] pp. 239–40; [1896] vol. 1, pp. 367–8. Rayleigh also explained the air currents observed by Vincenz Dvořák in 1876 in Kundt's tubes.

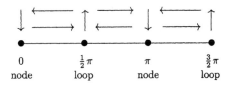

Fig. 5.14.   Motion of air near a vibrating plate. From Rayleigh [1883a] p. 250.

resulting motion is confined near to the plate in a layer of thickness $(v/f)^{1/2}$, where v is the kinematic viscosity and f is the frequency of the oscillations. This layer includes a periodic array of vortices, as shown in Fig. 5.14. The peripheral velocity of the vortices is of order v_0^2/V, where v_0 is the maximum velocity of the particles of the plate, and V is the celerity of the two progressive waves of which the standing wave motion of the plate is a superposition. As Rayleigh emphasized, this vortical velocity does not depend on the value of the viscosity v: 'We cannot, therefore, avoid considering this motion by supposing the coefficient of viscosity to be very small, the maintenance of the vortices becoming easier in the same proportion as the forces tending to produce the vortical motion diminish.'[98]

Rayleigh anticipated a similar singularity of the zero-viscosity limit in the case of plane parallel flow. This view agreed with Reynolds's assertion that intense shear near the walls caused the instability observed in pipe-flow experiments. As we will see in Chapter 7, in 1921 Ludwig Prandtl described a destabilizing mechanism for plane Poiseuille flow. In 1929 and 1947, his disciple Walter Tollmien proved the correctness of this intuition. In 1924, Heisenberg independently derived the instability of plane Poiseuille flow, through a method of approximation whose validity could only be established much later by Chia Chiao Lin and others. For circular pipes, the flow is probably stable at any Reynolds number, although a complete proof is still lacking. The latter problem is mathematically similar to plane Couette flow, for which a rigorous proof of stability is now available. Nineteenth-century experts on fluid mechanics did not possess the mathematical techniques that have proven necessary even in the simplest problems of viscous-flow stability. Yet they could anticipate various causes of instability, such as finite disturbances, intense shear in boundary layers, and irregularity of walls.[99]

5.5.6   *Reynolds's energetic approach*

Reynolds offered a last nineteenth-century approach to parallel-flow instability in a memoir of 1894. His reasoning was based on an equation he derived for the variation in time of the energy of the eddying motion. He thereby assumed the existence of a macroscopic averaging scale for which the mean motion no longer involved turbulent eddying. Under this assumption, the energy of the eddying motion is borrowed inertially from the energy of the mean motion and damped by viscous forces. As a stability criterion, Reynolds required the dominance of the damping term of his eddying-energy equation over the inertial term

---

[98]Rayleigh [1883a] p. 246.

[99]For modern knowledge regarding the stability of parallel flow, cf. Drazin and Reid [1981] pp. 212–13 (plane Couette flow), 221 (plane Poiseuille flow), 219 (Poiseuille flow in a circular pipe); also Lin [1966] pp. 11–14.

for any choice of the eddying motion. By laborious calculations he estimated the corresponding Reynolds number in the case of flow between two fixed parallel plates.[100]

Reynolds's method can at best yield a value of the Reynolds number below which the motion must be stable. It does not allow one to determine the Reynolds number from which certain perturbations (not necessarily of the random eddying kind) will grow. The general idea of studying the evolution of the energy of a perturbation of the laminar motion has nevertheless seduced later students of hydrodynamic instability, including Hendrik Lorentz, William Orr, Theodor von Kármán, and Ludwig Prandtl. In some cases, as the Prandtl–Tollmien boundary-layer instability, it provides some physical understanding of the mechanism of instability.[101]

Thomson's, Rayleigh's, and Reynolds's mathematical studies of parallel flow show how impenetrable the caprices of fluid motion could be to the elite of nineteenth-century mathematical physics. Where stability was hoped for, for instance in Kelvin's vortex rings, it turned out to be highly improbable. Where instability was observed, for instance in Reynolds's pipes, it turned out to be very hard to prove. The first failure threatened the British hope of basing the entirety of physics on the perfect liquid. The second stood in the way of concrete applications of fluid dynamics to hydraulic or aerodynamic processes. Yet the few mathematical successes obtained in simple, idealized cases, together with inspired guesses on general fluid behavior, opened a few paths of the modern theory of hydrodynamic instability.

In general, the nineteenth-century concern with hydrodynamic stability or instability led to well-defined, clearly-stated questions on the stability of the solutions of the fundamental hydrodynamic equations (Euler's and Navier's). Most answers to these questions were tentative, controversial, or plainly wrong. The subject that Rayleigh judged 'second to none in scientific as well as practical interest' remained utterly confused. Apart from the Helmholtz–Kelvin instability and Rayleigh's inflection theorem, the theoretical yield was rather modest. There was Stokes's vague, unproved instability of divergent flows, Thomson's unproved instability of vortex rings, the hanging question of the formation of discontinuity surfaces, and two illusory proofs of stability for simple cases of parallel viscous flow.[102]

The situation could be compared to number theory, which is reputed for the contrast between the simple statements of some of its problems and the enormous difficulty of their solution. The parallel becomes even closer if we consider that some nineteenth-century problems of hydrodynamic stability, for example the stability of viscous flow in circular pipes or the stability of viscous flow past obstacles, are yet to be solved, and that the few available answers to such questions were obtained at the price of considerable mathematical efforts. This long persistence of basic questions of fluid mechanics is the more striking because in physics questions tend to change faster than their answers.

In number theory, failed demonstrations of famous conjectures sometimes brought forth novel styles of reasoning, interesting side problems, and even new branches of

---

[100]Reynolds [1895]. See Chapter 6, pp. 259–62. What I here call 'mean motion' corresponds to Reynolds's 'mean-mean-motion'.

[101]Cf. Lin [1966] pp. 59–63.   [102]Rayleigh [1884*b*] p. 344.

mathematics. Something similar happened in the history of hydrodynamic stability, though to a less spectacular extent. Stokes's and Helmholtz's surfaces of discontinuity were used to solve the old problem of the *vena contracta* and to determine the shape of liquid jets. They also permitted Rayleigh's solution (1876) of d'Alembert's paradox, and inspired some aspects of Ludwig Prandtl's boundary-layer theory (1904). Rayleigh's formulation of the stability problem in terms of the real or imaginary character of the frequency of characteristic perturbation modes inaugurated the hydrodynamic application of what is now called the method of normal modes.[103]

As a last important example of fruitful groping, Stokes, Thomson, and Rayleigh all emphasized that the zero-viscosity limit of viscous-fluid behavior could be singular. Stokes regarded this singularity as a symptom of the instability of inviscid, divergent flows; Thomson regarded it as an indication that the formation of unstable states of parallel motion required finite viscosity; Rayleigh regarded it as a clue to why some states of parallel motion were stable for zero viscosity and unstable for a small, finite viscosity. Rayleigh even anticipated the modern concept of boundary-layer instability:[104]

> But the impression upon my mind is that the motions calculated above for an absolutely inviscid liquid may be found inapplicable to a viscid liquid of vanishing viscosity, and that a more complete treatment might even yet indicate instability, perhaps of a local character, in the immediate neighbourhood of the walls, 'when the viscosity is very small.'

In the absence of a mathematical proof, such utterances are of dubious value. Rayleigh himself warned that 'speculations on such a subject in advance of definite arguments are not worth much.' Many years later, Garrett Birkhoff reflected that speculations were especially fragile on systems like fluids that have infinitely many degrees of freedom. Yet, by imagining odd, singular behaviors, the pioneers of hydrodynamic instability avoided the temptation to discard the foundation of the field, the Navier–Stokes equation; and they sometimes indicated fertile directions of research.[105]

Early struggles with hydrodynamic stability are not only interesting for the clues they give on the later development of this topic; they also reveal fine stylistic differences among leaders of nineteenth-century physics. Due to the lack of rigorous mathematical solutions for the outstanding problems of fluid dynamics, these physicists had to rely on subtle, individual combinations of intuition, past experience or experiment, and improvised mathematics. They ascribed different roles to idealizations such as inviscidity, rigid walls, or infinitely-sharp edges. For instance, Helmholtz and Stokes believed that the perfect liquid provided a correct intuition of low-viscosity liquid behavior, if only discontinuity surfaces were admitted. Thomson denied that, and reserved the perfect liquid (without discontinuity) for his sub-dynamics of the universe. As the means to exclude rigorously one of these two views were lacking, the protagonists preserved their colorful identities.

[103]Kirchhoff [1869]; Rayleigh [1876b]; Prandtl [1905]. Cf. Drazin and Reid [1981] pp. 10–11. The method of normal modes originated in Lagrange's study of the stability of the solar system and in Clerk Maxwell's study of the stability of Saturn's rings.

[104]Rayleigh [1892] p. 577.   [105]*Ibid.* p. 576; Birkhoff [1950].

6

TURBULENCE

> If the velocities [of water in rivers] remained constant in each point of the traversed space, the surface of the liquid would look like a plate of ice and the herbs growing at the bottom would be equally motionless. Far from that, the stream presents incessant agitation and tumultuous, disordered movements, so that the velocities change in an abrupt and most diverse manner from one point to another and from one instant to the next. As noted by Leonardo da Vinci, Venturi, and especially Poncelet, one can perceive eddies, large and small, with a vertical mobile axis. One can also see, at the surface, *bouillons*, or eddies with a nearly horizontal axis, that constantly surge from the bottom and thus form genuine ruptures, with the intertwining and mixing motions that M. Boileau observed in his experiments.[1] (Adhémar Barré de Saint-Venant, 1872)

Hard to gain though it may be, any understanding of hydrodynamic instabilities is of a negative kind. Namely, it only tells us when and why the rigorous solutions of the hydrodynamic equations under given boundary conditions fail to represent natural flows. It does not tell us much on the sort of motion into which the unstable system settles after perturbation. From common observations, everyone knows the great complexity of this motion. The capricious eddying of water behind obstacles, or the hesitating, convoluted rise of smoke from a fire have indeed inspired poets with metaphors for the unpredictability of human life.

William Thomson began using the term 'turbulent' in the 1880s to characterize such irregular motions, as opposed to the 'laminar' flows in which successive fluid layers glide smoothly over each other. Much earlier, in 1822, Navier opposed 'linear' to 'nonlinear' flow, and from the 1830s Saint-Venant opposed 'tumultuous' to 'regular' flow. The unpredictability captured in this terminology has long deterred the theorists of fluid motion. Yet, the intellectual mastery of some aspects of turbulent flow has proved possible.[2]

Turbulence studies began in the nineteenth century with what French engineers called *eaux courantes*, or open-channel flow. Pipe flow and fluid resistance took second priority, perhaps because turbulence is less visible in pipes and more heterogeneous around an obstacle, but also because in those years French engineers were busy building new canals and improving the navigability of rivers. In 1822, just after proposing his equations for viscous-fluid motion, Navier recognized its impotence for describing the 'nonlinear' flows encountered in hydraulics. In the 1830s and 1840s, Saint-Venant suggested that the same

---

[1] Saint-Venant [1872] p. 650

[2] Cf. Thomson [1887*e*], [1894]. Navier's and Saint-Venant's contributions are discussed later on pp. 229–31. The opposition between turbulent and laminar flow is used here as roughly as it was in the nineteenth century, with no consideration of intermediate, oscillatory forms of motion.

equation could be applied to the large-scale average of a tumultuous flow if the viscosity parameter was made to depend on the circumstances of the flow. In the 1870s, his disciple Joseph Boussinesq implemented this approach in a monumental *Théorie des eaux courantes*.

These early quantitative and statistical theories of turbulent flow are described in Sections 6.2 and 6.3. Section 6.1 does not deal with turbulence *per se*, but with anterior studies of open channels in the 1820s and 1830s, mainly the problem of backwaters that largely motivated Saint-Venant's and Boussinesq's work. The authors of these studies did not calculate from the fundamental equations of hydrodynamics. Instead, they developed a semi-empirical approach that combined a parallel-slice idealization of the flow, mechanical principles, and some experimental input for wall friction. They ignored the turbulent character of the motion. In contrast, Saint-Venant argued that insights into the nature of turbulence would permit more fundamental solutions of hydraulic problems.

The mathematical theory of open-channel flow was a mostly French topic, usually avoided by British engineers. There was a significant exception, the brother James of William Thomson, who kept up with literature on this subject and agreed with his French counterparts that turbulence played a significant role in determining the flow pattern. While helping James explain an anomaly of the velocity profile, in 1887 William Thomson discovered that the turbulent fluid had effective rigidity and could thus propagate large-scale transverse vibrations. For a short, exhilarating time, he believed to have found the key to the perfect-fluid theory of the luminiferous ether. George Francis FitzGerald, who similarly dreamt of a 'vortex-sponge' theory of the ether, extended Thomson's speculation with much enthusiasm. These theories are described in Section 6.4.

Thomson's and FitzGerald's ether theories, as for Saint-Venant's and Boussinesq's hydraulics, only involved developed turbulence. They did not require an understanding of the transition from laminar to turbulent flow. In 1839, the German hydraulician Gotthilf Hagen discovered the sudden character of this transition in the case of pipe flow. His original purpose was to provide engineers with more exact retardation formulas, so he did not dwell on this curious phenomenon. In contrast, Reynolds's hydrodynamic investigations of the 1880s, described in Section 6.5, the final section of this chapter, concerned this transition and its 'criterion'.

Problems of navigation, rather than hydraulics, motivated Reynolds's interest in turbulence. While reflecting on propellers, wakes, and sea waves, he surmised that most hydrodynamic paradoxes and anomalies resulted from our ignorance of invisible vortex motion. William Thomson and James Clerk Maxwell had already made vortices in a pervasive, ideal fluid responsible for the magnetic properties of the ether and for the stability of matter. Reynolds made their continual production the main cause of resistance and retardation in real fluids. To reveal the secrets of fluid motion, he only needed a few drops of ink.

A more surprising source of Reynolds's reflections on turbulent flow was the kinetic theory of gases. Following an investigation of William Crookes's radiometer and Thomas Graham's transpiration phenomena, Reynolds argued that the nature of the flow of a dilute gas depended on the 'dimensional properties of matter', specifically on the ratio between the dimensions of the flow (vane size or tube diameter) and the mean free path. Similarly, he expected the nature of the flow of a denser fluid to depend on the dimensional properties of the Navier–Stokes equation. This led him to the idea of a transition controlled by the Reynolds number, to the experimental verification and sharpening of this idea, and to his later kinetic–statistical theory of the turbulent transition.

6.1 Hydraulic phenomenology

6.1.1 *Hydraulics versus hydrodynamics*

As the rational hydrodynamics of d'Alembert and Euler proved inept at practical problems of hydraulics and navigation, empirical or semi-empirical methods began to thrive. When, in the 1770s, the minister Turgot consulted d'Alembert, the Marquis de Condorcet, and the abbot Charles Bossut about the project of an underground canal in Picardie, they performed towing experiments that showed, among other things, that the resistance increased with the narrowness of the canal. Bossut taught empirical hydrodynamics at the Ecole Royale du Génie de Mézière. The Ministry of War funded his numerous experiments on retardation in pipe and channel flow. The second edition of the resulting treatise, published in 1786/87, long remained a reference for hydraulic engineers.[3]

Bossut praised the 'very profound and very generous method' of his friend d'Alembert as well as the 'scope and generality' of Euler's contribution. However, he did not try to apply these theories in the real world:

> These great geometers seem to have exhausted the resources that can be drawn from
> analysis to determine the motion of fluids: their formulas are so complex, by the
> nature of things, that we may only regard them as geometrical truths, and not as
> symbols fit to paint the sensible image of the actual and physical motion of a fluid.

Bossut measured the loss of head in pipes and channels of various breadths, for which he provided a wealth of numerical tables and the inference that the loss was roughly proportional to the square of the average velocity.[4]

The other French master of late-eighteenth-century hydraulics, Pierre Du Buat, agreed with Bossut that urgent hydraulic problems could only be solved by the experimental method. Yet he had the more theoretical ambition of providing general formulas for pipe and channel flow, as well as a detailed discussion of the course of rivers, which was his main interest. He applied Newton's second law to the bulk motion of water, guessed the form of retarding forces by molecular intuition, and inferred relevant parameters from abundant measurements.[5]

Du Buat formulated the 'key to hydraulics' as the balance between the accelerating force (due to pressure gradient or gravity) of a fluid slice and its friction on the walls. For uniform, permanent flow in an open channel, this leads to the equation (in Prony's later notation)

$$\rho g S i = \chi F_U, \qquad (6.1)$$

---

[3]Cf. Dugas [1950] pp. 300–3, Rouse and Ince [1957] pp. 126–8, Redondi [1997].

[4]Bossut [1786/87] p. XV.

[5]Du Buat [1786]. Cf. Dugas [1950] pp. 303–5, Rouse and Ince [1957] pp. 129–34.

where ρ is the density of water, g is the acceleration of gravity, S is the normal fluid section, i is the slope of the bottom (the sine of the angle that it makes with a horizontal plane), χ is the wetted perimeter of the channel, F_U is the retarding force per unit length, and U is the average velocity.[6]

Du Buat's intuition of fluid tenacity and 'molecular gearing' yielded an intricate expression for the retarding force F_U, which reduces to the quadratic form bU^2 in most practical cases. Combined with the equilibrium condition (6.1), this form gives the formula named after Antoine Chézy, who proposed it first in an unpublished report on a canal planned to bring the waters of the River Yvette into Paris. In 1804, the director of the Ecole des Ponts et Chaussées, Gaspard de Prony, inferred from Couplet's, Bossut's, and Du Buat's retardation measurements and from Charles Coulomb's understanding of surface friction the perennial form[7]

$$F_U = aU + bU^2. \tag{6.2}$$

As well as his treatment of uniform permanent flow, Du Buat gave semi-empirical formulas for weirs and backwaters. One of his main concerns was the improvement of the navigability of rivers, then usually achieved by a series of weirs that elevated the water level. The weir formula gives the height of the water above a weir as a function of the river's discharge and the weir's width and height. Upstream from the weir, the water surface has a curved shape that would asymptotically reach the natural level of the river if no other weir interfered. This is the 'backwater' phenomenon which Du Buat improperly called *remou*. The navigability of a naturally shallow river is improved by weirs placed so that the depth of the backwater of the nth weir at the foot of the $(n-1)$th weir exceeds the minimal depth required for navigation. A lock on the side of each weir permits the passage of the boats.[8]

6.1.2 *Bélanger's backwater theory*

Du Buat contented himself with a circular-arc approximation of the backwater curve, arguing that the knowledge of the relevant differential equation would be of no practical help. Some forty years later, Jean-Baptiste Bélanger, an engineer with the Ponts et Chaussées, judged differently. Like many former polytechnicians, Bélanger had faith in the practical usefulness of higher mathematics. While working on canals and adjacent rivers, he sought a theory of non-uniform flow that would permit more rational designs. The Royal Academy of Metz had recently advertized a prize for 'determining the curve that running water forms upstream from a weir'.[9]

The only known open-channel formulas concerned uniform permanent flow, for which the section of the channel is uniform and the slope of the water surface is the same as the slope of the bottom. In his new theory, Bélanger admitted a slow variation of the section of the channel and slight differences between the surface and bottom slopes. For simplicity, he assumed that the velocity (vector) within a section of the stream was nearly uniform,

[6]Du Buat [1786] vol. 1, p. xvii.   [7]*Ibid.* p. 62. Cf. Rouse and Ince [1957] pp. 141–3.

[8]Du Buat [1786] vol. 1, chap. 4, pp. 205–17.

[9]Bélanger [1828] p. iii. Cf. Saint-Venant [1887c] pp. 154–7, Rouse and Ince [1957] pp. 148–9. On the prize, cf. Poncelet [1845] p. 510 n.

although he knew from Du Buat that the velocity increased with the distance from the bottom. This assumption agrees with a variable fluid section as long as the departure from uniformity remains small.[10]

Following Bélanger, take the s-axis parallel to the common velocity of the approximately-parallel water filaments, and the x-axis normal to this axis (see Fig. 6.1). Denote by γ the angle that the s-axis makes with the horizontal. A given particle of the fluid experiences three forces, namely, its weight, the pressure gradient, and a frictional force, which Bélanger took to be the same at every point of a fluid section.[11] Its velocity v varies with the section S, as follows from the constancy of the discharge $Q = vS$ (the volume of water crossing a section of the channel per unit time). Newton's second law, projected onto the s- and x-axes then gives

$$-\frac{\partial P}{\partial s} + \rho g \sin \gamma - \frac{\chi}{S} F_v = \rho \frac{dv}{dt}, \tag{6.3a}$$

$$-\frac{\partial P}{\partial x} - \rho g \cos \gamma = 0. \tag{6.3b}$$

With the origin of the x-axis at the bottom of the section, integration of eqn (6.3b) gives

$$P = (h - x)\rho g \cos \gamma + P_0, \tag{6.4}$$

where h is the depth measured in the direction of the x-axis and P_0 is the atmospheric pressure.[12]

So far, the slope of the s-axis could have any value between the slope of the surface and that of the bottom. Bélanger ultimately placed the s-axis at the bottom (running through

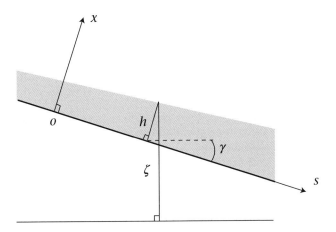

Fig. 6.1.   The geometrical parameters for slowly-varying flow in an open channel.

[10]Bélanger [1828] p. 5.

[11]This assumption is clearly valid in the uniform case, since the frictional force is then balanced by the gravitational force, which is a constant.

[12]Bélanger [1828] p. 8.

the lowest point of each section). For any particle of a given fluid section, the equation of motion (6.3a) then gives

$$-\frac{dh}{ds}\cos\gamma + \sin\gamma - \frac{\chi}{\rho Sg}F_v = \frac{1}{g}\frac{dv}{dt}.\tag{6.5}$$

Denoting by i the slope $\sin\gamma$ of the bottom, Bélanger obtained his backwater equation:

$$i\,ds - \sqrt{1-i^2}\,dh - \frac{\chi}{\rho Sg}(av + bv^2)\,ds + \frac{Q^2}{gS^3}\,dS = 0,\tag{6.6}$$

the last term of which comes from the identities $dv/dt = d(Q/S)/dt = -(Q/S^2)dS/dt$ and $dS/dt = (dS/ds)v = (dS/ds)Q/S$. This equation completely determines the back-water curve if the variation of the fluid section S with the depth h and the distance s is known.[13]

Bélanger provided a stepwise integration of this equation in the simple case of the horizontal aqueduct which had been built recently to bring the waters of the River Ourcq into Paris. In this case, the practical question was the height that the water must have at the beginning of the aqueduct for a given height at the end. Bélanger also gave a few examples of calculations of the backwaters before a weir, with the navigability of rivers in mind.[14]

6.1.3   Hydraulic jumps

In most practical cases, the values of the parameters in eqn (6.6) allow integration that does not conflict with the starting assumption of a slow variation of the depth h. However, Bélanger noticed the possibility of different behaviors. Consider the case of a straight canal with a wide rectangular section and with a purely quadratic friction. Denote by q the discharge per unit breadth, h_0 the depth $(bq^2/\rho gi)^{1/3}$ that the flow would have in the uniform case, and h_c the depth $(q^2/g)^{1/3}$. In these terms, eqn (6.6) takes the simple form

$$\frac{dh}{ds} = \frac{h^3 - h_0^3}{h^3 - h_c^3}\,\mathrm{tg}\gamma.\tag{6.7}$$

Although this equation can be integrated explicitly, the variations of h are more conveniently inferred from the sign that the derivative dh/ds takes according to the relative values of h, h_0, and h_c. In the frequent case of a swell ($h > h_0$) on a small-sloped bed ($h_0 > h_c$), the curve $h(s)$ is concave and has an upstream asymptote parallel to the bed and a horizontal downstream asymptote (see Fig. 6.2(a)). This means that the flow is asymptotically uniform in the upstream direction and then swells owing to a downstream cause, which could be a weir or the merging into a lake.[15]

---

[13]Bélanger [1828] p. 10.   [14]*Ibid.* pp. 11–28.
[15]Cf. Bresse [1860] pp. 218–30, Flamant [1891] pp. 237, 263–4, Forchheimer [1927] pp. 181–3.

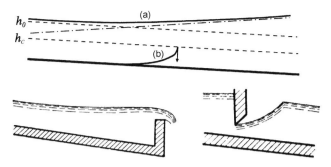

Fig. 6.2.   The backwater curves in the small-sloped case, and their concrete realizations according to Forchheimer [1927] p. 181.

In the case $h < h_c < h_0$, which would occur when water is forced through a sluice gate
into a small-sloped channel or when a high-sloped channel turns into a small-sloped one,
the depth increases in the downstream direction until it reaches the critical value h_c for
which the slope dh/ds becomes infinite (see Fig. 6.2(b)). The part of the curve close to this
critical point cannot be trusted, for it contradicts the approximation of parallel-slice flow.
Bélanger surmised that in this case the water level would suddenly increase to a value
higher than critical, and then again vary smoothly according to eqn (6.7). He identified this
behavior with the 'hydraulic jumps' that the Italian hydraulician Giorgio Bidone had
studied in the 1820s.[16]
   In order to determine the height of the jump, Bélanger appealed to the theorem of live
forces which his colleague Claude-Louis Navier and his friend Gaspard Coriolis had been
applying to the theory of machines. In Coriolis's statement of this principle, the variation
of live force of a mechanical system during a given time must be equal to the work of the
forces acting on the system during this time. Bélanger considered a portion of fluid
delimited by two planes perpendicular to the bottom and situated before and after the
jump. Denote by ζ and ζ' the surface heights (measured from a fixed horizontal plane) in
these planes, v and v' the corresponding fluid velocities, and z_G and z'_G the heights of their
gravity centers (see Fig. 6.3). During a time dt, the live force of the portion of fluid varies
by $(1/2)\rho(v'^2 - v^2)Qdt$. The work of the pressures acting on the sections during the same
time is $\rho g(\zeta - z_G - \zeta' + z'_G)Qdt$, because the pressures vary hydrostatically in the two
sections. The corresponding work of gravity is $\rho g(z_G - z'_G)Qdt$. Neglecting the work of
frictional forces, the theorem of live forces gives

$$\zeta' - \zeta = \frac{v^2}{2g} - \frac{v'^2}{2g}. \tag{6.8}$$

In the parlance of hydraulicians, the jump equals the decrease of the velocity head.[17]

[16]Bélanger [1828] pp. 29–31; Bidone [1820]. Cf. Rouse and Ince [1957] pp. 143–4, 149.

[17]Bélanger [1828] p. 32.

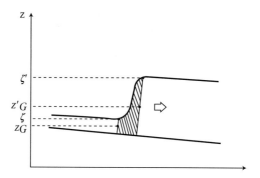

Fig. 6.3.   Geometrical parameters for a hydraulic jump.

6.1.4   *Backwater energetics*

As Bélanger later realized, this reasoning errs by ignoring the loss of live force that the abrupt change of motion implies.[18] For a gradually-varying flow, the balance of live force seems to apply to an infinitesimal change $d\zeta$ of surface height over the infinitesimal distance ds. Taking into account the work $-(\chi/S)F_v Q dt$ of frictional forces, this gives the equation

$$d\zeta + \frac{\chi}{S\rho g} F_v \, ds + d\left(\frac{v^2}{2g}\right) = 0, \qquad (6.9)$$

which is a simpler form of Bélanger's backwater equation, because of the relations $d\zeta = -i\,ds + dh\,\sqrt{1 - i^2}$ and $v = Q/S$ (see Fig. 6.1). Poncelet and Navier obtained the backwater equation in this manner, the former in the same year as Bélanger.[19]

   In 1836, Pierre Vaulthier, who only knew Bélanger's proof, obtained the Poncelet form of the backwater equation by suspiciously simple reasoning. The surface of the water in equilibrium, Vaulthier reasoned, is horizontal. Then the immediate cause of the flow in a channel must be the slope of the surface. In the uniform case, the descent $-d\zeta$ of the water surface is given empirically by the Prony formula $(\chi/\rho gS)(av + bv^2)\,ds$. For a frictionless fluid it should be equal to the variation $d(v^2/2g)$ of the velocity head, by analogy with frictionless fall on an inclined plane. Vaulthier simply added these two contributions to get the backwater equation. His main service to the subject was not this dubious proof, but his many applications of the backwater equation at a time when French engineers busied themselves with the improvement of rivers.[20]

[18]Bélanger [1841/42] pp. 94–6. Bélanger obtained the correct expression for the jump by equating the resultant of the pressures acting on the sides of the fluid portion to the variation of its momentum.

[19]Poncelet [1836] pp. 66n; Navier [1838] p. 190. Cf. Saint-Venant [1887c] p. 158. Bélanger's procedure of 1828 can also lead to the Poncelet form of the backwater equation if the *s*-axis is made parallel to the water surface.

[20]Vaulthier [1836] pp. 241–313. Cf. Saint-Venant [1887c] pp. 157–8.

Upon reading Vaulthier, Coriolis offered his own derivation of the backwater formula, which long remained the standard one. 'The question of the figure of back-waters', Coriolis declared, 'is the most important question that theoretical hydraulics presents to the engineer.' He meant that, whereas the known laws of efflux and uniform flow were essentially empirical, the backwater problem admitted a theoretical solution. He nonetheless disagreed with Bélanger's treatment, which he misread as an incorrect application of the theorem of live forces, and with Vaulthier's, which ignored the principles of mechanics. Even Poncelet's treatment fell short of satisfying Coriolis, for it maintained Bélanger's provisional assumption of uniform velocity in a given section of the flow.[21]

Avoiding the latter restriction (but still neglecting the curvature of the lines of flow), Coriolis expressed the variation dT of live force of a slice of fluid in the time dt and the work W_G of the gravity force and the work W_P of the pressure force in the same time as

$$dT = dt \, \Delta \int\int \rho v \frac{v^2}{2} \, dS,$$

$$W_G = -dt \, \Delta \int\int \rho v g z \, dS, \qquad (6.10)$$

$$W_P = -dt \, \Delta \int\int v[P_0 + \rho g(\zeta - z)] \, dS,$$

where Δ denotes the difference between the values that the expression following it takes on the two sides of the slice, z denotes the height of a point of the fluid section, and the integrals are performed over this section. For the work of frictional forces, Coriolis simply assumed an external friction with an effective sliding velocity equal to the average velocity. The resulting backwater formula is

$$I = -\frac{d\zeta}{ds} = \frac{\chi}{S} \frac{F_U}{\rho g} + \frac{d}{ds}\left(\alpha \frac{U^2}{2g}\right), \qquad (6.11)$$

where U denotes the average velocity of the fluid through a given section, and

$$\alpha = \frac{1}{S} \int\int \left(\frac{v}{U}\right)^3 dS. \qquad (6.12)$$

This equation only differs from Vaulthier's by the introduction of the coefficient α, which Coriolis determined from Du Buat's old measurements of the velocity profile.[22]

6.1.5 *Rivers and torrents*

There is more to say about the critical depth h_c. In 1851, Saint-Venant noted that the form of the backwater curves below and above the critical depth corresponded to tumultuous

[21]Coriolis [1836] p. 314. Cf. Saint-Venant [1887c] pp. 158–67, Rouse and Ince [1957] pp. 150–1.

[22]Coriolis [1836] p. 318.

and tranquil flows, respectively. The $h < h_c$ case defined '*torrents, the various parts of which seem to flow independently of each other and whose acquired velocity allows them to flow over small obstacles.*' The $h > h_c$ case defined '*rivers or quiet streams whose successive slices press on each other and move along together, so that they can only get over obstacles by means of the weight of the elevated water and so that every elevation in one part is felt in the upstream direction to a finite distance.*'[23]

For a rectangular canal of small slope, the condition $h \approx h_c$ for the possibility of a jump is equivalent to $U \approx \sqrt{gh}$. As Saint-Venant noted in 1870, this means that the velocity of the water is the same as the propagation velocity of a small swell as given by Lagrange. In a hydrostatic canal closed by two distant gates, with a rise of the water level obtained by constantly feeding water at one of the gates, the higher level propagates as a step along the canal with the Lagrangian celerity $\sqrt{gh}$, as indicated in Fig. 6.4. A small hydraulic jump can be obtained by superposing with this motion a constant flow at the velocity $-\sqrt{gh}$. As we saw in Chapter 2, Saint-Venant used this remarkable connection between jumps and waves to confirm his distinction between rivers and torrents. In a stream slower than $\sqrt{gh}$, the swells created by an obstacle must propagate in the upstream direction, so that water accumulates before it can pass the obstacle. This is the case of a river. In a stream faster than $\sqrt{gh}$, the water can pass obstacles without previous accumulation. This is the case of a torrent.[24]

With their sophisticated analysis of backwaters, hydraulic jumps, and critical depths, French hydraulicians could pride themselves on having transcended the more empirical approach of their predecessors. Yet they did not base their theories on the fundamental hydrodynamics of Euler and d'Alembert, which famously failed in most concrete problems. Following a via media between pure empiricism and fundamental deduction, they developed effective theories that exploited the principles of mechanics but required some experimental input and various theoretical idealizations.

The theorists of pipes and open channels obviously required empirical knowledge of the retarding action of the walls, and also of the transverse velocity profile in Coriolis's case. In a first idealization, they assumed the retarding effect of the walls to be transmitted uniformly to the inner filaments of the fluid by an unknown mechanism roughly independent of the varied character of the flow. Without knowledge of this mechanism, the velocity profile could not be derived. Most authors assumed approximate uniformity of

Fig. 6.4.   The progression (thin arrow) of a swell produced by feeding additional water (thick arrow) at one gate of a hydrostatic canal.

[23]Saint-Venant [1851*a*] p. 319.   [24]Saint-Venant [1870] pp. 186–95. See Chapter 2, p. 82.

the fluid velocity within a given section. Although Coriolis avoided this restriction, he still neglected the curvature of the fluid filaments and the resulting centrifugal force.

There was yet another simplification, so obvious that no one cared to mention it. The deductions of formulas for backwaters and hydraulic jumps all rested on the assumption that the only relevant motion was the average, macroscopic motion measured by standard gauging methods. Their authors must have been aware of the temporal and spatial irregularities constantly encountered in hydraulic experiments. However, they did not suspect that these irregularities could affect the average flow in pipes or channels of slowly-varying slope and section.

6.2 Saint-Venant on tumultuous waters

6.2.1 *Tumultuous flow*

Not all French engineer-mathematicians of this period confined themselves to a semi-empirical approach to hydraulics. An early exception was Navier, who in 1822 derived a differential equation of fluid motion based on a simple molecular assumption and intended to describe the behavior of real, viscous fluids. As we saw in Chapter 2, Navier quickly realized that his equation only applied to 'linear motions' (we would say laminar) and not to 'the more complex motions' occurring in typical hydraulic systems. We also saw that in 1834 his former student and admirer Saint-Venant had begun to develop an equally fundamental approach.[25]

Saint-Venant clearly distinguished two scales, namely, a larger scale at which the average velocity varies smoothly in space and time, and a smaller scale at which the motion can be highly irregular. In his derivation of the Navier–Stokes equation, he used volume elements that included 'the case when partial irregularities of the fluid motion force us to take faces of a certain extension so as to have regularly varying averages.' The effective viscosity parameter ε defined at this scale depended on the irregular motions at the smaller scale. Thus it could vary from one point of the fluid to another, and from one kind of flow to another.[26]

Among the irregularities of motion on which the variable ε depended, Saint-Venant included the undulations of molecular paths he had described in his memoir of 1834 as being caused by the sliding of successive fluid layers over each other. In a letter to Pierre Boileau of March 1846, he evoked the further possibility that 'the internal friction coefficient may vary with the general dimensions of the current and with the freedom that oblique motions and eddies thus have to develop and to disseminate live force, as you have very well said.'[27]

Leonardo da Vinci and Daniel Bernoulli had long ago noted the whirling motions induced by the sudden enlargement of a pipe or by obstacles. As we saw in Chapter 3, in 1799 the Italian hydraulician Giovanni Battista Venturi pleaded for a more realistic

---

[25]Navier [1823c] pp. 389–440.   [26]Saint-Venant [1843c] p. 1242n.

[27]Saint-Venant to Boileau, 29 Mar. 1846, Fond Saint-Venant, Ecole Polytechnique. Boileau ([1846] p. 215) had written: 'The viscosity of liquids seems to play [in the retardation of the upper fluid layers] a more important and more complex role than has been admitted by geometers, by giving birth to molecular motions oblique to the stream and by disseminating the live force of the fluid filaments ... in a manner related to their mutual friction.'

conception of fluid motion in which 'the lateral communication of motion' and the resultant eddies played an essential role. According to Venturi, 'the eddies of the water in rivers are produced by motion, communicated from the more rapid parts of the stream to the lateral parts, which are less rapidly moved.' They contribute to the retardation of the flow:[28]

> One of the principal and most frequent causes of retardation in a river is produced by the eddies incessantly formed in the dilations of the bed, the cavities of the bottom, the inequalities of the banks, the bends or windings of its course, the criss-crossing currents, and the streams that intersect with different velocities.

In a study of pressure losses in the pipes of steam engines published in 1838, Saint-Venant emphasized the role of '*extraordinary friction*, usually called *loss of live force*, and determined by the eddying of fluids especially at points where the section of the flow suddenly increases.' The following year, the military engineer Jean-Victor Poncelet published the second edition of his celebrated course 'for the artists and workers' of Metz, in which he gave much importance to the whirls observed during the sudden alteration of a flow. These motions, 'much more complicated than one usually thought', involved pulsations, intermittence, and the conversion of large-scale motions to smaller-scale ones, perhaps thus cascading to the molecular level:[29]

> Careful observation of the facts justifies the belief that independently of the gyratory motions shared by a whole portion of the fluid mass, there are also secondary or less apparent motions that involve smaller groups of molecules and develop in the intervals between the former motions.... We may further assume without much risk that the motions of rotation or oscillation thus impressed on individual molecules or on the smallest groups of molecules are, in addition to adhesion and cohesion ... one of the most important causes of the loss of motion in fluids and especially of the resistance that their stream lines experience when gliding on each other or on the surface of solids.

Poncelet thus provided the mechanism through which Joule and others later interpreted the dissipation of macroscopic motion into heat. He even explained Brownian motion as a consequence of the ensuing molecular agitation, instead of the vitality of organic particles imagined by naturalists. Less speculatively, he regarded the formation of eddies as 'one of the means that nature uses to extinguish, or rather to dissimulate the live force in the sudden changes of motion of fluids, as the vibratory motion themselves are another cause of its dissipation, of its dissemination in solids.' He also believed that smaller-scale whirling largely contributed to the effective friction between fluid filaments.[30]

Saint-Venant approved these considerations and brought them to bear on hydraulic problems. In 1846, he examined Borda's old formula for the loss of head during a sudden enlargement of a pipe. 'The molecular gearing [*engrènement moléculaire*]', he wrote, 'creates whirls and other non-translatory motions indicated by D. Bernoulli and by M. Poncelet, which, after being conserved for some time in the fluid, end up being

---

[28]Venturi [1797] prop. X. Cf. Rouse and Ince [1957] pp. 134–7. See also Chapter 3, pp. 105–8.

[29]Saint-Venant [1838] p. 47; Poncelet [1839] pp. 529–30.

[30]*Ibid.*

dissipated under the effect of friction and extraordinary resistance.' He then offered a simple derivation of Borda's formula, based on the balance of live forces in a reference system moving at the final velocity of the fluid.[31]

In the same year, Saint-Venant also considered the old, difficult problem of fluid resistance. As we saw in Chapter 3, he related the resistance to the live force of the non-translatory fluid motions induced by the immersed body. When tumultuous, whirling motion occurs at the rear of the body, the resistance largely exceeds the value it would have for a perfectly smooth flow.[32]

6.2.2 The effective viscosity

In a memoir of 1851 on retardation formulas and backwater tables, Saint-Venant publicized the idea of small-scale tumultuous motions being responsible for the variable ε that he had championed since 1834:[33]

> Newton's hypothesis, as reproduced by MM. Navier and Poisson, consists in making internal friction proportional to the relative velocity of the filaments sliding on one another; if it can be approximately applied to the various points of the same fluid section, every known fact indicates that the proportionality *coefficient* must increase with the dimensions of transverse sections; which may be to some extent explained by noticing that the filaments do not proceed in parallel directions with a regular gradation of velocity, and that the *ruptures*, the whirls and other complex and oblique motions that must considerably influence the intensity of friction develop better and faster in large sections.

Saint-Venant found much evidence for this view in experimental studies of open-channel flow. Boileau's contribution has already been mentioned in Chapter 3. In 1868, the American hydraulicians Andrew Humphreys and Henry Abbot published the results of their measurements and observations on the Mississippi River. From these sources, Saint-Venant extracted the contrast between small-scale disorder and large-scale order that justified the effective ε approach:[34]

> Beyond this disorder [in the local fluid motion], as was especially noted by [Captain Boileau] and as has been observed in larger masses by American engineers [Humphreys and Abbot], a certain order is nevertheless observed; for the same particularities of the velocity of the fluid quickly repeat themselves everywhere, so that the motion, if determined by constant causes, settles up by *periodicity* [Humphreys and Abbot's 'river pulses'] ... and the effective velocities undergo complex but small oscillations around constant averages relative to each point. These *local* average velocities for fluid translation or transport are 'those measured by floats and other hydrometric instruments and they determine the flows to be computed.'

---

[31]Saint-Venant [1846a] p. 147. Saint-Venant's reasoning was a simplification of an earlier reasoning by Bélanger that combined momentum and live-force balance in the natural reference system. Rather than directly estimating the live force lost to whirls, Bélanger and Saint-Venant assumed the pressure on the walls of the expanding part of the pipe to vary hydrostatically.

[32]Saint-Venant [1846a] pp. 28, 72–8, 120–1.   [33]Saint-Venant [1851a] p. 229.

[34]Humphreys and Abbot [1868] pp. 165–94; Saint-Venant [1872] p. 650.

6.2.3   Darcy's and Bazin's measurements

In 1857, still another engineer at the Ponts et Chaussées, Henry Darcy, published system-atic hydraulic measurements. During the construction of a new water supply system for the fountains of Dijon, Darcy had found faults in Prony's old retardation formula. His appointment as head of the municipal water service of Paris gave him the opportunity to perform experiments on pipes of various diameters and wall structure. These were by far the most extensive and sophisticated measurements of this kind since Bossut's and Du Buat's. Unlike his predecessors, Darcy used graphical plots as well as the method of least squares to find the best-fitting resistance laws. His most important finding was that retardation depended on the roughness of the walls. Du Buat had excluded such an effect, because he believed that an adhering layer of fluid prepared a smooth surface for the gliding fluid.[35]

Darcy further demonstrated that the friction per unit surface depended on the diameter D of the tube according to an $\alpha + \beta/D$ law, in conformance with the natural expectation that the effect of roughness should be more important for small pipes than for large pipes. Although he did not offer any precise theory, he suggested that the roughness of the wall played an essential role in determining the nature of the flow. He rejected the usual parallel-filament picture, because asperities on the wall implied 'gyratory motion by molecular groupings' and a concomitant loss of live force for the progressive motion. Far from the walls, or for a smooth wall, he believed that undulations or oscillations of eventual fluid filaments would also trigger gyratory motion at the surface of mutual contact.[36]

In this view, as in Saint-Venant's, the velocity profile had special theoretical interest. Darcy measured it with unprecedented accuracy thanks to an improved Pitot tube, and found a semi-cubic velocity profile ($v \propto r^{3/2}$) for a circular section.[37] For this profile, the dynamical equilibrium of successive cylindrical layers of the fluid may be obtained by making the internal friction proportional to the square of the velocity gradient. Darcy favored this odd theoretical choice 'against the opinion of several eminent hydraulicians'. Among the eminences, he mentioned Saint-Venant for his notion of a Newtonian internal friction (proportional to the velocity gradient) with a coefficient depending on ruptures and whirling motions.[38]

After completing his work on pipes, Darcy turned to open channels. He planned experiments on the canal of Burgundy and its reaches, with generous funding by the

[35]Darcy [1856], [1857]; Du Buat [1786] vol. 1, p. 41: 'Considering how water itself prepares the surface on which it flows, we see that the difference of the matters of which the wall may be composed cannot have a truly sensible effect on the resistance.' Cf. Rouse and Ince [1957] pp. 169–173, Brown, Garbrecht, and Hager [2003].

[36]Darcy [1857] pp. vi, 10, 188–93, 202–19. That Prony had failed to detect this dependence in the data accumulated by Couplet, Bossut, and Du Buat did not worry Darcy, because the larger pipes of Couplet's experiments were older and therefore rougher than the smaller pipes of Bossut and Du Buat.

[37]Darcy [1857] p. 128. The original Pitot tube, invented by Henri de Pitot in 1732 (see Rouse and Ince [1957] pp. 115–16) was made of two parallel glass tubes, one being straight, and the other bent through a right angle at its lower end; the water-level difference in the two tubes after vertical immersion in a stream gives the pressure at their lower end.

[38]Darcy [1857] pp. 181–2. Bazin's assumption for the internal shear stress is equivalent to a constant mixing length in Prandtl's later theory of turbulent flow (see Chapter 7, pp. 297–9). It does not really contradict Saint-Venant's theory, for the effective viscosity ε may depend on the velocity gradient. Saint-Venant ([1869] p. 585) welcomed Bazin's suggestion, but approved Maurice Lévy's rejection of generalizations that contradicted the form $\varepsilon(\partial_i v_j + \partial_j v_i)$ of the stress system.

Ministry of Public Works. After his sudden death in 1858, his gifted disciple Henri Bazin completed the measurements and published them in 1865. As in the pipe case, Darcy and Bazin found that the roughness of the walls played an important role. They noted the pronounced irregularities of motion, 'the very sudden jolts' and 'ruptures' that André Baumgarten (from the Ponts et Chaussées) had long ago observed on the River Garonne with Reinhardt Woltman's velocity-measuring mill. They determined the velocity profile and found it to be quadratic in the case of a wide rectangular section, and cubic in the case of a semicircular section.[39]

Bazin also studied varied flow, thus partly confirming previous backwater and jump theories, but also pointing to discrepancies and new phenomena. For example, he described the undulations of the jumps occurring in small-sloped channels, and he found the height of the jumps to be generally smaller than the change in velocity head, owing to 'losses by tumultuous motion'. Lastly, Bazin provided much data on non-permanent flow, including solitary waves and tidal bores, for which no satisfactory theory yet existed. For about half a century, Darcy's and Bazin's measurements remained the most reliable hydraulic data in France and abroad. Their wealth of new regularities and phenomena defied theory, Bazin thought: 'Maybe such a delicate part of science must long remain in the realm of experiment.'[40]

6.3   Boussinesq on open channels

6.3.1   *Terrestrial physics*

Among those undeterred by this pessimistic forecast was Saint-Venant's protégé Joseph Boussinesq. Born in Hérault and from a family of small farmers, Boussinesq became a high-school *surveillant* and teacher. In this position he found time to take analysis and mechanics at the University of Montpellier and to study works of higher analysis by himself. He was mainly self-taught, which makes his writings sometimes difficult to penetrate. Saint-Venant noticed him in 1867, upon reading his memoir on anisotropy induced by compression. The two men had a common interest in applied mechanics and also in religion. Their correspondence covers both topics with equal prolixity. Saint-Venant knew that a religious provincial without higher academic training had little chance to gain recognition. With his usual generosity, he supported Boussinesq so efficiently as to win him a chair at the University of Lille in 1873, the chair of experimental and physical mechanics at the Sorbonne in 1885, and election to the Academy of Sciences in 1886.[41]

Boussinesq defined 'the aim of his life' as 'the study of mathematics as they came alive in Creation, or, if you prefer, the study of the traces left in nature by the geometer who organized her when he produced her.' He shared Saint-Venant's dislike for the abstractions of rational mechanics, and worked hard to develop a 'physical mechanics' based on a more realistic, molecular conception of matter. According to Saint-Venant, the latter

---

[39]Darcy and Bazin [1865a] pp. 23–5; Baumgarten [1847]. These velocity profiles, together with the finite slip at the walls, can now be seen as approximations of the logarithmic profiles given by turbulent boundary-layer theory, cf. Prandtl [1933] p. 833n.

[40]Darcy and Bazin [1865a] pp. 30–7, [1865b].

[41]Cf. Le Tourneur [1954], Picard [1933], Douysset [undated], Blaquière [1931].

mechanics, developed in the 1820s and 1830s by Poisson, Navier, Coriolis, Poncelet, Bélanger, and himself, later met considerable skepticism from the powerful Joseph Bertrand and other academicians, who pursued 'an absolute and immediate rigor that prohibited any application of analysis to phenomena, even in celestial mechanics.'[42]

Saint-Venant presented Boussinesq as the savior of the true, physical, 'terrestrial', or 'intimate' mechanics:[43]

> The sterility that seemed to strike [this mechanics] some twenty years ago and the resulting skepticism in some excellent minds only derived from the manner of formulating questions; one tried to solve the general problem of each science, dealing with it in all generality and all inextricable complication, whereas what needed to be done was to seek a simplifying cause to permit simple, approximate laws or give a handle for successive approximations, without which celestial mechanics itself could not have been built.

Laplace and Poisson excepted, the promoters of physical mechanics had all been engineers trained at the Ecole Polytechnique and one of the *Ecoles d'application*. Boussinesq was not. Saint-Venant nevertheless found him to possess a strong sense of the concrete:[44]

> Very well versed in analysis ... his positive spirit yet never rests in abstractions; even though he never went through any engineering school, he has passion for the real, the concrete as presented by our terrestrial world, which is certainly more difficult to know than the planetary world. The general law of the latter is already known, whereas the study of the other requires that a law be extracted for each subject, and a constant effort to discern the relative magnitudes and orders of approximation.

6.3.2 *Eaux courantes*

Boussinesq's biographers report that his interest in hydraulics started early, with the contemplation of whirls and waves during walks along the River Hérault with his high-school teacher. Some of his early works, published in the *Journal de mathématiques pures et appliquées*, bore on internal fluid friction. They display the combination of analytical power and concrete sense praised by Saint-Venant. For example, Boussinesq solved the problem of the Poiseuille flow in pipes of elliptical, rectangular, and triangular sections, and applied this kind of flow to the phenomena of infiltration and transpiration (in Graham's sense). Before William Thomson, who is usually credited for this discovery, he derived the helicoidal nature of the flow in a curved tube or channel, and used it to explain the evolution of meanders in rivers.[45]

In these studies Boussinesq confined himself to laminar flow in small sections. He nonetheless gave the most striking illustration of the failure of this assumption in hydraulic pipes and channels: regular flow at ordinary temperature in a semicircular channel would yield, for a diameter of one meter and a slope of 10^{-4}, the absurdly high velocity of 187 m/s

---

[42]Boussinesq to Saint-Venant, quoted in Picard [1933] p. 14; Boussinesq to Saint-Venant, 21 Apr. 1876, Bibliothèque de l'Institut.

[43]Saint-Venant [1876].   [44]Saint-Venant [1880] p. 23.

[45]Cf. Douysset [undated] (Hérault), Boussinesq [1883].

for the central fluid filament. In harmony with his mentor's view, Boussinesq concluded that in this case irregular transverse motions induced a much higher effective velocity than occurs in regular flow. Aware of Darcy's and Bazin's relevant measurements, he began a thorough investigation of open-channel flow that took the whirling agitation of the water into account.[46]

Boussinesq borrowed from Saint-Venant the fundamental idea of a large-scale, effective viscosity that depended on the 'intensity of the whirling agitation'. Saint-Venant judged the distribution and effect of this agitation to be 'a hopeless enigma'. Similarly, Bazin concluded his hydraulic research with the pessimistic words:

> The question grows more complex and obscure just when new, numerous, and precise experiments would be expected to throw a brighter light … We do not yet have reasonable notions about the internal motions of fluids and the mutual actions of their molecules.

Boussinesq cut the Gordian knot by guessing the form of the effective viscosity and verifying observable consequences of the guess.[47]

According to Boussinesq's intuition, whirling originates in the macroscopic fluid slide on an unavoidably irregular wall and then propagates through the rest of the fluid, with an intensity depending globally on the breadth of the flow and locally on the distance from the wall. The effective viscosity then results from the additional momentum exchange that local agitation implies between successive fluid filaments. In the case of a wide rectangular channel, Boussinesq assumed an effective viscosity proportional to the sliding velocity v_0 on the walls and to the depth h of the water. In the case of a circular channel, he took it to be proportional to the sliding velocity on the walls, to the radius R of the channel, and to the ratio r / R of the distance from the axis of the channel to its radius.[48]

Boussinesq substituted these values into the Navier–Stokes equation, which he carefully rederived for large-scale, secular motions, in a manner reminiscent of Saint-Venant's reasoning of 1843. For the boundary conditions, Boussinesq assumed a friction proportional to the square of the velocity slip on the bottom and a vanishing stress component on the surface. He then derived and discussed relations between surface slope and discharge in the following cases of increasing difficulty: permanent and uniform, permanent and slowly varying, permanent and quickly varying, and non-permanent. The consideration of the last two cases was new, except for some anticipations by Saint-Venant. The consideration of the first two cases led to important corrections to Bélanger's and Coriolis's theories.[49]

For a better control of approximations, Boussinesq generally started from the Navier–Stokes equation in an adequate coordinate system and integrated over a fluid section. The following is a sketch of simpler but less rigorous deductions based on a momentum

---

[46]Boussinesq [1868].

[47]Boussinesq [1870], [1871*b*], [1872*c*]; Saint-Venant [1872] p. 774; Darcy and Bazin [1865*a*] p. 30.

[48]Boussinesq [1877] pp. 45–51.   [49]Cf. Saint-Venant [1873].

balance for fluid elements or slices. Although Boussinesq knew of such deductions, Saint-Venant was the first to publish them in full in a suggestive memoir of 1887.[50]

6.3.3 Coriolis revisited

In the permanent, uniform case, the main novelty of Boussinesq's approach was the determination of the velocity profile. For a wide rectangular channel, the viscosity ε is a constant which is proportional to the depth and to the sliding velocity at the bottom. Balancing the weight of a truncated fluid filament with the stresses acting on its lower and upper sides (the pressures on the right and left sections cancel each other), Boussinesq found that

$$-\varepsilon \frac{d^2 v}{dx^2} = \rho g i, \tag{6.13}$$

where x is the distance of the fluid filament from the surface and i is the slope. This implies a quadratic velocity profile, in conformance with Bazin's (still partial) measurements (see Fig. 6.5).[51]

Denote by v_0 the velocity on the bottom, and h the distance between the bottom and the surface. The expression $\varepsilon = \rho g A h v_0$ for the viscosity, the formula $F = \rho g B v_0^2$ for the friction on the bottom, the second-order equation (6.13), and the boundary conditions $v'(0) = 0$ and, $-\varepsilon v'(h) = F$ lead to the formula

$$\frac{v}{v_0} = 1 + \frac{B}{2A}\left(1 - \frac{x^2}{h^2}\right). \tag{6.14}$$

Hence the average velocity U is proportional to the velocity v_0, which is itself related to the slope i by the formula $\rho g h i = F = \rho g B v_0^2$ which expresses the balance between the weight of a normal slice of fluid and the external friction F. In conclusion, the Chézy formula (6.1) still holds, despite the transverse variation of the velocity.[52]

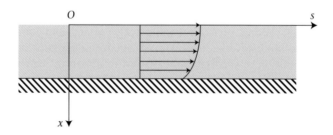

Fig. 6.5.   Boussinesq's velocity profile (horizontal arrows) for permanent flow in a wide rectangular channel.

[50]Saint-Venant [1887c] pp. 168–76. See also the simplified deductions for the case of a broad, rectangular channel in Boussinesq [1897].

[51]Boussinesq [1877] p. 72.   [52]Ibid. p. 73.

In the permanent, slowly-varying case, consider a slice ds of fluid perpendicular to the s-axis, which is chosen to be parallel to the surface. The s-component of the weight of this slice is $\rho g S I\, ds$, where S is the fluid section and I is the slope of the surface. The pressures on both sides mutually cancel (up to a higher-order term), since their transverse variation is hydrostatic. The frictional force is $-ds \int F_{\bar{v}}\, d\chi$, where $F_{\bar{v}}$ denotes the wall friction per unit surface area for the shift velocity $\bar{v}$ and the integration runs over the wetted perimeter χ (along which $\bar{v}$ may vary). The momentum variation per unit time is $-d \iint v(\rho v\, dS)$, where the integration is performed over the section S. Denoting by η the number such that

$$\iint v^2\, dS = (1+\eta)SU^2,\tag{6.15}$$

and neglecting the variation of this number with s, the dynamical equilibrium of the slice reduces to

$$I = \frac{1}{S\rho g}\int F_v\, d\chi + (1+\eta)\frac{d}{ds}\left(\frac{U^2}{2g}\right).\tag{6.16}$$

This equation differs from Coriolis's equation (6.11) in two ways. Firstly, the coefficient $1+\eta$ differs from Coriolis's α. Secondly, the velocity $\bar{v}$ in the frictional force differs from the average velocity U.[53]

The extent of the latter difference can be derived from the velocity profile. By a reasoning similar to that used in the uniform case, for a wide rectangular channel this profile satisfies the equation

$$-\varepsilon\frac{d^2v}{dx^2} = \rho g I + \rho\dot{v},\tag{6.17}$$

where $\dot{v}$ denotes the acceleration of a fluid particle that results from the variation of the fluid section. This acceleration is given by $-v^2 h^{-1} dh/ds$ in the approximation for which the slope of the fluid filaments varies linearly between the bottom and the surface. Since it is not uniform in a given section of the fluid, a distortion of the velocity profile results. Boussinesq determined this distortion approximately by replacing the acceleration with the value it would have if the velocity profile were still parabolic. The resulting difference between Boussinesq's and Coriolis's frictional terms has the same form as the inertial term, so that eqn (6.16) can be rewritten as[54]

$$I = \frac{\chi}{S\rho g}F_U + (1+\eta+\beta)\frac{d}{ds}\left(\frac{U^2}{2g}\right).\tag{6.18}$$

This equation has exactly the same form as Coriolis's, but with a new value for the numerical coefficient of the gradient of the velocity head. As Boussinesq emphasized, this similarity of form should not be mistaken for a justification of Coriolis's method. On the

---

[53]*Ibid.* p. 66.   [54]Boussinesq [1877] pp. 92, 112.

contrary, the new derivation makes it clear that Coriolis erred in equating the work of pressure, gravity, and frictional forces to the increase of *large-scale* kinetic energy; whenever the fluid section changes, part of the energy of the large-scale motion goes to smaller-scale, eddying motion. Coriolis also underestimated the error he committed in assuming the same expression for the work of frictional forces in the uniform and varying cases.[55]

Boussinesq then approached the more delicate problem of quickly-varying permanent motion. In this case, a non-negligible centrifugal force acts on the fluid filaments. To first order in the filaments' curvature, and for a wide rectangular channel, the resulting equation is

$$I = \frac{\chi F_U}{S\rho g} + (1 + \eta + \beta)\frac{\mathrm{d}}{\mathrm{d}s}\left(\frac{U^2}{2g}\right) - h^2\left[\frac{1}{3}\frac{\mathrm{d}^3}{\mathrm{d}s^3}\left(\frac{U^2}{2g}\right) + \frac{1}{2}\frac{U^2}{gh}\frac{\mathrm{d}^2 i}{\mathrm{d}s^2}\right]. \qquad (6.19)$$

Boussinesq deduced the profile of a hydraulic jump and confirmed Bazin's distinction between two sorts of jumps with or without long-range oscillations. He also refined Saint-Venant's distinction between torrents and rivers, now introducing an intermediate category of 'moderate torrents' for which jumps can occur, but only long and wavy ones. Lastly, Boussinesq obtained equations for non-permanent flow, and used them to discuss waves, river tides, tidal bores (*mascarets*), and floods.[56]

6.3.4 *Praise and neglect*

Boussinesq's theory of open channels appeared in several papers between 1870 and 1872, and formed, together with an extension to pipe flow, the substance of a long essay submitted in 1872 to the Académie des Sciences and published five years later in its *Mémoires des savants étrangers* together with Saint-Venant's laudatory report. Saint-Venant had himself introduced an equation for non-permanent flow in 1871 and had integrated it in a simple case of fluvial tide, though only for negligible curvature. Also, he had obtained the equation for rapidly-varying permanent flow as early as 1851. After his death, Boussinesq discovered and published the relevant manuscript as a homage to his mentor's modesty. Probably not to discourage his beloved disciple, Saint-Venant had hidden his priority on this important aspect of the theory of open channels.[57]

In his report on Boussinesq's essay, Saint-Venant emphasized its recourse to subtle methods of approximation and its agreement with Darcy's and Bazin's data:

> These numerous results of a high analysis, founded on a detailed discussion and on judicious comparisons of quantities of various orders of smallness, sometimes to be kept, sometimes to be neglected or abstracted, and their constant conformity with the results obtained by the most careful experimenters and observers, appear most remarkable to me.

---

[55]Boussinesq [1877] pp. 112–13.

[56]*Ibid.* pp. 193, 196–217, 242–529. There is also a fourth part which includes notes on efflux, weirs, turns, and some effects of capillarity.

[57]Boussinesq [1870], [1871*b*], [1872*c*], [1877]; Boussinesq's introduction to Saint-Venant [1887*b*] pp. 5–6 (hidden priority); Saint-Venant [1873], [1871*a*], [1851*b*].

Elsewhere, Saint-Venant praised 'the man who possessed intuition no less than high calculus, who knew how to invent new integrals for the needs of this *intimate* mechanics born in our century from our France, and pertaining to the real things of the terrestrial world that we inhabit.'[58]

The reception of Boussinesq's theory was not as warm as Saint-Venant's comments would suggest. Most hydraulicians could not follow its analytical sophistication and regarded the practical circumstances of hydraulics as too complex for precise quantitative analysis. For example, in his authoritative treatise of 1882, the French hydraulician Michel Graeff condemned the application of backwater theory to the navigability of rivers because of the complex and variable shape of real beds. After praising Saint-Venant's and Boussinesq's analytical skills, he put forward his own empirical methods. British and American hydraulicians ignored the higher French theories. Some German-language hydraulicians picked up on Boussinesq's theory. In his major treatise of 1914, the Austrian Philipp Forchheimer gave detailed accounts of this and earlier French theories, which Saint-Venant's disciple Alfred Flamant had made more accessible in his clear, pedagogical treatise of 1891. In 1920, Josef Koženy improved Boussinesq's theory of pipe flow by introducing a new form for the effective viscosity in circular pipes. Effective viscosity parameters remain a practically important approach to turbulent flow both in hydraulics and in atmospheric physics.[59]

For physicists, the most important aspect of Boussinesq's essay was the nonlinear theory of waves discussed in Chapter 2. However, this theory was confined to waves on calm water, for which the effects of internal friction are negligible. In general, Boussinesq's results depended little on his specific treatment of turbulence and internal friction. His equation for slowly-varying flow had exactly the same form as Coriolis's old equation, even though Coriolis's proof turned out to be unacceptable. For the curvature-dependent terms, Boussinesq mainly used an approximation in which the lack of uniformity of the velocity in a given fluid section was irrelevant. Consequently, Boussinesq's readers could accept and rederive his final equations for the shape of the water surface without paying attention to his innovative treatment of turbulence.[60]

6.4 The turbulent ether

6.4.1 *A hydraulic anomaly*

French hydraulics attracted little attention from Britons, who preferred more empirical methods. The Glasgow engineer James Thomson was in this respect atypical. In 1878, his reading of Boileau, Darcy, and Bazin prompted him to reflect on an anomaly emphasized by Captain Boileau, namely that the maximum velocity in a fluid section lay somewhat below the water surface, at variance with the 'laminar theory' based on mutual friction between successive fluid layers. In this context, Boileau introduced the 'oblique motions'

---

[58]Saint-Venant [1873] p. xxi; [1880] p. 16.

[59]Graeff [1882], vol. 2, p. 130; Forchheimer [1914] (re-edited in 1924 and 1930); Flamant [1891]; Koženy [1920] p. 31.

[60]In his *Hydrodynamics* (Lamb [1895], [1932] for the sixth edition), Horace Lamb only referred to Boussinesq for his derivation of the profile of solitary waves (Boussinesq [1871*a*]) and for his calculations of laminar flow in pipes of various sections (Boussinesq [1868]).

that met Saint-Venant's approval in the earlier cited letter. James Thomson similarly assumed transverse, 'commingling' motion of the water:[61]

> The laminar theory constitutes a very good representation of the viscid mode of motion; but it offers a very fallacious view of the motion in the flow of water in ordinary cases in which the inertia of the various parts of the fluid is not subordinated to the restraints of viscosity ... [In these cases], indefinite increase of velocity of the water situated in the interior of the current is prevented by continual transverse flows thereto, and commingling therewith, of portions of water already retarded through their having been lately in close proximity to the resisting channel face.

To explain Boileau's maximum-velocity anomaly, Thomson assumed that 'deadened' water from the bottom of the channel reached the fluid surface and slowed the laminar motion there. Gravel, mud, and weeds at the bottom, when present, were a possible cause of this effect. However, Thomson believed it also occurred for a smooth bed. After consultation with his brother William, he assumed a thin layer of dead water at the bottom, on which the next layer slid with a finite velocity. This motion, being unstable according to Helmholtz and William Thomson (Helmholtz–Kelvin instability), led to turbulent transverse fluxes.[62]

6.4.2   Turbulent rigidity

These reflections prompted William Thomson to investigate the distribution of the average velocity in the open channel. He was 'surprised to discover the seeming possibility of a law of propagation as of distortional waves in an elastic solid.' In a paper of 1887, he applied a mixture of line, surface, and volume averages, as well as Fourier analysis, to the plane, laminar disturbance of homogeneous, isotropic, turbulent flow. The following is a generalization of his arguments to an arbitrary disturbance, with the benefit of modern tensor notation.[63]

Thomson conceived the flow as a superposition of a large-scale regular ('laminar') component **u** and a small-scale turbulent component **v** assumed to be homogeneous and isotropic, in a sense that will become clear. The total flow obeys Euler's equation

$$\partial_t(u_i + v_i) = -\partial_i P - \partial_j[(u_i + v_i)(u_j + v_j)] \qquad (6.20)$$

in tensor notation and for unit density. The partial flows obey the incompressibility equations $\partial_i u_i = 0$ and $\partial_i v_i = 0$. Taking the average of Euler's equation over a volume that is small at the laminar scale and large at the turbulent scale, we obtain

$$\partial_t u_i = -\partial_i \bar{P} - \partial_j(u_i u_j) - \partial_j(\overline{v_i v_j}). \qquad (6.21)$$

The last term corresponds to a stress system $\overline{v_i v_j}$ (the Reynolds stress).[64]

---

[61]J. Thomson [1878] pp. 114, 117, 120 (quote); Boileau [1846]; Saint-Venant to Boileau, 29 Mar. 1846 (cited earlier on p. 229). J. Thomson's paper probably inaugurated the hydrodynamic meaning of 'laminar'. Boileau, unlike Thomson, made viscosity responsible for the oblique movements, see footnote 27.

[62]J. Thomson [1878] pp. 121, 124.   [63]W. Thomson [1887e] p. 314.

[64]*Ibid.* See later on p. 261 for the Reynolds stress. Thomson did not reason in terms of stresses. The last equation is a generalization of his equation (34).

When there is no laminar motion ($\mathbf{u} = \mathbf{0}$), the homogeneity and isotropy of the turbulent motion implies that

$$\overline{v_i v_j} = \alpha^2 \delta_{ij}, \tag{6.22}$$

where α represents the constant intensity of the turbulence. Now suppose a laminar motion to begin. Subtracting eqn (6.21) from eqn (6.20) and multiplying by v_j, we obtain

$$\partial_t(v_i v_j) = -v_j[\partial_i(P - \overline{P}) + \partial_k(u_i v_k + u_k v_i + v_i v_k - \overline{v_i v_k})] + (i \leftrightarrow j). \tag{6.23}$$

The large-scale average of the second of the resulting terms is

$$-\overline{v_j \partial_k(u_i v_k)} = -\overline{v_j v_k} \partial_k u_i = -\alpha^2 \delta_{jk} \partial_k u_i = -\alpha^2 \partial_j u_i. \tag{6.24}$$

That of the third term vanishes after symmetrizing with respect to the indices i and j:

$$\overline{v_j \partial_k(u_k v_i)} + (i \leftrightarrow j) = u_k \partial_k(\overline{v_i v_j}) = 0. \tag{6.25}$$

The averages of the fourth and fifth terms vanish because they contain odd-degree products of the components of the initial turbulent velocity, which is isotropic. The most delicate part is the evaluation of the first term.

The incompressibility conditions together with eqns (6.20)–(6.22) yield

$$\Delta(P - \overline{P}) = -2\partial_i u_j \partial_j v_i. \tag{6.26}$$

Consequently, the first term of eqn (6.23) is

$$-v_j \partial_i(P - \overline{P}) = 2v_j \partial_i \Delta^{-1}(\partial_k u_l \partial_l v_k). \tag{6.27}$$

Since the spatial variation of $\mathbf{u}$ is much slower than that of $\mathbf{v}$, this may be rewritten as

$$-v_j \partial_i(P - \overline{P}) = 2\partial_k u_l(v_j \Delta^{-1} \partial_i \partial_l v_k). \tag{6.28}$$

Owing to the isotropy of the original turbulent flow, averaging leads to[65]

$$\overline{-v_j \partial_i(P - \overline{P})} = 2\partial_k u_l \overline{v_j v_l \Delta^{-1} \partial_i \partial_l v_k} = \frac{2}{3}\alpha^2 \partial_j u_i - \frac{1}{3}\alpha^2 \partial_i u_j. \tag{6.29}$$

Altogether, we have

$$\partial_t(\overline{v_i v_j}) = -\frac{2}{3}\alpha^2(\partial_i u_j + \partial_j u_i). \tag{6.30}$$

Strictly speaking, this expression only holds at the beginning of the laminar motion $\mathbf{u}$. However, when this motion is small and periodic, eqn (6.30) remains valid at any time if the induced anisotropy of the turbulent motion stays small. Combining this equation with the average Euler equation (6.21) yields, to first order in $\mathbf{u}$,

[65]According to Thomson ([1887e] p. 316), isotropy and incompressibility lead to the relations $\overline{v_x \Delta^{-1} \partial_x^2 v_x} = \overline{v_x \Delta^{-1} \partial_y^2 v_x} = \overline{v_x \Delta^{-1} \partial_z^2 v_x} = \alpha^2/3$, $\overline{v_x \Delta^{-1} \partial_x \partial_y v_y} = (1/2)\overline{v_x \Delta^{-1} \partial_x(\partial_y v_y + \partial_z v_z)} = -(1/2)\overline{v_x \Delta^{-1} \partial_x^2 v_x}$ $= -\alpha^2/6$, to similar relations for similar terms, and to the vanishing of other kinds of terms.

$$\partial_t^2 \mathbf{u} - \frac{2}{3}\alpha^2 \Delta \mathbf{u} = \mathbf{0}, \tag{6.31}$$

which means that transverse waves propagate in the turbulent liquid with a velocity proportional to the average velocity of the turbulent motion.[66]

6.4.3   *The vortex sponge*

In Thomson's eyes, this result meant much more than a hydrodynamic curiosity. He announced it as 'something seemingly towards a solution (many times tried for within the last twenty years) of the problem to construct, by giving vortex motion to an incompressible fluid, a medium which shall transmit waves of laminar motion as the luminiferous ether transmits waves of light.' In 1847, Thomson had written to Stokes:

> I perceived a fine instance of elasticity in an incompressible liquid, in a very simple observation made at Paris, on a cup of thick 'chocolat au lait'. When I made the liquid revolve in the cup, by stirring it, and then took out the spoon, the twisting motion (in eddies, and in the general variation of angular vel[ocity] on acc[ount] of the action of the spoon overcoming the inertia of the liquid, and the fric[tion] at the sides) in becoming effaced, always gave rise to several oscillations so that before the liquid began to move as a rigid body, it performed oscillations like an elastic (incompressible) solid.

In the 1850s, after Thomson developed the mechanical theory of heat, he became convinced that every physical phenomenon could be reduced to pure motion. In particular, he believed that the rigidity of solid bodies or of the ether derived from the centrifugal inertia of internal, rotary motions.[67]

In this state of mind, Thomson could turn his brother's hydraulic problem into a hope for ether theory. His formulation was nevertheless cautious in that he only claimed 'something seemingly towards a solution' of the ether problem. He feared that the condition of persistent randomness of the turbulent flow could be 'vitiated by a rearrangement of vortices'. At the end of his paper, he showed that a symmetrical distribution of vortex rings satisfied the condition, but only if (as he now doubted) the vortex rings were themselves stable. 'I am thus driven to admit', he concluded, 'that the most favourable verdict I can ask for the propagation of laminar waves through turbulently moving inviscid liquid is the Scottish verdict of *not proven*.'[68]

The Irish ether theorist George Francis FitzGerald embraced Thomson's deduction without the worries about vitiating effects. Two years earlier, while reviewing various mechanical theories of the ether, he had introduced the 'vortex-sponge theory' of the ether that became his foremost philosophical project:

> It seems certain that the only way in which a perfect liquid can become everywhere endowed with properties analogous to rigidity is by being everywhere in motion. The

---

[66]Equations (6.30) and (6.31) generalize the equations (50) and (51) of Thomson [1887*e*].

[67]*Ibid.* p. 308; Thomson to Stokes, 20 Oct. [1847], in Wilson [1990]. Cf. Smith and Wise [1989] chap. 12. The relevance of Thomson's casual observation is questionable as hot chocolate is hardly a perfect liquid.

[68]Thomson [1887*e*] pp. 317, 320. On the instability of vortex motion, see Chapter 5, pp. 191–5.

most general supposition of this kind would be, that it was what Sir William Thomson
has called a vortex sponge, i.e. everywhere endowed with vortex motion, but with this
motion so mixed up as to have within any sensible volume an equal amount of vortex
motion in all directions. There are many ways in which this supposition seems to be in
accordance with what we know of the properties of the ether.

Thomson arrived at the vortex sponge in the 1870s while considering the degradation of a
cylindrical vortex, in connection with his speculations on vortex atoms. With little proof
and strong faith, FitzGerald made this residue of Thomson's matter theory the primitive
material of the universe.[69]

Impressed by Thomson's deduction of laminar motion in a turbulent fluid, FitzGerald
tried to interpret it in electromagnetic terms. This was part of the Maxwellian endeavor to
find a mechanical medium whose equations of motion would correspond to Maxwell's
equations. FitzGerald had already done so on the basis of James MacCullagh's rotation-
ally-elastic medium. The turbulent liquid, or vortex sponge, was a more appealing candi-
date, for it derived elastic behavior from pure motion. In a first attempt published in 1889,
FitzGerald identified the Reynolds stress system $\overline{v_i v_j}$ with Maxwell's electromagnetic
stress system ($E_i E_j + H_i H_j$ for the off-diagonal elements), and the fluid velocity with the
electromagnetic momentum flux $\mathbf{E} \times \mathbf{H}$ (the vectors $\mathbf{E}$ and $\mathbf{H}$ denote the electric and
magnetic field vectors, respectively). This works in the case of plane disturbances, the
only ones considered by Thomson and FitzGerald.[70]

At the close of the century, FitzGerald still believed in the vortex sponge as the ultimate
basis for a theory of ether and matter. He reasserted his conviction that Thomson's
vitiating rearrangement was improbable. He upheld the electromagnetic interpretation,
though in a different guise; he now compared the linearized, plane-disturbance counterparts of eqns (6.21) and (6.30) directly with Maxwell's equations, so that the large-scale
velocity $\mathbf{u}$ corresponded to the electric field vector and the Reynolds stress $\overline{v_i v_j}$ to the
magnetic field.[71]

FitzGerald's analogies between the electromagnetic ether and a turbulent liquid fail for
arbitrary, non-plane disturbances. Whether a better analogy of the same kind can be
found is an open question. From a historical point of view, it matters only that the
turbulent perfect liquid nourished the hopes of a few British ether theorists, in spite of
or because of its inexhaustible complexity.

6.5 Reynolds's criterion

Girard's discharge experiments with narrow tubes in the 1810s showed that, for a given
head, the character of pipe flow depended on the diameter of the pipe. For small diameters, the flow was 'linear', that is, divisible into straight or slightly-curved filaments;

---

[69]FitzGerald [1885] p. 154. Cf. Hunt [1991] pp. 96–107. On the origin of the vortex sponge, see Chapter 5,
pp. 194–5.

[70]FitzGerald [1889]. With the extension provided above, FitzGerald's analogy is easily seen to fail in the
general case.

[71]See FitzGerald [1899a], and [1899b] for a more picturesque expression of the same analogy in terms of
spiraling vortex filaments.

the efflux was proportional to the head and it depended strongly on temperature. For the large diameters encountered in hydraulics, the flow was 'complicated', that is, the fluid particles followed tortuous, variable paths; the efflux was roughly proportional to the square root of the head. Girard, Navier, Poiseuille, Darcy, and Bazin all knew of this difference. Saint-Venant and Boussinesq first took it into account in theories of hydraulic flow. Yet none of these investigators examined the transition between the two kinds of flow. Their lack of interest in this question is not surprising: they believed the transition to depend on accidental circumstances (entrance effect, irregularities of the walls, etc.), and they probably expected it to be gradual, in analogy with Coulomb's study of fluid friction.[72]

6.5.1 *Hagen's transition*

In 1839, the German hydraulician Gotthilf Hagen accidentally observed the transition while experimenting with small pipes. In order to enhance the effect of wall friction, he used hydraulically unrealistic diameters, of the order of a millimeter. While varying the head of water h, he observed a sudden change in the efflux for velocities larger than a certain (small) fraction of $\sqrt{2gh}$. He also noted[73]

> an essential change of appearances themselves when this limit was crossed ... which was very clearly marked in all series of observations: when I let the water flow freely in the air, for smaller pressure heads the jet had a permanent shape, and near the pipe it looked like a solid piece of glass; but as soon as the velocity, by stronger pressure, exceeded the given limit, the jet started to fluctuate and the efflux was no longer uniform and occurred by pulses.

As we saw in Chapter 3, he carefully established that below the turbulence threshold the loss of head per unit length of the tube was proportional to Q/R^4, where Q denotes the efflux and R is the radius of the tube (Hagen–Poiseuille law). In the turbulent case, he found the loss of head to be much larger and roughly proportional to the square of the discharge, in conformance with previous hydraulic knowledge. However, he focused on the regular case, in his view the only one suited to precision measurement. For turbulent flow he believed that 'the water [in the tube] lacked the tension [*Spannung*] necessary for the transmission of pressure', so that the conditions of motion were inherently underdetermined.[74]

Some fifteen years later, Hagen carefully studied the effect of temperature on the flow in his small pipes. He corroborated his earlier observation that the loss of head depended strongly on temperature in the non-turbulent case, whereas it did not in the turbulent case. He also observed that an increase in temperature could induce a transition from non-turbulent to turbulent flow. In his interpretation, the temperature increase implies a diminution of the internal friction of the fluid, which in turn causes the tension of the fluid to vanish at some point and fluctuations to occur. Then part of the pressure head is

---

[72]Girard [1816]. See Chapter 3, pp. 104–6.

[73]Hagen [1839] p. 424. Cf. Schiller [1933] pp. 83–4, Rouse and Ince [1957] pp. 157–61. Hagen's criterion for the transition agrees with Reynolds's, because below it, in the laminar mode, the viscous force is comparable to the pressure head.

[74]Hagen [1839] p. 442.

lost in the production of 'internal motions, which are induced by the smallest irregularities of the walls, or perhaps during the entrance into the tube.'[75]

Hagen insisted on the existence of these internal motions besides the motion measured by hydraulicians:[76]

> Being reckoned from the efflux, the velocity [of the water] is only measured along the axis of the pipe and does not take into account internal motions and eddies. Hence it does not represent the total motion of the water; it only represents the part of the motion that corresponds to the progression of the whole mass. Special observations, which I performed with glass tubes, show the two kinds of motion very clearly. As I let saw dust enter these tubes along with the water, I observed that for small pressures the saw dust propagated only in the direction of the tube, whereas for strong pressures it shot from one side to another and often assumed an eddying motion.

Non-German physicists and hydraulicians long remained unaware of Hagen's remarkable observations. Poiseuille is usually credited for the laminar discharge law, and Reynolds for the discovery of the suddenness of the transition to turbulent flow and for its criterion. The latter attribution does not misrepresent history as much as the former, because Hagen's research focused on the laws of capillary flow, not on the turbulent transition. The study of this transition remained wide open.

6.5.2 An eccentric philosopher–engineer

Born in Belfast in 1842, Osborne Reynolds entered Queen's College at Cambridge University after completing an apprenticeship in mechanical engineering. He graduated seventh wrangler in 1867. The following year he obtained the new chair of engineering at Owens College, Manchester. Reynolds belonged to a new kind of British engineering professors and consultants, well versed in higher mathematics and familiar with recent advances in fundamental physics. Throughout his life he maintained a double interest in practical and philosophical questions.[77]

Reynolds's teaching and research styles seemed highly idiosyncratic to his students and colleagues. His scientific papers were written in an unusually informal, concrete, and seemingly naive language. At first reading they often seem obscure, but tend to make more sense after careful study. They rely on astute analogies with previously known phenomena rather than deductive reasoning. Even though some of these analogies later proved superficial or misleading, in most cases Reynolds gained valuable insight from them. As he did not bother scanning older literature on his subject, he often duplicated previously known results. Perhaps for the same reason, some of his output was brilliantly original.[78]

---

[75]Hagen [1854] p. 81. Hagen believed that the maximum translatory velocity of the fluid, being superior to its average velocity, could exceed the value for which the pressure becomes negative according to Bernoulli's law. On his establishing of the Poiseuille law, see Chapter 3, pp. 140–1.

[76]Hagen [1854] pp. 80–1.

[77]Cf. Lamb [1913] pp. xv–xxi, Gibson [1946], Allen [1970].

[78]Reynolds's duplicate discoveries include the deflection of sound in a velocity gradient (known to Stokes), the law of discharge of gases under pressure (known to Saint-Venant), and the internal cohesion of liquids (known to Laplace); cf. Allen [1970] pp. 26–7, and J. J. Thomson, quoted *ibid.* p. 25: 'When he took up a problem, he did not begin by making a bibliography and reading the literature about the subject, but thought it out for himself from

Reynolds did his famous work on the transition between laminar and turbulent flow in the 1880s. Two themes of his earlier research conditioned his approach. The first was the importance of eddying motion in fluids, and the second was the dimensional properties of matter related to its molecular structure. A paper of 1874 on steam boilers brought the two themes together. Reynolds puzzled over the rapidity of the transfer of heat through the surface of the boiler. His interest in this problem was not solely practical:[79]

> The rapidity with which heat will pass from one fluid to another, through an intervening plate of metal, is a matter of such practical importance that I need not apologize for introducing it here. Besides its practical value, it also forms a subject of very great philosophical interest, being intimately connected with, if it does not form part of, molecular philosophy.

As an admirer of James Joule and James Clerk Maxwell, Reynolds had a deep interest in the kinetic molecular theory of heat and the resulting insights into transfer phenomena. In the boiler case, he concluded that ordinary diffusion bound to invisible molecular agitation did not suffice to explain the observed heat transfer. Adding to this process 'the eddies caused by visible motion which mixes the fluid up and continually brings fresh particles into contact with the surface', he deduced the form $A + Bv$ of the total heat transfer rate, where v denotes the velocity of the water along the walls of the boiler. He also noted the analogy with the Prony form $av + bv^2$ of fluid resistance in pipes.[80]

6.5.3   Revealing vortices

In the same year, Reynolds encountered eddying fluid motion while investigating the racing of the engines of steamers. As British seamen had learned at their expense, when the rotational velocity of the propeller becomes too large, its propelling action as well as its counteracting torque on the engine's axis suddenly diminish. A damaging racing of the engine follows. Reynolds explained this behavior by a clever analogy with efflux from a vase. The velocity of the water expelled by the propeller in its rotation, he reasoned, cannot exceed the velocity of efflux through an opening of the same breadth as its own. For a velocity higher than this critical velocity, a vacuum should be created around the propeller, or air should be sucked in if the propeller breaks the water surface.[81]

According to this theory, a deeper immersion of the propeller should retard the racing (for the efflux velocity depends on the head of water) and the injection of air next to it should lower the critical velocity (for the efflux velocity into air is smaller than that into a vacuum). While verifying the second prediction, Reynolds found out that air did not rise in bubbles from the screw, but followed it in a long horizontal tail. Suspecting some peculiarity of the motion of the water behind the screw, he injected dye instead of air and observed

---

the beginning before reading what had been written about it. There is, I think, a good deal to be said for this method. Many people's minds are more alert when they are thinking than when they are reading, and less liable to accept a plausible hypothesis which will not bear criticism.'

[79]Reynolds [1874c] p. 81. Cf. Silver [1970].

[80]Reynolds [1874c] p. 82. Reynold's argument is strikingly similar to that found in Saint-Venant [1838] on the additional retardation caused by eddy formation in the pipes of steam engines (see earlier on p. 230).

[81]Reynolds [1874a]. Reynolds later elaborated on cavitation in fluids. Cavitation in turbines was already known to Euler, cf. Ackeret [1957] p. L.

a complex vortex pattern in the trail. Similar experiments with a vane moving obliquely through water displayed vortex bands issuing from the angles of the vane.

From these observations, Reynolds inferred that hidden vortex motion played 'a systematic part in almost every form of fluid motion'. These considerations came well after Helmholtz's famous paper of 1858 on the theory of vortex motion, and after Thomson's, Tait's, and Maxwell's involvement in a vortex theory of matter. Reynolds convinced himself, in conversations with William Froude and William Thomson, that his forerunners had only seen the tip of the iceberg. The following year he brought invisible vortex formation to bear on a phenomenon well known to sailors, namely, the power that rain has to calm the sea. Letting a drop of water fall on calm water covered by a thin layer of dye, he observed the formation of a vortex ring at the surface followed by a downward vertical motion. When the drops of rain fall on agitated water, he reasoned, part of the momentum of this agitation is carried away by the induced vortices, so that the agitation gradually diminishes.[82]

Reynolds made vortex motion the subject of a popular conference at the Royal Institution in 1877. He began by promising a revelation:

> In this room, you are accustomed to have set before you the latest triumphs of mind over matter, the secrets last wrested from nature from the gigantic efforts of reason, imagination, and the most skillful manipulation. To-night, however, after you have seen what I shall endeavour to show you, I think you will readily admit that for once the case is reversed, and that the triumph rests with nature, in having for so long concealed what has been eagerly sought, and what is at last found to have been so thinly covered.

He went on with the failure of hydrodynamics to account for the actual motion of fluids and propounded that this failure was due to the lack of empirical knowledge of their internal motions.[83]

Reynolds then recalled casual observations of vortex rings above chimneys, from the mouth of a smoker, or from Tait's smoke box. These rings had only been studied 'for their own sake, and for such light as they might throw on the constitution of matter.' To Reynolds's knowledge, no one had understood their essential role in fluid motion. This he could reveal 'by the simple process of colouring water', which he had first applied to elucidate the motion of water behind a propeller or oblique vane. By the same means, he studied the vortex rings formed behind a disc moved flatly through water. The resistance to the disc's motion appeared to be caused by the continual production and release of such rings.[84]

Reynolds emphasized that 'imagination or reason had failed to show' such forms of fluid motion. Everyone knew the impotence of rational hydrodynamics, but 'it would seem that a certain pride in mathematics has prevented those engaged in these investigations from availing themselves of methods which might reflect on the infallibility of reason.' Only with hints from colored water could mathematicians proceed further:[85]

> Now that we can see what we are about, mathematics can be most usefully applied; and it is expected that when these facts come to be considered by those best able to do

---

[82]Reynolds [1877a] p. 188, [1875].   [83]Reynolds [1877a] p. 184.
[84]*Ibid.* pp. 187, 191.   [85]*Ibid.* pp. 185, 191.

so, the theory of fluid motion will be placed on the same footing as the other branches of applied mechanics.

6.5.4   *The dimensional properties of matter*

Meanwhile, Reynolds became interested in what he called 'the dimensional properties of matter'. The context was an attempt to explain the working of William Crookes's radiometer by evaporation from the black side of its vanes. According to the kinetic theory of gases, Reynolds reasoned, the ejection of a molecule from the surface of a vane implies a recoil of this vane with a momentum opposite to that of the molecule. Through specific experiments, he verified that the evaporation of a liquid caused a pressure on its surface. In an appendix to the ensuing paper, he noted that Crookes's effect could also be explained by surface heating: the adsorbed molecules of the residual gas leave the dark side of a vane at a higher velocity than those of the silvery side.[86]

As Reynolds came to realize, this simple and still popular explanation of the radiometer leads to a serious paradox. Consider two infinite parallel plates immersed in a gas, with different temperatures of their facing sides and equal temperature of their external sides. If Reynolds's simple theory held, the forces acting on the two plates should be different, since the temperature asymmetries are different for the two plates. However, the equality of action and reaction implies that these two forces should be equal and opposite. In order to escape from this paradox, Reynolds introduced a finite extension of the plates. With the military analogy of two batteries of guns, he explained how the oblique shots near the corners of the plates disturbed the balance of forces, because for them the recoil of a gun on one plate was not necessarily compensated by the impact of a bullet on the other.[87]

For a consideration of this sort to work in the radiometer case, the mean free path of the molecules has to be of the same order as the breadth of the vanes. This explains why a radiometer only works for a very small pressure of the residual gas. Reynolds further surmised that a radiometer with very tiny vanes would work at ordinary pressures. Such an experiment being practically impossible, he turned to the 'inverse phenomenon', that is, the gas motion induced by the temperature gradient of the vanes. This suggested a thermal counterpart to the 'transpiration' of a gas through a porous plug that Thomas Graham had investigated half a century earlier. Reynolds justified 'thermal transpiration' theoretically through an extension of Maxwell's kinetic theory of gases that included stresses in thermally heterogeneous gases, and experimentally by extending Graham's experiments to temperature gradients.[88]

Graham had noted that the law of transpiration became different for very small pores, but fell short of a theoretical conclusion. In contrast, Reynolds argued that the change of law occurred when the size of the pores became of the order of the mean free path. In the kinetic molecular conception of a gas, transpiration should only depend on the ratio between these two quantities. Reynolds therefore expected the transpiration curves to be homothetic whenever the product of the density of the gas and the diameter of the pores was the same. This he verified by means of a logarithmic plot of his data. He believed that

---

[86]Reynolds [1874*b*], [1876]. On early theories of the radiometer, cf. Everitt [1974] pp. 224–5.

[87]Reynolds [1879] pp. 304–5 (paradox), 306–9 (gun batteries).

[88]*Ibid.* p. 261.

he had thus reached an 'absolute experimental demonstration that gas possesses a hetero-geneous structure.'[89]

6.5.5   *Guessing the criterion*

In his works on transpiration and kinetic theory, Reynolds wrote as a Cambridge-trained natural philosopher. He even enjoyed criticism from his hero James Clerk Maxwell, who had his own theory of stresses in rarefied gases. Yet the engineer was always alive in Reynolds. While philosophizing on the distinction between capillary and ballistic flow, he remembered the existence of a similar distinction between capillary and hydraulic flow, or between 'direct' and 'sinuous' motion in his terms. He called for a similar dimensional analysis in this case:[90]

> As there is no such thing as absolute space and time recognised in mechanical philosophy, to suppose that the character of motion of fluids in any way depended on absolute size or absolute velocity, would be to suppose such motion without the pale of the laws of motion. If then fluids in their motion are subject to these laws, what appears to be the dependence of the character of the motion on the absolute size of the tube, and on the absolute velocity of the immersed body, must in reality be a dependence on the size of the tube as compared with the size of some other object, and on the velocity of the body as compared with some 'other velocity'.

In the case of pipe flow, the relevant theory was hydrodynamics based on the Navier–Stokes equation (with uniform viscosity). By the 1870s, the validity of this equation for capillary and small-scale motion was established, and its failure for larger-scale motion was blamed on unknown or uncontrollable circumstances of the motion rather than on the form of the equation itself. In his Royal Institution lecture on vortex motion, Reynolds criticized the deductive approach to hydrodynamics, but not its fundamental equation. As he remembered in 1883, 'the equations of [fluid] motion had been subjected to such close scrutiny, particularly by Professor Stokes, that there was small chance of discovering anything new or faulty in them.' Moreover, Reynolds knew that in the case of gases Maxwell had provided a kinetic-theoretical derivation of the Navier–Stokes equation.[91]

Reynolds noted that, after the elimination of pressure from this equation (by taking its curl), the time derivative of the vorticity $\boldsymbol{\omega} = \nabla \times \mathbf{v}$ had two terms:

$$\frac{\partial \boldsymbol{\omega}}{\partial t} = \nabla \times (\mathbf{v} \times \boldsymbol{\omega}) + \frac{\mu}{\rho} \Delta \boldsymbol{\omega}. \tag{6.32}$$

If the motion depends on a single velocity parameter U and on a single linear parameter L, then the first term has the 'factor' U^2/L^2 and the second $(\mu/\rho)(U/L^3)$. Hence Reynolds

---

[89]*Ibid.* p. 259.

[90]Reynolds [1883] p. 53. The present reconstruction agrees with Reynold's statement, *ibid.* p. 54: 'It is always difficult to trace the dependence of one idea on another. But it may be noticed that no idea of dimensional properties, as indicated by the dependence of the character of motion on the size of the tube and the velocity of the fluid, occurred to me until after the completion of my investigation of the law of transpiration of gases, in which was established the dependence of the law of transpiration on the relation between the size of the channel and the mean range of the gaseous molecules.'

[91]Reynolds [1883] pp. 54–5. On Stokes's opinion, cf., e.g., Stokes [1850*b*].

concluded that 'the birth of eddies depend[ed] on some definite value of $[LU\rho/\mu]$.' This dimensionless number is what Arnold Sommerfeld later called the Reynolds number.[92]

6.5.6   *Testing the criterion*

Reynolds tested his theory with two kinds of experiment. In the first, performed in 1880, he injected ink at the center of the conical entrance of a glass tube in a tank of still, isothermal water. He controlled the water flow by a valve at the lower end of the tube. A preliminary small-scale trial failed to show the transition because the maximal velocity of the flow was too small. Reynolds therefore ordered the larger apparatus of Fig. 6.6. For a slow flow, the injected ink formed a steady band along the axis of the tube (see Fig. 6.7(a)). At a certain value of the velocity, and at some distance from the entrance, 'the colour band appeared to expand and mix with the water so as to fill the remainder of the pipe with a coloured cloud' (see Fig. 6.7(b)). By moving the eye so as to follow the motion of the water, or under a flash of light, the cloud appeared to be made of two or three waves followed by distinct eddies (see Fig. 6.7(c)).[93]

   To test his criterion for the transition, Reynolds varied the diameter of the tube and the temperature of the water. The latter amounted to a variation of viscosity, calculable through a formula by Poiseuille. The experiments proved to be difficult, because the transition occurred suddenly and the slightest disturbance of the water entering the tube lowered the critical velocity. Reynolds had to wait several hours to reach sufficient equilibrium before each run. Instead of heating up the water, he cooled it with ice so as to minimize convection currents. He thus managed to establish that the critical velocity was proportional to the diameter of the tube and inversely proportional to the viscosity. This confirmed the theoretical criterion, with a Reynolds number of 6415 and the radius of the tube as the characteristic length.[94]

6.5.7   *The two kinds of instability*

The simple reasoning behind this criterion suggested that the instability of the fluid did not depend on the size of the disturbances, since it made the relative size of the two terms of eqn (6.32) depend only on the breadth and the velocity of the flow. The observations on pipe flow sharply contradicted this expectation. The water appeared to be in an unstable state with respect to finite disturbances well before the critical point was reached. Reynolds verified this condition by artificially inducing a finite disturbance with a wire placed in the tube (see Fig. 6.7(d)). He also noted that, for narrow tubes and velocities slightly above the critical point, the eddying occurred in a series of distinct 'flashes' (see Fig. 6.7(e)). As Reynolds put it,

> the critical velocity was very sensitive to disturbance in the water before entering the tubes ... This showed that the steady motion was unstable for large disturbances

---

[92]*Ibid.* p. 55; Reynolds to Stokes, 25 Apr. 1883, in Stokes [1907], vol. 1, pp. 232–3; Sommerfeld [1908] p. 599.

[93]Reynolds [1883] pp. 59–61, 68–77, 72 (quote).

[94]Reynolds [1883] pp. 44, 73–5. Later investigators found much higher critical numbers. According to Drazin and Reid ([1981] p. 216), the Poiseuille flow in a cylindrical pipe is probably stable (no rigorous proof exists), but there is an instability owing to the boundary layer that forms at the entrance of the pipe.

Fig. 6.6.   Reynolds' apparatus for studying the turbulent transition of the flow of water in a tube (from
         Reynolds [1883] p. 71). Water from the tank enters the horizontal glass tube through the conical funnel. The
         valve with the long handle on the right controls the flux, whose value is inferred from the lowering of the
         floater. Ink from the flask is injected continuously in the middle of the entrance of the tube.

  long before the critical velocity was reached, a fact which agreed with the full-blown
  manner in which the eddies appeared.

Indeed, the latter observation indicated a kind of snowball effect, a higher instability
induced by the disturbances that appeared at the critical velocity.[95]

  Reynolds found this result the more surprising because other kinds of flow agreed with
the expected behavior, namely, the transition from direct to sinuous motion independent
of the size of disturbances, and the gradual divergence from direct motion above the
critical velocity. Reynolds knew this from casual observations of crossing streams of water

[95]Reynolds [1883] pp. 61 (quote), 76–7.

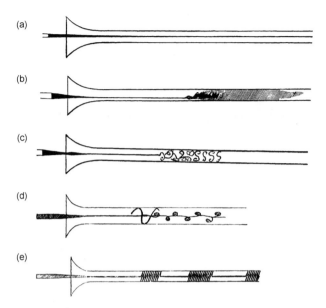

Fig. 6.7.   Reynolds' drawings of kinds of flow in a tube, as indicated by the ink jet method (from Reynolds
          [1883] pp. 59–60, 76–77): (a) 'direct' flow, (b) 'sinuous' flow, (c) the same observed with a flash of light
          (d) the disturbance of direct flow by a wire, and (e) intermittent 'flashes' of eddying.

(see Fig. 6.8). He studied the corresponding type of instability with a clever device in which
he made two fluids of different density slide over each other with a gradually increasing
velocity (see Fig. 6.9). In this 'very pretty experiment', small waves appeared beyond a
well-defined critical velocity and then grew until they curled and broke, 'the one fluid
winding itself into the other in regular eddies'. In modern terms, what he observed was a
simple spectrum of perturbations followed by developed turbulence.[96]

6.5.8   *Private calculations*

Having empirically established the existence of two different kinds of instability, Reynolds
was 'anxious to obtain a fuller explanation ... from the equations of motion.' He first
studied the stability of steady solutions of Euler's equation for frictionless fluids and found
that parallel flow in one direction was stable, while parallel flow in opposite directions was
unstable. Since Reynolds assumed that viscosity could have only a stabilizing effect, he
could not explain the observed instability of pipe flow. After a long period of puzzlement, at
the end of 1882 he attempted a similar study in the more difficult case of the Navier–Stokes
equation. He then found that the boundary condition for viscous fluids (vanishing velocity
at the walls) implied instability for sufficiently small values of the viscosity. The transition
between stability and instability still depended on the value of the Reynolds number.[97]

[96]*Ibid.* pp. 56, 61 (pretty), 62 (eddies).

[97]*Ibid.* pp. 62–3. See Chapter 5, p. 215. The boundary-layer instability of the *plane* Poiseuille flow was
suggested by Prandtl in 1821, and proved by Walter Tollmien in 1829 (cf. Drazin and Reid [1981] p. 216, this
book, Chapter 7, pp. 294–6). It seems doubtful that Reynolds anticipated this difficult analysis.

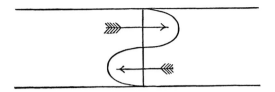

Fig. 6.8.   A case of unstable parallel flow (Reynolds [1883] p. 56).

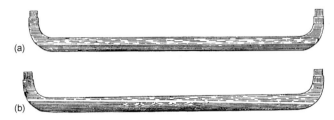

Fig. 6.9.   Reynolds's device for studying the instability of the sliding of two fluids over one another (Reynolds
[1883] pp. 61–2). (a) In the original configuration the higher-density colored fluid and the lower-density fluid
rest horizontally in two superposed layers. (b) A double, conflicting flow is obtained by inclining the tube.
Above a certain velocity, waves appear on the separating surface.

Reynolds never published the relevant calculations. In his memoir of 1883 he only gave the
results, together with a few empirical confirmations. In the course of experiments performed
in 1876 to study the calming effect of oil on wind waves, he had incidentally observed another
effect of the wind, namely, the formation of eddies beneath the oiled water surface. In the
light of his new theory, he argued that the stiffness of the oil film introduced a new boundary
condition on the water surface and thus destabilized the parallel flow beneath it.[98]

As for the lowering of the critical velocity under finite disturbances in the case of pipe
flow, Reynolds explained that 'as long as the motion was steady, the instability depended
upon the boundary action alone, but once eddies were introduced, the stability would be
broken down.' He thereby meant that the introduction of an eddy changed the distribution
of velocity and thus induced an instability of the frictionless kind (inflection in the velocity
profile). The latter instability could overcome a much higher viscous damping than the
boundary-based instability.[99]

6.5.9 *Pipe discharge*

As a corollary, there should be a value of the Reynolds number below which instability
with respect to finite disturbances disappears. Reynolds inferred the existence of a second
critical velocity of pipe flow, 'which would be the velocity at which previously existing
eddies would die out, and the motion become steady as the water proceeded along

[98]Reynolds [1883] pp. 58–9, 63.

[99]Reynolds [1883] p. 63.

the tube.' The method of colored bands could not test this conjecture, since the diffusion of the ink in turbulent water is irreversible. Reynolds therefore appealed to the law of discharge, the study of which he had so far avoided because of its greater experimental difficulty.[100]

For temperature stability and to assure a wide range of pressure, Reynolds used the water from the Manchester main. He measured the fall of pressure within pipes of various diameters using a differential pressure gauge, and the corresponding discharge by means of a special weir gauge (see Fig. 6.10). Since the pressure from the main could vary during a run, Reynolds's assistant kept it constant with an additional valve and manometer. The pipes were fed through a T-shaped connection that caused considerable disturbance of the entering water. Reynolds placed the differential pressure gauge away from this connection, in order to give time for the regularization of the flow in the subcritical case. He conducted the experiments in the workshop of Owens College, 'which offered considerable facilities owing to arrangements for supplying and measuring the water used in experimental turbines', and with the help of a Mr Forster, a skillful and clever technician.[101]

As Reynolds expected the character of the flow to be the same for equal values of the number $DU\rho/\mu$, he used the 'method of logarithmic homologues' that had served him so well in his transpiration studies. He plotted the logarithm of the pressure slope i (the fall of pressure head per unit length) versus the logarithm of the velocity U (see Fig. 6.11). The curves had a well-defined critical point, corresponding to a Reynolds number of 1015. They were composed of two straight lines, save for a small curved portion around the critical point. They could be very accurately superposed through a shift of the ordinates by $\ln(D^3/\nu^3)$ and of the abscissas by $\ln(D/\nu)$, where ν is the ratio μ/ρ.[102]

Accordingly, the relation between pressure slope and velocity has the general form

$$\frac{D^3 i}{\nu^2} = f\left(\frac{DU}{\nu}\right). \tag{6.33}$$

Up to the critical point, the function f is linear, in conformance with Poiseuille's law. A little beyond the critical point, Reynolds's discharge law takes the form

$$\frac{D^3 i}{\nu^2} \propto \left(\frac{DU}{\nu}\right)^\alpha, \tag{6.34}$$

with $\alpha = 1.753$. Reynolds confined his measurements to pipes of relatively small sections. For larger pipes, he relied on Darcy's raw data. The law (6.34) still held, but with an exponent depending slightly on the roughness of the pipe's walls.[103]

Unknown to Reynolds, German and French hydraulicians had already suggested fractional-power discharge laws. For example, in 1851 Saint-Venant favored such a law over Prony's $aU + bU^2$ law to permit the use of logarithms in backwater calculations.

[100]Reynolds [1883] pp. 64.   [101]Ibid. pp. 78–85 (quote on p. 79).

[102]Ibid. pp. 93–4. The expressions for the abscissa and ordinate shifts result from the validity of the Poiseuille law below the critical point.

[103]Reynolds [1883] pp. 94–7 (formula), 98–105 (on Darcy).

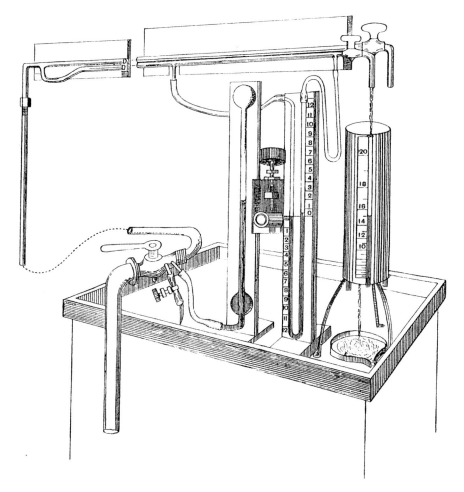

Fig. 6.10. Reynolds's apparatus for studying the loss of charge in a tube as a function of the discharge rate (Reynolds [1883] p. 79). Water from the main is fed at a constant rate into the lower horizontal tube, to which a differential manometer (the vertical U-shaped tube) is connected. The discharge rate is measured with the cylindrical weir gauge on the right-hand side.

Reynolds's motivation, as well as the special form he gave to the discharge law, were nonetheless new. They reflected the dimensional properties of his stability criterion.[104]

6.5.10 *The Reynolds legend*

In retrospect, Reynolds's achievement seems enormous. He gave a precise characterization of the transition to turbulence for pipe flow with respect to eddying and the discharge law, he defined the two relevant critical points and their relation to the Reynolds number of the

[104]Saint-Venant [1851*a*]; also Woltman [1791–1799], and Hagen [1854].

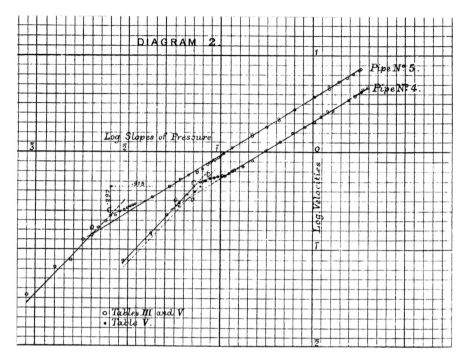

Fig. 6.11.   Reynolds's log-log plot of head loss versus fluid velocity (Reynolds [1883] p. 94).

flow, he introduced the distinction between two kinds of hydrodynamic instability, and he gave the now canonical form of the turbulent discharge law in a smooth pipe. However, some of the discoveries usually credited to him were not his. Hagen already knew of the correlation between eddying and the discharge law, as well as the suddenness of the transition to turbulent flow. Saint-Venant, Hagen, and earlier German hydraulicians had used discharge laws with fractional powers.[105]

Reynolds did not provide the dimensional analysis for which he is most famous. He did not discuss the scaling invariance properties of the Navier–Stokes equation, even though his stability criterion and his discharge laws reflect this invariance. When Stokes read Reynolds's theoretical derivation of the criterion (article 6 of the 1883 memoir), he saw in it an argument of dynamical similarity. Reynolds mildly protested: 'I had no intention whatever of laying down the conditions of dynamical similarity, although I now see that Art. 6 not only bears this construction but really fails to express what I meant.' What he truly meant was a comparison of the magnitude of viscous and inertial forces.[106]

Stokes had noted the similarity property of the Navier–Stokes equation in the context of his pendulum studies. There he found that the equation was invariant under a change of space scale (L), time scale (T), viscosity (μ), and density (ρ) if the ratio $(L^2/T)/(\mu/\rho)$

[105]For blasting criticism of Reynolds's priority claims and concomitant praise of Saint-Venant and Boussinesq, see Knibbs [1897].

[106]Reynolds to Stokes, 25 Apr. 1883, in Stokes [1907], vol. 1, p. 232.

remained the same. This implies the similarity of flows related by the same change. From this property, Stokes inferred that the mass correction of the pendulum, already known to depend on the density of the fluid around it, also had to depend on the pendulum's period.[107]

Stokes expected the similarity to hold only in the case of regular motions for which the Navier–Stokes equation effectively determined the flow. This explains his assessment of Reynolds's contribution:[108]

> Professor Reynolds has shown that the same conditions of similarity hold good, as to the average effect, even when the motion is of the eddying kind; and moreover that if in one system the motion is on the border between steady and eddying, on another system it will also be on the border, provided the system satisfies the above conditions of dynamical as well as geometrical similarity. This is a matter of great practical importance, because the resistance to the flow of water in channels and conduits usually depends mainly on the formation of eddies; and though we cannot determine mathematically the actual resistance, yet the application of the above proposition leads to a formula for the flow, in which there is a most material reduction in the number of constants for the determination of which we are obliged to have recourse to experiment.

Through this influential reading of Reynolds's contribution, Reynolds found himself credited for strategies that he did not use. Helmholtz was the first physicist, in 1860, to apply the similarity condition of the Navier–Stokes equation to pipe flow. He showed that a flow known to be 'linear' for certain values D of the pipe's diameter, v of the fluid's velocity, and $\Delta P / \Delta L$ of the pressure gradient, would also be linear for the values nD, v/n, and $\Delta P/\Delta Ln^3$ of these quantities. In 1873, the Prussian Ministry asked him to investigate the problem of the steering of balloons. Since the aerial motion around the rudder was too complex to be calculated, he decided to lay out the conditions for the rational use of small-scale experiments. As in Stokes's reinterpretation of Reynolds's memoir, Helmholtz assumed the similarity condition to extend to turbulent motion.[109]

The use of the similarity condition to restrict the form of the resistance law was Lord Rayleigh's invention. Reynolds discovered the restricted form empirically, due to the logarithmic plotting of his data. Rayleigh reasoned by dimensional analysis. If the density ρ, the kinematic viscosity v, the velocity U, and the diameter D entirely determine the flow, then dimensional homogeneity requires that the resistance $\Delta P/\Delta L$ (pressure fall per unit length) should have the form $(\rho U^2/D)\psi(DU/v)$ or $(\rho v^2/D^3)\phi(DU/v)$. This is to be compared with Reynolds's relation (6.33), which implies the form $(g\rho v^2/D^3)f(DU/v)$ for the resistance (since $\Delta P/\Delta L = i\rho g$). The occurrence of the acceleration of gravity in the latter formula confirms Reynolds's neglect of dimensional analysis. In contrast, Rayleigh had been using dimensional analysis since the beginning of his career. He never missed an opportunity to denounce its neglect and to emphasize the severe constraint it

---

[107]Stokes [1850a] p. 17. Stokes's analysis did not include the $(\mathbf{v} \cdot \nabla)\mathbf{v}$ term.

[108]Stokes's statement in presenting the Royal Medal to Reynolds, in Stokes [1907] vol. 1, pp. 233–4.

[109]Helmholtz and Piotrowski [1860] p. 173; Helmholtz [1873]. In the same period William Froude applied a different similitude argument to naval construction, see later on pp. 278–9. Bertrand [1848] had discussed mechanical similarity in general.

placed on the laws of fluid resistance. As we will see in the next chapter, dimensional and similitude arguments gradually pervaded fluid mechanics.[110]

6.5.11   *Color bands, magic cubes, and military discipline*

The color-band experiments on the two manners of fluid motion lent themselves to brilliant public shows. Reynolds did not miss this opportunity. In 1884, at the Royal Institution, he claimed a partial verification of his earlier prediction that 'the method of coloured bands would reveal clues to those mysteries of fluid motion that had baffled philosophy.' Besides the transition from direct to sinuous flow in straight pipes, he showed instabilities occurring in jets and in diverging pipes. John Tyndall had already displayed the sensitivity of smoke jets to sound in the same room. Helmholtz had enunciated the theoretical instability of the implied surfaces of discontinuity. Reynolds showed the transitory formation of a vortex ring when the flow came on gradually, the stabilizing effect of viscosity for slow motion, and the existence of a critical velocity for which the jet became unstable.[111]

Reynolds also offered a few spectacular analogies for the effects of hidden motion. At the beginning of his lectures, he showed the oscillations of two visually identical cubes suspended on springs. One cube obeyed the laws of mechanics, while the other apparently violated them. Reynolds then revealed to his audience that a gyrostat had been mounted inside the second cube. In a humorous allusion to Newton's apple, he argued that the laws of mechanics would have been as impenetrable as the laws of fluid motion if apples had hidden internal motion.[112]

To illustrate the dependence of the manner of fluid motion on viscosity, size, and velocity, Reynolds resorted to an analogy with a disciplined troop. If the discipline is respected, then the motion of the troop should be regular. Yet the march may turn into a whirling, struggling mob under an external disturbance. This instability clearly increases with the velocity of the march, with the size of the troop, and with the difficulty of the maneuver ordered; it decreases if the discipline is reinforced. In a real fluid, discipline corresponds to viscosity, and the difficulty of the maneuver to the boundary conditions.[113]

In 1893, Reynolds gave his last color-band show at the Royal Institution. This time he distinguished between two kinds of internal fluid motion. Firstly, wave-like motions are possible (in an incompressible fluid) when the fluid has a free surface. In this case, 'the colour bands, however much they may be distorted, cannot be relatively displaced, twisted, or curled up, and in this case motion in water once set up continues almost without resistance.' In the other class of internal motion, the colored band becomes thoroughly mixed with the rest of the water. As in cooking, wool spinning, and metal rolling, the mixing process involves 'folding, piling, and wrapping, by which the attenu-

---

[110]Rayleigh [1892]. See also Rayleigh [1904], [1909] (plate resistance), [1915*b*] (many examples). Cf. Rott [1992], Roche [1998] pp. 208–9. Although similarity arguments (Helmholtz) and dimensional analysis (Rayleigh) are not quite the same, they are intimately related to each other: in one case the scale of the actual system is changed, in the other the scale of units is changed.

[111]Reynolds [1884] pp. 154 (quote), 158–9 (jets).

[112]*Ibid.* p. 154.   [113]*Ibid.* pp. 155–6.

ated layers are brought together.' Such mixing occurs when waves break at the free fluid surface, or when the fluid slides over a solid surface with sufficient velocity. It explains the ease with which stirring brings uniformity in a liquid, compared to the slowness of molecular diffusion.[114]

Lastly, Reynolds showed the precise mechanism by which the motion of a spoon or vane achieved mixing, namely, the creation of a spiraling vortex sheet behind the vane, together with wave motion outside the sheet. The theory of the teacup was not the sole ambition of this display:[115]

> We can hope to interpret the parallel of the vortex wrapped up in the wave, as applied to the wind of heaven, and the grand phenomenon of the clouds, as well as those things which directly concern us, such as the resistance of ships.

6.5.12   *The philosophy of scales*

Reynolds intended his experiments, especially the color-band shows, to suggest new theoretical insights. Conversely, his experimental work depended on theoretical analysis. As we saw, his original analysis of pipe discharge depended on his theoretical understanding of the instabilities that led to turbulent motion. In his memoir of 1883, he mentioned that he had spent many months studying the infinitesimal perturbation of stationary solutions to the Euler and the Navier–Stokes equations. Yet he never published his calculations, perhaps because two giants of British hydrodynamics, Lord Kelvin and Lord Rayleigh, controlled the field.[116]

Reynolds's idea that the dimensionless number UD/ν determined the character of the fluid motion also derived from a theoretical consideration, namely, comparing the inertial and viscous terms in the vorticity equation. In 1894, Reynolds published a purely theoretical memoir on this criterion. Instead of the vorticity equation, he now considered the energy distribution at various scales. His inspiration came from an analogy with the kinetic theory of gases and also from Stokes's analysis of dissipation in a viscous fluid. In 1850, after being converted to Joule's concept of heat as a kind of motion, Stokes explained that the internal work of viscous stresses implied a continual conversion of the observable fluid motion into the microscopic motion called heat.[117]

Stokes derived the rate of this dissipation from the Navier–Stokes equation. In a reformulation inspired by Maxwell's identification of stress with momentum flux, Reynolds wrote this equation as

$$\rho \frac{\partial v_j}{\partial t} = \partial_i \pi_{ij},\tag{6.35}$$

with

$$\pi_{ij} = p_{ij} - \rho v_i v_j.\tag{6.36}$$

---

[114]Reynolds [1893] pp. 529, 531.   [115]*Ibid.* p. 534.

[116]Reynolds [1883] p. 63. Cf. earlier pp. 253–4 Chapter 5, p. 211.

[117]Reynolds [1895]; Stokes [1850*b*] pp. 67–70.

The part p_{ij} of this tensor gives the viscous stresses and corresponds to molecular momentum transfer. The rest corresponds to molar momentum transfer.[118]

The convective derivative of the macroscopic kinetic energy density $(1/2)\rho v^2$ follows:

$$\left(\frac{\partial}{\partial t} + \mathbf{v} \cdot \nabla\right)\left(\frac{1}{2}\rho v^2\right) = \partial_i(v_j p_{ij}) - p_{ij}\partial_i v_j. \tag{6.37}$$

After integration over an internal portion of the fluid, the first term yields the work of the pressures acting on the portion's surface. For an incompressible fluid, the second term should therefore represent the '*vis viva* consumed by internal friction' and 'converted into heat'. The expression

$$p_{ij} = \mu(\partial_i v_j + \partial_j v_i) - P\delta_{ij} \tag{6.38}$$

of the stress system implies that this term is always negative, in conformance with its dissipative character.[119]

As Reynolds correctly pointed out, this derivation assumes the possibility of discriminating between small-scale motion to be identified with heat, and larger-scale motion to be described by the Navier–Stokes equation. Since the distinction involves some arbitrariness, Stokes's interpretation of the dissipative term in his energy balance is not entirely compelling. Part of the small-scale internal motion produced by dissipation might be different from heat.[120]

Reynolds's lengthy discussion of this point boils down to the simple requirement that the average velocity **v** in the Navier–Stokes equation should vary very slowly at the scale of the averaging length and time. Concretely, if **v**(**r**) represents the spatial average of the lower-scale velocity in a domain of radius R centered on the point **r**, then the variation of the function **v**(**r**) should be very small over the length R. As Reynolds emphasized, this property depends on the nature of the mechanical system. Reynolds hoped that the elucidation of this 'discriminative cause' would clarify the 'hitherto obscure ... connection between thermodynamics and the principles of mechanics.'[121]

6.5.13   *Proving the criterion*

If the requirement of smooth averages is met at one scale, then it will be for any neighboring scale for which the variation of averages remains slow. This defines a certain scale band of consistent averaging. The equation of motion is approximately the same for any scale in this band, and implies the same dissipation rate. The Navier–Stokes equation corresponds to a first scale band. From hydraulics and from previous observations of

---

[118]Reynolds [1895] p. 544. Of course, Reynolds did not use the tensor notation. FitzGerald ([1889] p. 254) introduced the momentum-transfer form of the Euler equation.

[119]Stokes [1850b] p. 70; Reynolds [1895] p. 545.

[120]*Ibid.* p. 546.

[121]*Ibid.* pp. 537–44, 547–50, 560 (quote). Similarly, Burbury's and Boltzmann's contemporary notion of molecular chaos allowed for equations of large-scale evolution (the Boltzmann equation) that no longer referred to the low-scale, molecular dynamics.

turbulent flows, Reynolds assumed a similar band to exist at a much higher scale, of the same order of magnitude as the 'pulse' and spatial 'period' of turbulence.[122]

This scale is also the averaging scale that Saint-Venant and Boussinesq had introduced in their theories. Reynolds, who had not read them, called the corresponding average motion the 'mean-mean-motion', for it could be obtained by averaging the 'mean-motion' of the Navier–Stokes equation, which itself comes from averaging the molecular motion. Whereas Saint-Venant determined the equation of mean-mean-motion by a symmetry argument, Reynolds relied on an analogy with the kinetic theory of gases. In 1866, Maxwell derived the Navier–Stokes equation for the mean-motion by averaging over molecular processes. Similarly, Reynolds obtained his equation of mean-mean-motion by averaging the Navier–Stokes equation.[123]

Specifically, Reynolds divided the mean-motion $\mathbf{v}$ into the mean-mean-motion $\bar{\mathbf{v}}$ and the 'relative-mean-motion' $\mathbf{v}'$. Averaging the form (6.35) of the Navier–Stokes equation then leads to

$$\rho \frac{\partial \bar{v}_j}{\partial t} = \partial_i(\bar{p}_{ij} - \rho\bar{v}_i\bar{v}_j - \rho\overline{v'_i v'_j}). \tag{6.39}$$

This equation has the same form as the Navier–Stokes equation, save for the additional stress system $-\rho\overline{v'_i v'_j}$. The latter stress corresponds to momentum transfer between consecutive layers of the fluid through the relative-mean-motion, whereas the viscous stress $\bar{p}_{ij}$ corresponds to momentum transfer through molecular motion. For large Reynolds numbers, the viscous stress is negligible.[124]

Having no way to determine the statistical behavior of the relative-mean-motion, Reynolds could not reach a more determinate form of the equation of mean-mean-motion. He could, however, discuss energy transfers in a manner similar to Stokes. In the counterpart of the earlier balance (6.37), the additional stress system $-\rho\overline{v'_i v'_j}$ implies an additional term $-\rho\overline{v'_i v'_j}\partial_i\bar{v}_j$, corresponding to the conversion rate of mean-mean-motion energy into relative-mean-motion energy.[125]

Another equation can be written for the convective variation of the kinetic energy of the relative-mean-motion:

$$\left(\frac{\partial}{\partial t} + \mathbf{v}\cdot\nabla\right)\left(\frac{1}{2}\rho\overline{v'2}\right) = \partial_i[\overline{v'_j(p'_{ij} - \rho v'_i v'_j)}] - \overline{p'_{ij}\partial_i v'_j} + \rho\overline{v'_i v'_j}\partial_i\bar{v}_j, \tag{6.40}$$

with $p'_{ij} = p_{ij} - \bar{p}_{ij}$ for the viscous stress of the relative-mean-motion. Once integrated over a fluid portion, at the borders of which the relative-mean-motion vanishes, this equation means that the variation in the energy of the eddying motion in this portion is the sum of two terms, namely, a negative one corresponding to the conversion of eddying motion into heat, and a positive one corresponding to the conversion of mean-mean-motion into eddying motion.[126]

---

[122]Reynolds [1895] pp. 538, 551–3.   [123]*Ibid.* pp. 553–4.   [124]*Ibid.* p. 554.

[125]Reynolds [1895] p. 555.   [126]*Ibid.* p. 556.

Reynolds applied the latter balance to the determination of the Reynolds number below which turbulent flow becomes impossible, namely, the damping term dominates in the above equation for any relative-mean-motion that is compatible with the Navier–Stokes equation. With much laborious calculation, the best Reynolds could do was to determine a lower limit (517) of this number for the flow between two fixed parallel plates.[127]

Reynolds's memoir should not be judged only for the 'determination of the criterion' promised in its title. It contained some basic ideas for a statistical approach to turbulence, namely, the mean-mean-motion, the turbulent stress, and the cascading of energy from one scale to another. Unknown to Reynolds, Saint-Venant had already introduced such notions. However, the purposes and manners of reasoning were different. Saint-Venant aimed at a fundamental foundation of *hydraulics* and offered new effective strategies for solving hydraulic problems. Reynolds had a more philosophical ambition: he wanted to reform *hydrodynamics* by specifying the conditions for turbulent flow and subjecting this kind of flow to methods similar to those of Maxwell's kinetic theory of gases.[128]

Although Reynolds's contribution to our understanding of turbulence is the most memorable, it was not the only nineteenth-century achievement on this arduous subject. For the investigators honored in this chapter, it was clear that many of the flows occurring in nature and in hydraulic systems had the unpredictable, multi-scale whirling character that we now call turbulent. Saint-Venant and Poncelet ascribed an essential role to this property in large-scale transfer phenomena. Saint-Venant, Thomson, Boussinesq, and Reynolds inaugurated statistical approaches to turbulent flow, based on effective viscosity and kinetic–molecular analogy. Hagen and Reynolds provided fine descriptions of the transition from laminar to turbulent flow. In this context, Reynolds introduced important dimensional considerations and the dimensionless number that relates viscous flows at different scales.

This early history of turbulence offers a rich sample of the strategies that physicists may deploy when confronted with the complex, highly-irregular behavior of a dynamical system. Most primitively, they may completely ignore the complexity and reason on an ersatz system that obeys simple mechanical laws. This is what early backwater theorists such as Bélanger did with a certain amount of success. In the more refined strategy inaugurated by Saint-Venant, they may try to average out the irregularities, to seek general relations between the averages based on symmetry considerations, conservation laws, and dimensional analysis, and to determine the leftover parameters empirically. At a still more advanced stage, they may investigate the stochastic processes that relate the macroscopic parameters to the microscopic dynamics. This is what Reynolds began to do by analogy with the kinetic theory of gases.

These achievements guided later theories of turbulence, for instance Ludwig Prandtl's in the 1920s and Geoffrey Taylor's in the 1930s. As the latter theories, especially Prandtl's,

---

[127]Reynolds [1895] pp. 557–77.

[128]Reynolds was not only interested in the philosophical aspects of hydrodynamics. In 1886, he gave an influential theory of lubrication based on the Navier–Stokes equation. Although this theory involves laminar flow only, he mentioned the existence of 'molar viscosity' in the turbulent case ([1886] pp. 236–8).

were the first to provide reliable guidance for hydraulic and aeronautic engineers, and because their authors founded important schools of pure and applied fluid mechanics, they are often regarded as the true starting-point of the scientific study of turbulence. In the next chapter, we will see how much twentieth-century success in this field and in the related resistance problem capitalized on insights from the previous century.[129]

---

[129]For a historically sensitive review of the kinetic and statistical approaches to turbulence, cf. Farge and Guyon [1999]. On Taylor, cf. Battimelli [1990]; Batchelor [1996].

7

DRAG AND LIFT

> When the complete mathematical problem looks hopeless, it is recommended to enquire what happens when one essential parameter of the problem reaches the limit zero.[1] (Ludwig Prandtl, 1948)

The problem of the forces that a fluid exerts on a solid body challenged hydrodynamics since d'Alembert's foundational *Essai* of 1752. It only found a quantitative, wide-ranging solution in the first third of the twentieth century, after a few partial successes in the nineteenth century and earlier. The first section of this chapter is devoted to these older attempts, including Newton's molecular-impact theory, Rayleigh's dead-water theory, and eddy-resistance theories by Poncelet and Saint-Venant. None of these theories truly succeeded in making quantitative predictions, and they all lacked a solid conceptual basis. Newton's theory artificially neglected the mutual action of fluid molecules, Rayleigh's implied an absurdly large wake, and Saint-Venant's required some observational input. Yet they all contained important elements of the modern understanding of fluid resistance. Newton understood how a similitude argument constrained inertial resistance to be quadratic. Rayleigh's theory foreshadowed the separation process now admitted for non-streamlined flow. Saint-Venant correctly described the eddy resistance resulting from the instability of separated flow.

Section 7.2 is devoted to ship resistance. The development of steam navigation in the Victorian empire motivated the efforts of a few learned engineers to reflect on the optimal shape of ship hulls. John Scott Russell saw how to minimize wave resistance. William Rankine clearly distinguished skin friction, large-eddy resistance, and wave resistance. Lastly, and most importantly, William Froude expressed the conditions for a rational use of models and developed the relevant experimental techniques. In his analysis of skin friction, he finely described what Prandtl later called a turbulent boundary layer. His and Rankine's insights into the mechanisms of high-Reynolds-number resistance nevertheless remained qualitative. The means were still lacking to turn them into efficient computational schemes.

In Section 7.3, we will step into the twentieth century and follow the successful development of Ludwig Prandtl's boundary-layer theory of fluid resistance at high Reynolds numbers. Contrary to a well-spread myth, this theory was not suddenly born out of a Prandtl paper of 1904. Prandtl benefited from the rich conceptual resources of earlier hydrodynamics, and many years elapsed before the aims expressed in this paper were truly reached. The boundary-layer concept only became practically useful around 1930, after Prandtl and Theodore von Kármán understood the role of turbulence in these layers.

---

[1]Prandtl [1948] p. 1606.

Building on Boussinesq's and Reynolds' intuition of eddy viscosity, they discovered the logarithmic velocity profile on which the modern understanding of hydraulic pipe retardation and turbulent boundary-layer theory is based. One important aspect of this evolution was the pervasive use of similitude and dimensional arguments that Stokes, Helmholtz, Froude, and Rayleigh pioneered in the previous century.

Section 7.4, the final section of this Chapter, will take us into the air with the first successful theories of a peculiar, useful form of fluid resistance, namely aerodynamic lift. Whereas the leaders of late-nineteenth-century fluid mechanics adhered to concepts of fluid resistance that made it very difficult for birds and planes to fly, a British automobile engineer, Frederick Lanchester, and a young German mathematician, Wilhelm Kutta, independently arrived at the circulatory flow that occurs around any properly working wing. In Russia, Nikolai Joukowski clarified the relation between circulation and lift, and greatly generalized Kutta's two-dimensional theory. Amidst the German war effort, Prandtl capitalized on these analytical considerations and on Lanchester's intuitions to develop modern wing theory in three dimensions. Together with boundary-layer theory, this achievement crowned nineteenth-century efforts at reconciling theoretical and real flows. These efforts, recounted in the previous chapters, indeed brought many of the concepts that permitted Prandtl's breakthroughs: vortex structures, discontinuity surfaces, viscous stress, parallel-flow instability, eddy viscosity, conformal transformations, and similitude.

7.1 Tentative theories

7.1.1 *Early views on fluid resistance*

The earliest quantitative studies of fluid resistance occurred in the seventeenth century, when Edme Mariotte and Christiaan Huygens investigated the impact of a fluid jet on a plate, with waterwheels and windmills in mind. They found the impulsion on the plate to be proportional to the density of the fluid and to the squared velocity of the jet. Huygens explained this result by analogy with the pressure that the water exerts on the bottom of the vessel from which the jet issues. In modern terms, he obtained the form $\frac{1}{2}\rho v^2 S$ for the impulsion, where ρ is the fluid's density, v is its velocity, and S is the section of the jet. Closer to our point, Mariotte believed the resistance encountered by a solid body moving through a fluid to result from the impact of fluid veins acting like the jets on which he experimented.[2]

In his *Principia*, Isaac Newton obtained a similar formula while trying to show that Descartes' matter-filled space led to an absurd resistance to planetary motion. He first used a similitude theorem to show that the resistance was necessarily proportional to the fluid density, to the square of the linear dimensions of the body, and to the square of its velocity. According to this influential theorem (anticipated by Mariotte), for any possible motion of a mechanical system, the similar motion obtained by uniformly rescaling all

---

[2]Cf. Saint-Venant [1887*b*] pp. 12–15, Rouse and Ince [1957] pp. 65–7. These researches originated in an Academic commission of 1668 mainly devoted to the verification of Torricelli's law of efflux. Mariotte and Huygens, who belonged to this commission, were unaware of the *vena contracta* (the contraction of the fluid vein near to the opening on the vase) first described in the second edition of Newton's *Principia*. Cf. Blay [1992] pp. 339–42.

lengths and velocities is also a possible motion if all forces acting in the system are simultaneously rescaled as the inverse of a length times the square of a velocity.[3]

Newton next assumed that the particles of the fluid were individually deflected by the solid surface, and identified the force acting on this surface with the destroyed momentum (see Fig. 7.1). For normal impact, the resulting force on the surface element dS is $\rho v^2 dS$. When the incoming flow makes an angle θ with the surface element, the particles' flux and the destroyed momentum of each particles are both multiplied by $\sin \theta$. The resulting force is normal to the surface and has the intensity $\rho v^2 dS \sin^2 \theta$. This implies the existence of a prow shape for which the resistance is a minimum for a given maximal breadth, a result Newton believed to be relevant to navigation. Newton next integrated the differential pressure over the front half of a sphere and compared the result with his own experiments on metal spheres dropped from the dome of Saint Paul's Cathedral in London. As the resistance turned out to be smaller than expected, he offered another theory based on efflux through a partially-obstructed opening. Assuming the 'cataract' flow of Fig. 7.2, he equated the resistance to the weight of the static pyramid of fluid above the obstacle.[4]

This odd theory was soon forgotten, except for the dubious implication that the largest transverse section of the body determines the fluid resistance. The discrete-impact theory survived for about two centuries in engineering quarters, for it gave the correct dependence

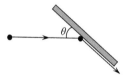

Fig. 7.1.   The Newtonian deflection of a fluid particle against an inclined surface element.

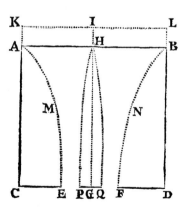

Fig. 7.2.   Newton's cataract (AEPH and BFQH) for the flow through an opening (EF) partially obstructed by the disc PQ. From Newton [1713]: book 2, sect. 7, prop. 36.

[3]Newton [1687], [1713], book 2, sect. 7, prop. 32 (similitude theorem), 33 (general form of resistance).

[4]*Ibid.* prop. 34 (implicit $\sin^2 \theta$ law), 35 (resistance of globe), 36–9 (cataracts), 40 (experiments). Cf. Saint-Venant [1887*b*] pp. 15–19, G. Smith [1998].

on density and velocity in the simplest manner. Yet careful resistance measurements disproved it several times in the second half of the eighteenth century. In the 1760s, Jean-Charles de Borda found that the resistance more nearly depended on the sine of the inclination of the surface elements of the body. In the 1770s, Charles Bossut included two other factors, namely, the dimensions of the towing tank, and the formation of waves in the case of partially-immersed bodies. In the 1880s, Pierre Du Buat discovered that the form of the rear of the body largely controlled the resistance. He also found that a 'negative pressure' (a pressure inferior to that of the undisturbed fluid) occurred in this region and contributed to the resistance. In the 1890s, the British Colonel Mark Beaufoy identified fluid friction along the walls of the body as an important contribution to the resistance. Through precise experiments with plates towed edgewise, he showed that this frictional effect depended on the power 1.8 of the velocity. In conformance with Borda's remark, Samuel Vince proved that the resistance offered by a slightly-inclined plate varied as the sine of the inclination, rather than the Newtonian squared sine.[5]

This impressive rise of empirical knowledge went along with a distrust of higher hydrodynamic theory. In 1768, d'Alembert admitted that he did not know how to avoid the vanishing resistance predicted by his theory. Two years earlier, Borda deduced the absence of resistance in a very general manner based on the principle of live forces: he showed that no work is needed to pull a body uniformly through a still fluid, because the live force of the fluid motion around the body remains globally unchanged in this process. Even earlier, in 1745, Euler had shown that momentum balance along tubes of flow yielded zero resistance unless the contribution of the rear part of the tubes was ignored. In a memoir of 1760 on fluid resistance, he explored the latter assumption, thus effectively returning to the Newtonian theory of resistance and giving up recourse to his own hydrodynamic equations.[6]

In summary, at the beginning of the nineteenth century the best-founded hydrodynamic theories, namely those of d'Alembert and Euler, led to a vanishing fluid resistance. Newton's old theory accounted for a finite resistance proportional to density and squared velocity, but failed in any other respect.

7.1.2 *Discontinuity surfaces*

A first way to avoid d'Alembert's paradox was to introduce viscosity, as Navier first did in 1822. Twenty years later, Stokes successfully obtained the viscous damping of pendulum oscillations and the linear resistance formula for slowly-moving, small spheres such as the droplets of clouds. He knew, however, that the most common kinds of fluid resistance eluded this theory, since their dependence on velocity was quadratic instead of linear. One of Stokes's early suggestions for solving this difficulty was that a dead-water region circumscribed by surfaces of discontinuity occurred in the wake of bodies traveling through an inviscid fluid. Such discontinuities were indeed compatible with Euler's equations.[7]

In 1868, Helmholtz independently introduced surfaces of discontinuity (*Trennungs-fläche*) in order to explain jet formation within a fluid. He believed that such surfaces

---

[5]Cf. Saint-Venant [1887*b*] pp. 27, 37–9, Rouse and Ince [1957] pp. 128–9, 133–4.

[6]Euler [1745], [1760b]. Cf. Saint-Venant [1887*b*] pp. 9–11, 21–37, Truesdell [1954] p. XL.

[7]See Chapter 3.

were formed whenever the pressure of the hypothetical irrotational flow became negative, typically near a sharp edge of a solid wall. He interpreted them as infinitely-thin vortex sheets and used his vortex theorems of 1858 to show their tendency to spirally unroll under an infinitesimal perturbation. In geometrically simple cases of two-dimensional flow, he introduced the velocity potential φ and the stream function ψ in the manner of d'Alembert, Lagrange, and Stokes, and managed to determine the form of the discontinuity surfaces by seeking a holomorphic function $\varphi + i\psi$ that satisfied the required boundary conditions in the plane of the complex variable $x + iy$.[8]

Kirchhoff and Rayleigh soon applied this technique to the motion around a flat plate (a segment in two dimensions, see Fig. 7.3) to derive the resistance formula

$$R = \frac{\pi \sin \theta}{4 + \pi \sin \theta} \rho v^2 S, \tag{7.1}$$

where ρ is the fluid density, v is the constant fluid velocity far from the plate, S is the surface of the plate, and θ is its inclination. Whereas the normal direction of the resistance, and its dependence on density and velocity agree with Newton's theory, the angular dependence does not and fits experiments much better, as Rayleigh judged on the basis of Vince's old data. Helmholtz agreed that the formation of surfaces of discontinuity was the main source of resistance in any large-scale (high-Reynolds-number) motion, although the instability of these surfaces cast doubt on any quantitative use of them.[9]

The recourse to surfaces of discontinuity and dead water was a controversial issue. William Thomson strictly rejected them, even though he was the British physicist closest to Helmholtz. According to Thomson, surfaces of discontinuity could never be formed in a perfect liquid, because Lagrange's and other theorems forbade the creation of vorticity; and the discontinuous state of motion, with its infinitely-long dead-water wake, was patently absurd. In his view, the true cause of any apparent departure from potential flow was viscosity or cavitation. In the viscous case, the intense shear of the flow past an edge induced the production of a series of vortices that roughly imitated Helmholtz's

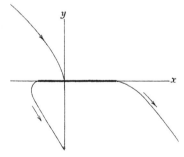

Fig. 7.3.   Flow around an inclined plate (thick line) according to Kirchhoff [1869] p. 425. The two lines of
current from the edges of the plate delimit the dead-water region.

[8]Helmholtz [1868b]. See Chapter 4, pp. 163–5.

[9]Kirchhoff [1869]; Rayleigh [1876b]; Helmholtz [1873]. See Chapter 4, pp. 164–5.

vortex sheets. In his playful correspondence with Thomson on this matter, Stokes valiantly defended the discontinuity surfaces, arguing that none of Helmholtz's theorems forbade the growth of surfaces of discontinuity from a germ on the wall surface, and that the zero-viscosity limit of viscous flow led to surfaces of discontinuity whenever the lines of the Eulerian flow diverged too much (typically, behind a body with a bluff rear).[10]

In 1898, Stokes's focus on the zero-viscosity limit of viscous flow prompted Kelvin to reflect on the nature of the flow around a globe. He imagined that the motion of the globe was impulsively started from rest and then kept uniform. In the first instant, he reasoned, the induced motion is very similar to the potential flow ruled by Euler's equation, with a finite sliding velocity on the walls. The corresponding infinitesimal layer of vorticity is then subjected to three effects, namely, viscous diffusion, convection through the impressed flow, and shear instability. The competition between the first two effects leads to deviations from the potential flow confined within an adjacent fluid layer that grows downstream:

> If the velocity is sufficiently great, the motion of the fluid at small distances from its surface all round will always be very nearly the same as if the fluid were inviscid, and the difference will be smaller near the front part than near the rear of the globe.

Shear instability within the boundary layer, Thomson went on, induces a trail of turbulent motion behind the globe. Stokes agreed with this scenario, except that, in his view, the trail commenced with a surface of sudden slip whose instability led to the observed turbulence.[11]

These private considerations did not foster any quantitative estimate of resistance. In Britain, the Rayleigh–Kirchhoff dead-water theory remained the only published, quantitative estimate of resistance in a fluid of small viscosity. Before the end of the century, Samuel Langley's and William Henry Dines's resistance measurements showed that this theory failed by at least a factor of three in the case of a blade moving parallel to itself. Helmholtz's surfaces of discontinuity nonetheless enjoyed some popularity among physicists and mathematicians. Horace Lamb and Alfred Basset devoted to them long sections of their widely-used treatises. In Germany they graced the lectures of the influential Munich professor August Föppl (published in 1899), who even declared:[12]

> The consideration of separation surfaces represents the first and the most important step toward a theory that better accounts for the facts ... Helmholtz's doctrine of fluid jets is therefore to be regarded as a remarkable progress of hydrodynamics, even though it leaves much to be desired with regard to physical exactness.

In 1901, the Italian mathematician Tullio Levi-Civita made discontinuity surfaces responsible for the fluid resistance of bodies of any shape (see Fig. 7.4). Like Stokes, he traced the failure of early ideal-fluid theories of resistance to an unwarranted assumption:

> The [usual] analytical formulation of the problem [of the flow of an ideal fluid around a solid body] introduces some elements, seemingly innocuous but much more remote

---

[10]See Chapter 5, pp. 197–207.

[11]Thomson to Stokes, 27 Dec. 1898, *ST*; Stokes to Thomson, 30 Dec. 1898, *ST*. See Chapter 5, pp. 204–6. The modern reader may recognize a boundary layer with separation.

[12]Langley [1891]; Dines [1891]; Föppl [1899] p. 396.

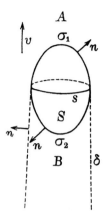

Fig. 7.4.   The discontinuity surface (δ) and dead-water region (B) for the flow past a bluff body moving at the velocity v through a perfect liquid. From Levi-Civita [1907] p. 131.

from reality than is the character of the perfect fluid. Such is, in my opinion, the hypothesis of the continuity of the movement of the fluid in the entire space around the body.

Again like Stokes, Levi-Civita assumed the formation of discontinuity surfaces even for round bodies. He then showed that a dead-water wake generally implied a resistance proportional to the fluid density and to the squared velocity of the body. However, he did not address the question of the location of the curve s from which the discontinuity surface departed. Nor did he know, in 1901, how to solve the system of equations that determine the shape of the discontinuity surface for a given curve s. He accomplished this difficult task in a larger memoir of 1907 through a broad extension of Helmholtz's complex-plane method.[13]

In summary, at the turn of the century, discontinuity surfaces remained the main analytical approach to the resistance problem for a slightly-viscous fluid. Yet they had well-identified shortcomings, namely: they led to utterly unstable and physically impossible motions, they gave smaller resistances than in reality, and they were essentially indeterminate in the case of smoothly-shaped bodies. For a Kelvin, these defects were fatal. For a Rayleigh, a Föppl, or a Levi-Civita, discontinuity surfaces marked a significant step toward a successful theory of resistance.

7.1.3   *Turbulent wakes and shear layers*

Another way to solve d'Alembert's paradox was to assume some instability of the laminar flow of a slightly-viscous fluid that prompted turbulent eddying in the rear of the body. Stokes first suggested this option in 1843. Poncelet and Saint-Venant made it the basis of quantitative resistance estimates. In 1846, Saint-Venant placed the fixed body within the current of a cylindrical pipe (see Fig. 7.5) which was so wide as to leave the resistance

---

[13]Levi-Civita [1901] p. 130, [1907].

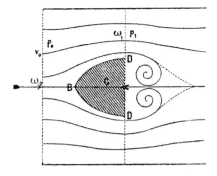

Fig. 7.5.   Flow around a truncated body (BDD) placed within a cylindrical pipe. From Saint-Venant [1887b] p. 89.

unchanged. The momentum which the incompressible fluid conveys to the body in a unit time is equal to the difference $P_0 S - P_1 S$ between the pressures on the faces of a column of fluid extending far before and after the body, because the momentum of the fluid column remains unchanged. The work $(P_0 S - P_1 S) v_0$ of these pressures in a unit time is equal the live force of the 'non-translatory motions' generated in the fluid. Hence the resistance is given by this live force divided by the original velocity of the fluid. In the non-translatory motion, Saint-Venant included both the small-scale motions that are a direct consequence of viscous stress and the 'tumultuous', whirling motions observed at the rear of bluff bodies.[14]

Saint-Venant then improved on a method invented by Poncelet to estimate the magnitude of the resistance, and based on the assumption that the wall pressure behind a separation point does not differ much from the value that Bernoulli's law gives it in the most contracted section of the flow. In the simple case of the truncated body of Fig. 7.5, the pressure at the rear is thus made equal to the pressure P_1 in the section ω_1. Saint-Venant further included a stress acting tangentially to the walls of the body, mostly due to eddy viscosity in the case of a turbulent incoming flow. In conformance with the standard treatment of pipe and channel retardation, he assumed a large-scale sliding velocity of the fluid along the walls, and made the friction proportional to this velocity squared.[15]

In the case of a plate parallel to the flow, for which wall friction is clearly the only cause of resistance, Saint-Venant compared Beaufoy's and Du Buat's measurements with the then-accepted friction coefficient in cylindrical pipes. The result indicated that the sliding velocity along the plate had to be smaller than the velocity of the incoming flow. Saint-Venant explained this difference as a retarding effect of eddy viscosity for the large-scale flow next to the slide on the walls (see Fig. 7.6).[16]

Although Saint-Venant's disciple Boussinesq did not address the resistance problem *per se*, he abundantly developed Saint-Venant's idea of eddy viscosity in pipe and open-channel flow, with velocity profiles that had finite slide on the walls and parabolic (for

---

[14]Saint-Venant [1846b]. See Chapter 3, pp. 134–5.

[15]*Ibid.* pp. 28, 72–8, 120–1; Saint-Venant [1887b] pp. 56–192.

[16]*Ibid.* (from a MS of 1847) pp. 116–49. The modern reader may recognize here a turbulent boundary layer, although Saint-Venant neglected any variation of this layer along the wall (in conformance with his assumption of a quadratic dependence on velocity).

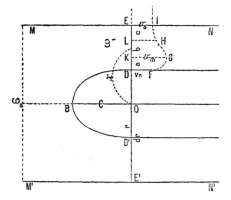

Fig. 7.6.   Velocity profile (FGHI, with the velocities DF, KG, LH, EI) for the flow between a body (DBD') and a cylindric wall (MM'NN'). From Saint-Venant [1887b] p. 132.

rectangular channels) or cubic (for circular pipes) increase from the walls.[17] In 1880, Boussinesq discussed the more academic question of the role played by viscosity and adherence to the walls at the beginning of a laminar flow. He did this in reaction to Jacques Bresse's erroneous extension to viscous fluids of Lagrange's theorem, according to which a velocity potential exists for any fluid motion started from rest.[18]

With Boussinesq, consider the simple case of a constant, uniform, and horizontal accelerating force ρk applied at time zero and onward to the entire mass of a viscous fluid resting over the horizontal plane $z = 0$. The resulting flow is obviously parallel to the plane, and its velocity u vanishes on the plane at any time. The Navier-Stokes equation for a kinematic viscosity ν gives

$$\frac{\partial u}{\partial t} = k + \nu \frac{\partial^2 u}{\partial z^2}, \tag{7.2}$$

in which Boussinesq immediately recognized Fourier's equation for the diffusion of heat. The relevant solution is

$$u = kt\left[1 - \frac{2}{\sqrt{\pi}} \int_\alpha^\infty \left(1 - \frac{\alpha^2}{\beta^2}\right) e^{-\beta^2}\, d\beta\right], \tag{7.3}$$

with $\alpha = z/2\sqrt{\nu t}$. Consequently, the retarding effect of the wall is only sensible in a layer whose thickness is comparable to $\sqrt{\nu t}$.[19]

In a second note, Boussinesq insisted that wall stress played an essential role in determining the nature of the motion in any hydraulic problem and that his simple calculation revealed the general mechanism through which rotational motion began in any flow:[20]

---

[17]These profiles agreed with Darcy's and Bazin's measurements. They can now be seen as approximations to the logarithmic profiles given by turbulent boundary-layer theory (cf. Prandtl [1933] p. 833n).

[18]Boussinesq [1880a].

[19]Stokes had already treated a similar problem in his pendulum memoir of 1850 (see later on pp. 290–1).

[20]Boussinesq [1880b] p. 967.

The retarding influence of a wall will first only be sensible in the vicinity of this wall. Hence some time will elapse before the similar influences of the other walls reach this region, and it will therefore be permitted to evaluate the velocity variation at the beginning of motion as if ... the wall under consideration had infinite breadth and the fluid mass had infinite thickness ... Hence, [my previous calculation] *most simply expresses what happens at the beginning of any flow, and demonstrates the general mechanism, abstracted from accessory complications.*

Ten years later, Boussinesq examined a similar question in an attempt to correct for entrance effects in some of Poiseuille's experiments, namely: how does the velocity profile of a viscous fluid entering a capillary tube evolve toward the uniform, parabolic profile? Assuming a rectangular profile at the entrance of the tube, he showed that an annular layer of retarded fluid grew from the walls until it reached the central part of the tube. Through an approximate solution of the Navier–Stokes equation, he found that the departure from the steady profile varied as $e^{-16\nu x/UR^2}$, where x is the distance from the entrance, R is the radius of the tube, U is the average velocity, and ν is the kinematic viscosity. Consequently, the retarded layer reaches the thickness R for a distance of the order $x = UR^2/\nu$, which means that the thickness grows with x as $\sqrt{\nu x/U}$.[21]

To summarize these French contributions, Saint-Venant's semi-empirical approach to hydraulic questions led to a well-defined strategy to take into account the turbulent character of the fluid motion in the resistance problem and in similar problems of retardation. Most essential were the recourse to momentum and energy balance in astutely-chosen spatial domains, and the concept of effective stress depending on eddy viscosity. Saint-Venant and Boussinesq thus made sense of a large number of hydraulic measurements. Their theories nonetheless lacked predictive power, for they involved adjustable *functions* giving the distribution of the turbulent eddying in the fluid. This objection does not apply to laminar flows, in which case Boussinesq obtained the beginning of the motion by purely deductive means. He did not extend these insights to the resistance problem, for he lived in a world of rivers and canals rather than ships or airplanes.

7.2 Ship resistance

Until the 1830s, the form of ship hulls was usually decided according to conservative and empirical principles. Naval architects distrusted theory—for good reason, as we may retrospectively judge. Contemporary hydraulics and hydrodynamics yielded an about even share of correct and incorrect ideas. True were the mostly quadratic dependence of resistance on velocity, Bossut's wave contribution, and Beaufoy's skin friction. Wrong were the concept of a bow resistance resulting from the impact of repelled water, the proportionality of the resistance with the mid-ship section, and the notion of a solid of least resistance. The latter ideas were dangerously stamped with Newton's authority. Pierre Bouguer enshrined them in his widely-used *Traité du navire* (1746).[22]

For most kinds of commercial ships, the resistance of water was only a minor consideration among others that determined the preferred form of the hull. The required amount of

---

[21]Boussinesq [1890], [1891]. The modern reader recognizes Prandtl's law for the growth of a laminar boundary layer.

[22]Cf. Wright [1983] chap. 2.

wood, the weight and volume of the intended cargo, and the stability at sea were most important. A better understanding of ship resistance only began to matter with the development of steam-powered, high-speed navigation in the first decades of the nineteenth century. One of the most important duties of the British Association for the Advancement of Science, founded in 1831, was precisely to favor the scientific study of navigation. A series of expert committees were formed to study ship resistance, stability, and propulsion.[23]

Scott Russell steered a couple of these committees in the 1830s, and thus promoted his ideas on the contribution of wave formation to ship resistance. The hollow lines of the bow, and the proportions he gave to the rest of the hull were meant to minimize wave formation. Although they improved on more conservative designs, they rested on a fragile theoretical basis. Russell understood little of the principles of mechanics, and reasoned mostly through intuition, analogy, and empirical induction. Where we would see energy wasted through the constant emission of periodic waves, he instead saw a conflict between the 'bow wave' (surge of water) and the progression of the ship.[24]

7.2.1 *Rankine's friction layer and stream lines*

The first British theorist of ship resistance who knew enough fluid mechanics was the Glasgow engineering professor William Macquorn Rankine. Educated at the University of Edinburgh, experienced in railways and hydraulic engineering, and a major contributor to the new mechanical theory of heat, Rankine best embodied a rising engineering science that profited from the fundamental theories of physics. His first considerations of ship resistance derived from his friendship with the Scottish shipbuilder James Napier who, in 1857, asked him for advice about the engine power necessary to propel a ship of given shape and size. Apparently, Napier did not trust the 'Admiralty formula' that had so far been used for this purpose:

$$\Pi = CSV^3, \tag{7.4}$$

where Π is the power, S is the mid-ship section, V is the velocity, and C is an empirical constant.[25]

Rankine communicated his own formula privately to Napier, and, 'for the sake of record', as an anagram in the August 1858 issue of the *Philosophical Magazine*. He proudly announced:[26]

> In the course of last year there were communicated to me *in confidence* the results of a great body of experimentation on the engine power required to propel steam-ships of various sizes and figures at various speeds. From these results I deduced a general formula for the resistance of ships having such figures as usually occur in steamers, which on the 23rd of December, 1857, I communicated to the owner of the experimental data; and he has since applied it to practice with complete success.

Five years later, Rankine revealed his secret theory to the learned public. For ships designed according to Russell's wave principle, Rankine reasoned, the main cause of resistance had to be 'skin friction', that is, the force exerted tangentially by the water sliding

---

[23]Cf. Wright [1983] Chap. 3.   [24]See Chapter 2, p. 51.
[25]Cf. Wright [1983] pp. 89, 106–19.   [26]Rankine [1858] p. 238.

on the hull. In approximate conformance with Beaufoy's old measurements and by analogy with pipe retardation, Rankine took this force to be proportional to the fluid density and to the square of the sliding velocity. Unlike Beaufoy, however, he did not assume this velocity to be constant along the hull. Instead, as 'the only assumption', he propounded 'that the agitation in the water caused by the friction on the ship's bottom extends only to a layer of water which is very thin as compared with the dimensions of the ship.' Beyond that layer, he assumed a smooth motion obeying Euler's equation. In the case of Russell's trochoidal lines, Rankine determined the corresponding flow in Gerstner's manner and thus obtained the sliding velocity v as a function of the curvilinear abscissa s along the hull.[27]

Rankine then equated the propelling engine power to the work done by the frictional forces:

$$P = f\rho V^3 G \int \left(\frac{v}{V}\right)^3 ds \approx f\rho V^3 GL\left(1 + \frac{\pi^2 B^2}{L_1^2}\right), \tag{7.5},$$

where f is a friction coefficient borrowed from Julius Weisbach's pipe-retardation formula, ρ is the density of water, G is the mean girth of the ship, V is its velocity, B is its greatest breadth, and L_1 is the length between the bow and the stern. The parenthesis or 'augmented length' factor contains the effect of the curvature of the hull. Rankine further obtained the resistance as the ratio P/V. The direct summation of the longitudinal component of the frictional force leads to a smaller result. Rankine attributed this discrepancy to a reaction of the hull on the water that slightly deformed the lines of flow and thus lowered the pressure on the stern.[28]

In 1864, Rankine clearly distinguished three contributions to ship resistance:

- a blunt stern leading to large eddies;
- a front surge leading to surface waves;
- frictional eddies.

For a fair-shaped ship at moderate velocity, this last cause was the only important one. Rankine described the relevant process in a manner probably reminiscent of Darcy:[29]

> The resistance due to frictional eddies ... is a combination of the direct and indirect effects of the adhesion between the skin of the ship and the particles of water which glide over it; which adhesion, together with the stiffness [viscosity] of the water, occasions the production of a vast number of small whirls or eddies in the layer of water immediately adjoining the ship's surface. The velocity with which the particles of water whirl in these eddies bears some fixed proportion to that with which these particles glide over the ship's surface: hence the actual energy of the whirling motion impressed on a given mass of water at the expense of the propelling power of the ship, being proportional to the square of the velocity of the whirling motion, is proportional to the square of the velocity of gliding.

[27]Rankine [1862] p. 23. Rankine identified the flow in a horizontal plane with the vertical section of a Gerstner wave (see Chapter 2, pp. 73–5). This kind of flow is not irrotational, unlike those later favored by Rankine.

[28]Rankine [1862] p. 24, [1863] p. 137, and [1864] p. 323 for the indirect effect of viscosity.

[29]Rankine [1864] p. 322. For Darcy's and Du Buat's similar ideas, see Chapter 6, pp. 224, 234.

In 1871, Rankine refined his description:

> It is well known through observation: that the friction between a ship and the water
> acts by producing a great number of very small eddies in a thin layer of water close to
> the skin of the vessel, and also an advancing motion in that layer of water; that this
> *frictional layer* (as it may be called) is of insensible thickness at the cutwater, and
> gradually increases in thickness towards the stern, by communication of the com-
> bined whirling and progressive motion to successive streams of particles; and that,
> finally the various elementary streams of which the frictional layer is composed,
> uniting at the stern of the ship, form the *wake*—that is, a steady or nearly steady
> current, full of small eddies, which follow the ship, but at a speed relatively to still
> water which is less than the speed of the ship.

From this picture, Rankine derived the equality of the resistance with the momentum flux
in the wake. If V is the velocity of the ship, A is the area of a section of the wake, and U
is its average velocity, Rankine reasoned, then the mass of water fed into the wake is
$\rho A(V - U)$ per unit time, and its momentum is $\rho A(V - U)U$. Rankine chose $U = V/2$,
which gives the smallest wake section, $4R/\rho V^2$, for a given resistance R.[30]

As the dominance of skin resistance depended on the fairness of the ship's shape,
Rankine wondered whether Russell's wave lines were the only ones that prevented wave
formation. In order to answer this question, he considered simple, two-dimensional,
potential flows obtained by superposing the flows defined by two opposite foci (a source
and a sink) and a uniform flow directed along the lines joining the two foci. Figure 7.7
represents the lines of flow that asymptotically merge with the uniform flow. Rankine
called them 'oögenous neoïds', for they correspond to the potential flow around a solid

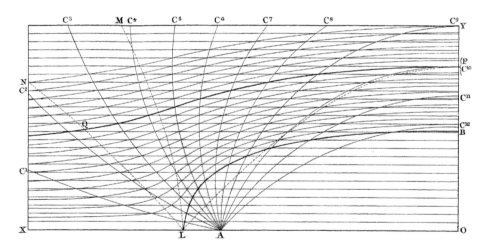

Fig. 7.7.   Rankine's bifocal lines of flow. A is one of the foci (the other being its mirror image through OY), LB
is the limiting oval, and PQ is the 'lissenoïd' or line of minimal vertical disturbance. The lines AC are
construction lines. From Rankine [1865] plate.

[30]Rankine [1871] pp. 300–1.

limited by the central oval. Yet he did not regard the oval as a plausible ship shape, since the resulting flow implies an abrupt vertical disturbance (through Bernoulli's law, the local depression of the free surface varies as the square of the fluid velocity). Rather, he selected the stream lines for which the velocity differs the least from that of the neighboring lines. As these 'lissenoïds' unfortunately have parallel asymptotes, Rankine cut them off at the point of slowest gliding and completed them with a plausible, edge-shaped stern and bow. The resulting shapes resembled Scott Russell's wave lines.[31]

When Rankine published these considerations, in 1864, he seems to have believed that laminar flow was only possible around special, simple solids such as those given by the bifocal method. In reality, Laplace's equation for the velocity potential admits a solution that meets the boundary conditions for a body of any shape. Being close to William Thomson, to whom potential theory had no secret, Rankine could not remain long in error. In a note of 1870 he explained:

> Although every surface is a possible stream-line, the surface of a ship is not even approximately an actual stream-line surface unless it is such that she does not drag along with her a mass of eddies of such volume and shape as to cause the actual tracks of the particles of water to differ materially in form from those which would be described in the absence of eddies.

Being now aware of William Froude's water-bird profiles (to be discussed shortly), Rankine added two more foci to his previous scheme and obtained the 'cynoid' lines. As the number of foci was in principle unlimited, there seemed to be no limit to the variety of imaginable ship shapes. While gaining generality, Rankine's method lost predictive power.[32]

7.2.2   Froude's models

Even though Rankine's contributions marked a significant progress in the understanding of ship resistance, they turned out to be of little value in the computation of resistance, or so it appears from the report of the British Association Committee on 'Resistance of water' that Russell, Napier, Rankine, and Froude directed from 1863 to 1866. This failure probably motivated William Froude's experiments of 1865–1867 on models. This country gentleman worked as a railway engineer until 1845 and retired at the early age of thirty-five to look after his ailing father. He had an elementary knowledge of mathematics, but a very good understanding of the laws of mechanics. In 1856, his former employer and Chief Engineer of the Great Eastern, Isambard Brunel, asked him for help in the study of wave-induced ship rolling. Froude's outstanding contributions to this subject won him the favors of the British Association, the ear of the Admiralty, and a membership of the Institution of Naval Architects. Worth noting are his consideration of skin friction as one of the damping factors of the rolling motion, and his use of similitude conditions to exploit rolling measurements carried out on a model of the Great Eastern.[33]

---

[31]Rankine [1865].   [32]Rankine [1871] p. 267n (note dated Dec. 1870).

[33]For biographical information, cf. Abell [1933], Brown [1992]. For a penetrating analysis of his works, cf. Wright [1983] chaps 6, 7.

In the absence of a priori means to determine the most advantageous ship shapes, Froude consulted experiments. As full-scale trials excluded any radically innovative shape, he built models of small dimensions and towed them in Dartmouth harbor. Unlike previous ship-model experimenters, Froude understood that the rational use of models to derive the behavior of full-scale ships required adequate scaling rules. On the basis of the expression $\sqrt{g\lambda/2\pi}$ for the celerity of a deep-water wave of length λ, he argued that similar wave patterns for models at different scales required a towing velocity proportional to the square of the dimensions of the model. Assuming that the total resistance varied as the square of the velocity and the square of a 'ruling dimension', he further expected this resistance to vary as the cube of the dimensions of the model.[34]

Such were the first scaling rules explained by Froude in an unpublished report to the Admiralty of April 1868. In an improved report of December 1868, he recognized that the wave component of the resistance did not generally vary as the squared velocity, but nonetheless varied as the cube of the dimensions of the ship. The advocated reason for this simple law was that the height of the waves as well as their length and breadth varied as the linear dimensions of the ship, so that their energy varied as the cube of these dimensions. As for the skin resistance, Froude believed that pipe-retardation measurements sufficiently proved its quadratic form. In his opinion, Beaufoy's 1.8 exponent probably resulted from experimental errors—a view that Froude revised a couple of years later.[35]

Froude did not explicitly introduce the 'Froude number' of modern navigation theory. Nor did he reason from fundamental principles or equations. Newton had briefly done so in the section of the *Principia* devoted to fluid resistance, and Joseph Bertrand had given the general similitude conditions of rational mechanics in 1848. Most relevantly, the director of the Ecole d'Application du Génie Maritime in Lorient, Ferdinand Reech, derived the similitude conditions for ship models in 1844 and included them in his lectures on mechanics. Although Froude seems to have been unaware of Reech's reasoning, its generality and rigor deserve a few lines.[36]

In any mechanical system, Reech reasoned, the equations of equilibrium are unchanged under global change of the length scale, as long as this change affects all forces in the same proportion. Taking d'Alembert's principle into account, the equations of motion are unchanged through a change of the length and velocity scales, if this change affects all forces, including the inertial forces, in the same proportion. Denote by α and β the factors by which the lengths and velocities are respectively multiplied. Then inertial and gravitational forces are multiplied by $\alpha^2\beta^2$ and α^3, respectively. For a system in which the acting forces are inertial and gravitational, the equations of motion will be invariant if $\alpha^2\beta^2 = \alpha^3$, or $\alpha = \beta^2$; so velocities must vary as the square root of a length in order that both kind of forces vary as the cube of a length. As Reech further noted, atmospheric pressure and viscosity forces behave differently, so that the rule no longer applies in problems for which these forces are

---

[34]Unpublished report sent to Edward Reed, discussed in Wright [1983] p. 210. Rankine ([1862] p. 28) had earlier expressed the condition that 'the velocities of the model and of the ship should be proportional to the square roots of their linear dimensions' in order that the wave effects should be comparable.

[35]Froude [1868].

[36]Reech [1844] p. 166, [1852] pp. 265–75. On the broader history of similitude and models, cf. Wright [1983] chap. 8, [1992].

not negligible. With this restriction, Reech concluded that 'Newton's theorem of similitude would always be the best and often the unique foundation of many practical applications [of mechanics]', and contrasted this power of similitude arguments with the meager yield of higher theories based on Euler's or Navier's 'special equations'.[37]

In order to verify his (or Reech's) scaling rules, in 1867 Froude built a series of models at different scales (3, 6, and 12 feet) for two shapes, namely, a wave-line shape he called Raven, and a water-bird shape he called Swan (see Fig. 7.8). The results confirmed his expectations, although modern analysis of his data has shown enormous errors (up to 50%!), probably due to a flawed dynamometer. He also concluded that the odd water-bird shape was superior to the wave-line shape at high velocities. This finding justified the need for further model experiments in which a large variety of unusual shapes could be tested. Froude soon planned the construction of a towing tank that would permit sufficient precision in such experiments.[38]

The project required important funds, which Froude secured from the Admiralty. The chief constructor of the Navy, Edward Reed, approved Froude's exploratory approach to ship form, in part because for iron-clad warships the high cost of iron excluded the slender forms recommended by Russell. Civil naval engineers were far less enthusiastic. In a British Association report of 1869, the Principal of the Royal School of Naval Architecture and Marine Engineers, Charles Merrifield, pointed to Reech's similitude conditions and held ignorance of these conditions responsible for the past failures of the models approach. Yet his general distrust of theory prompted him to recommend a new series of full-scale experiments in the name of the BA committee.[39]

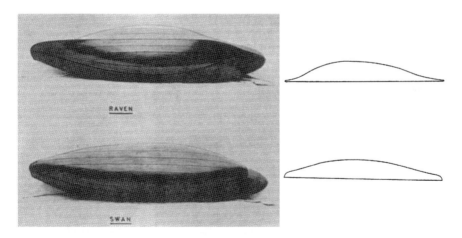

Fig. 7.8. Froude's Raven and Swan models. From Froude [1957] p. 132 (photos), [1869b] (half water lines).

[37]Reech [1852] p. 274.

[38]Cf. Wright [1983] pp. 131–6.

[39]Merrifield [1869] pp. 24–5. Merrifield had read Eugène Flachat's treatise on navigation, which reproduced writings by Siméon Bourgois and Stanislas Dupuy de Lôme, including discussions of Reech's similitude conditions (Flachat [1866] vol. 1, pp. 165n–167n, 214n). Bourgois noted that the frictional resistance measured by Beaufoy did not meet these conditions, so that their application to the total resistance could only be approximate (*ibid.* p. 167n).

Froude, who belonged to this committee, defended his own model approach in a long appendix to the report. He did the same at the Institution of Naval Architects, where Russell cited his own past failure to exploit model data and ironically questioned the future of this approach:

> You will have on the small scale a series of beautiful, interesting little experiments, which I am sure will afford Mr. Froude infinite pleasure in making them, as they did to me, and will afford you infinite pleasure in the hearing of them; but which are quite remote from any practical results upon the large scale.

Froude's defense brought forward the similitude conditions, the need to explore unusual shapes, the practical impossibility of predicting wave resistance, the agreement of his views with Rankine's earlier theories, and the success of his preliminary experiments on Raven and Swan. He conceded difficulties with small-scale towing, especially in the achievement of uniform speed, but felt able to surmount them. Merrifield, whose own full-scale towing project had just been rejected by the Admiralty, rejoiced magnanimously over the support given to 'a man of proven ability' (Froude).[40]

Froude built a 25-foot long, 33-foot wide, 10-foot deep tank in Chelton Cross, near his home town of Torquay. In his first experiments in this tank, reported in 1872, he towed a plate edgewise through the water, with the skin friction of ship hulls in mind. Like Rankine, he believed that the resistance of any fair-shaped ship was mainly due to skin friction, and therefore computable if the laws of this sort of resistance were known. However, he doubted the correctness of Rankine's and others' assumptions about these laws. His suspicion derived from his involvement in a water-main problem in Torquay around 1869. After a few tests, he had determined that the deplored loss of head was not due to obstructions, but to the roughness of the oxidized internal surface of the pipe. Scraping solved the problem.[41]

While pondering on the effect of roughness—which he wrongly believed to be unknown to hydraulicians—Froude came to question Beaufoy's and Rankine's assumption that the friction on a plate moving edgewise was uniform along it. In a memoir of 1869, he explained why it should not be so:[42]

> It is certain that the anterior portions of the surface, in rubbing against the particles which it passes, and experiencing resistance from them, *must* impress on them an equivalent force in the direction of the motion, and *must* impart to them some velocity in that direction. Thus, though it may be in some sense asserted that the anterior portions of the plane rub against the contiguous particles with the entire velocity of the plane, since these particles are undisturbed, this cannot be truly asserted of the posterior portions of the plane, since the particles against which these rub have already received a velocity conformable to that of the plane; and a 'state of motion' will be thus produced in the contiguous particles involving a widening body of fluid, and with increasing velocity imparted to it, as we recede foot by foot sternward along the plane; forming in fact a 'current', created and left

[40]Froude [1869*b*]; Merrifield [1870] pp. 82 (Russell's comment), 87–90 (Froude's defense), 80–1 (Merrifield).

[41]See R. Froude [1869].

[42]W. Froude [1869*a*] p. 212. Darcy had earlier emphasized the role of roughness, see Chapter 6, p. 232. Froude's description anticipates three features of modern boundary-layer theory, namely, the growth of the layer, the gradual decrease of wall stress along the wall, and the momentum balance.

behind, by the transit of the plane, such that if we could integrate the volume of current created in each unit of time, and the exact velocity possessed by each of its particles, the aggregate momentum must be precisely that which is due to the frictional resistance of the entire plane acting during that unit of time. Obviously the sternward portions of the plane moving forward in such a favouring current, must experience a less intense frictional resistance than the anterior portions.

With Rankine, Froude shared the idea of a growing layer of dragged fluid and the relation between wake momentum and resistance. Unlike Rankine, he did not regard the friction on the walls as being determined by the sliding velocity of the potential flow along the surface of the body. Instead, he made this friction depend on the normal gradient of the longitudinal fluid velocity. Whereas Rankine's sliding velocity is a constant along a plane, Froude's wall stress decreases along the plate owing to internal fluid friction. As Froude later wrote, 'it is the motion of the surface relative to contiguous particles, and not relative to distant ones, that governs the resistance.' Based on this idea, he indicated a way to compute the spreading of the motion from an infinite plate suddenly set in uniform motion in its own plane: he assumed the frictional force to be $\alpha(\partial u/\partial y)^2$ between consecutive layers of the fluid, and balanced the inertial force of each layer (of thickness η) with the difference $2\eta\alpha(\partial u/\partial y)(\partial^2 u/\partial y^2)$ between the frictional forces on its two faces. Froude next suggested that the solution would also apply to the case of a finite plane penetrating a still fluid with a constant velocity. However, he was reaching the limits of his mathematics.[43]

In his plank-towing experiments of 1872 and 1874, Froude verified that the resistance was not proportional to the length of the plank, and that it depended on the roughness of the surface, with a velocity exponent ranging between 1.83 for the smoothest surface (varnished) and 2.0 for the roughest one (sand-coated) in the case of the longest plank (50 feet). Whereas he did not comment on the relation between exponent and roughness, he gave the following discussion of the unexpectedly slow decrease of the friction a few feet behind the cutwater.[44]

Assuming an approximately linear transverse variation of the velocity in the current induced by the plank's motion on each face, denoting by H the thickness of this current at the end of the plank, U the velocity of the plank, and ρ the density of water, Froude estimated the momentum flux (per unit breadth) in the wake to be $\rho H U^2/3$. Equating this value to the measured resistance in Rankine's manner, he derived values of H that matched observations and that increased with the length of the plank. This growth of the favoring current suggested a rapid decrease of the friction with the distance from the cutwater, in contradiction with the measurements. As a solution to this paradox, Froude imagined that violent eddying in the boundary current fed undisturbed fluid particles from the outer margin of the current to the surface of the plank.[45]

[43]Froude [1874a] p. 253 (quote), [1869a] pp. 212–13 (computation). The modern reader may recognize Prandtl's assumption for the stress within a turbulent boundary layer in the case of a constant mixing length $\sqrt{\alpha/\rho}$ (in reality, the mixing length grows linearly with the distance from the wall), as well as Rayleigh's idea ([1911]) of connecting temporal and spatial growths of a boundary layer. See later on pp. 290–1, 297–9.

[44]Froude [1872], [1874a].

[45]Froude [1874a] p. 253. In conformance with Froude's view, Kármán's theory of 1921 yields a turbulent boundary layer growing as the power 4 / 5 of the distance from the cutwater, and a wall stress decreasing as the power $-1/5$ of this distance. See later on pp. 296–7.

In the same period, Froude performed full-scale experiments on HMS *Greyhound* and compared the results with measurements done on a model of this ship. This time Froude no longer assumed a quadratic form of the skin friction. Instead, he computed the skin resistance by extrapolation from his plank measurements. Then he subtracted this resistance from the measured total resistance, and applied his scaling rules to the remaining resistance. An impressive match between the model data and the full-scale data resulted. Froude concluded:[46]

> The experiments with the ship, when compared with those tried with her model, substantially verify the law of comparison which has been propounded by me as governing the relation between the resistances of ships and their models. This justifies the reliance I have placed on the method of investigating the effects of variation of form by trials with varied models—a method which, if trustworthy, is equally serviceable for testing abstract formulae, or for feeling the way towards perfection by a strictly inductive process.

Froude's main service to naval engineering was indeed the development of the rational use of models. As he showed, the proper exploitation of model data required the knowledge of the scaling laws for non-frictional resistance, and some understanding of the mechanism of skin friction. He modestly admitted to having borrowed most of his theoretical ideas from colleagues with higher mathematical skills (Rankine, Thomson, and Stokes): 'I am but insisting on views which the highest mathematicians of the day have established irrefutably; and my work has been to appreciate and adapt these views when presented to me.' Froude nevertheless grasped aspects of fluid motion that had eluded his predecessors. He understood that the variation of friction along a ship hull depended on an internal fluid-stress mechanism acting within a growing boundary layer of dragged fluid. Moreover, he foresaw the role of destructive wave interference in lowering the wave resistance of some ship shapes, such as the Swan of 1867, and he described important wave phenomena, such as group velocity and echelon waves, thus stimulating mathematical studies by Stokes, Rayleigh, and Kelvin.[47]

A last service of Froude was his simple, pedagogical explanation of the principles of ship resistance for lay audiences. Unconsciously imitating Euler, he derived d'Alembert's paradox through momentum balance along tubes of flow, thus condemning the fallacy of 'head resistance' that had so long impeded the progress of naval architecture. The true causes of ship resistance, he went on, were skin friction, wave emission, and large-eddy production. About each kind of resistance he had a simple wisdom to offer. Skin resistance is about the same on a ship hull as on a flat surface, wave resistance only counts at velocities for which the wavelength is comparable to the ship's dimensions, and eddy resistance essentially depends on the tendency of stream lines to separate from a blunt stern and thus to form a dead-water, eddying region: 'Blunt tails rather than blunt nose cause eddies.'[48]

---

[46]Froude [1874*b*] p. 59.

[47]Froude [1877*a*] p. 213, [1877*b*] (on ship waves). See Chapter 2, pp. 85–6.

[48]Froude [1875], [1877*a*] p. 205. Rayleigh ([1918] p. 553) claims to have obtained from Froude the idea (usually attributed to Prandtl) that separation is due to 'the loss of velocity near the walls in consequence of fluid friction, which is such that the fluid in question is unable to penetrate into what should be the region of higher pressure.' I have not been able to locate any statement of this sort in Froude's writings.

In summary, the development of steam-powered navigation prompted scientific studies of fluid resistance by Russell, Rankine, and Froude, spanning from the mid-1830s to the 1870s. These three investigators recognized, with increasing accuracy, the importance of wave resistance for partially-immersed bodies moving at sufficient speed. None of them, however, could theoretically predict the amount of this resistance. Froude remedied this weakness by a rational use of model measurements. Rankine and Froude recognized that the motion of water around a fair-shaped ship hull was mostly governed by the corresponding irrotational solution of Euler's equation, except for a layer of fluid adjacent to the hull, in which complex eddying motion occurred. Froude understood that the behavior of this layer and the resulting skin friction depended on internal friction within the layer. As he did not have the means to develop a quantitative theory of this behavior, he again relied on small-scale experiments, and extrapolated the results to large-scale skin friction.

This research better achieved its aim, namely the prediction of ship resistance, than the resistance theories discussed above and based on the concepts of discontinuity surfaces and eddy viscosity. The key to this empirical efficiency was not the elaboration of a quantitative, deductive theory. It was a qualitative understanding of the implied physical processes along with the rational exploitation of small-scale experiments.

7.3 Boundary layers

7.3.1 *Prandtl's Heidelberg paper*

After completing his engineering studies at the Technische Hochschule in Munich, Ludwig Prandtl obtained a doctorate in 1898 under Ludwig Föppl on the lateral instability of beams in bending. From Föppl he learnt a kind of engineering science that relied on higher mathematical skills, fundamental physical theory, and multifarious approximation strategies. He went on to work in the Maschinenfabrik Augsburg-Nürnberg, where he was asked to improve a suction device for the removal of shavings. While working on this project, he realized that the pressure rise expected in a sharply-divergent tube failed to occur because the lines of flow tended to separate from the walls—as Daniel Bernoulli had long ago noted in a similar hydraulic case. Prandtl later remembered this observation to have started the chain of reasoning that led him to the boundary-layer approach to resistance in slightly-viscous fluids.[49]

In the following years, Prandtl developed his resistance theory and tested it with a water tank of his own making, while teaching mechanics at the Technische Hochschule in Hannover. At the third international congress of mathematics held in Heidelberg in 1904, he had ten minutes to announce results that inspired much fruitful research in subsequent years. In the short, dense report published in 1905, he began with the 'unpleasant properties' of the Navier–Stokes equation

$$\rho\left[\frac{\partial \mathbf{v}}{\partial t} + (\mathbf{v} \cdot \nabla)\mathbf{v}\right] = -\nabla(P + V) + \mu\Delta\mathbf{v}. \tag{7.6}$$

Solutions were known, he noted, for the simpler equations obtained by omitting either the nonlinear term $\rho(\mathbf{v} \cdot \nabla)\mathbf{v}$ or the viscous term $\mu\Delta\mathbf{v}$. No non-trivial solution had yet been

---

[49]Prandtl [1948]. For biographical data, cf. Lienhard [1975], Rotta [1990].

found for the complete equation. For a slightly-viscous fluid such as water or air, a natural course was to omit the viscous term. Alas, the resulting solutions to resistance problems differed widely from the observed behavior. A different strategy was needed.[50]

Prandtl assumed that the viscous term $\mu \Delta \mathbf{v}$ could be neglected everywhere except in a 'boundary layer' (*Grenzschicht*) or 'transition layer' (*Übergangschicht*) of fluid near the solid walls on which the fluid adheres. This layer remains thin only if the path of fluid particles along the walls is not too long. Without proof, Prandtl further asserted that, if the viscosity μ is an infinitesimal of second order, then the width of the transition layer and the normal velocity within the layer are of first order, and the normal pressure gradient and the curvature of the lines of flow are negligible. He presumably reasoned as follows.[51]

For a two-dimensional flow, denote by δ the thickness of the transition layer, u the parallel velocity, v the normal velocity, x the curvilinear abscissa along the wall, and y a normal curvilinear coordinate. As long as the curvature radius of the surface is large compared to the thickness of the layer, the differential equations of the motion within the layer have the same form as if x and y were Cartesian coordinates. As the velocity within the layer varies much faster in the normal than in the parallel direction, $\partial^2 u / \partial x^2$ is negligible compared to $\partial^2 u / \partial y^2$, and the Navier–Stokes equation for u reads

$$\frac{\partial u}{\partial t} + u \frac{\partial u}{\partial x} + v \frac{\partial u}{\partial y} = -\frac{1}{\rho} \frac{\partial P}{\partial x} + v \frac{\partial^2 u}{\partial y^2}. \tag{7.7}$$

The continuity equation reads

$$\frac{\partial u}{\partial x} + \frac{\partial v}{\partial y} = 0. \tag{7.8}$$

In the zero-viscosity limit, and at a given fraction y/δ of the transition layer, the terms $u \partial u / \partial x$, $-(1/\rho) \partial P / \partial x$, and $\partial u / \partial x$ in these equations must remain finite; the term $v \partial^2 u / \partial y^2$ is of the order of v/δ^2, and the term $\partial v / \partial y$ is of the order of v/δ. Consequently, v is of the same order as δ, which is of the same order as $\sqrt{v}$, and all of the terms of eqn (7.7) are of the same order. The Navier–Stokes equation for v further implies that $\partial P / \partial y$ is negligible, because all other terms are of the same order as δ. Therefore, the term $-(1/\rho) \partial P / \partial x$ in eqn (7.7) may be regarded as a known function of x only that can be obtained by solving the Eulerian flow problem along the given solid body. Prandtl obtained the velocity profile of the boundary layer through the numerical, stepwise integration of eqns (7.7) and (7.8).[52]

The simplest case is that of a uniform flow of velocity U encountering a parallel, infinite blade (see Fig. 7.9) in the domain $x > 0, y = 0$. The corresponding Eulerian flow is strictly uniform, so that the pressure gradient vanishes. Prandtl asserted without proof that the velocity component u was a function of $y/\sqrt{x}$ only. Presumably, he guessed that, in the

---

[50]Prandtl [1905] pp. 575–6. Cf. Ackroyd *et al.* [2001] Chap. 9; Heidelberger [2006].

[51]*Ibid.* pp. 576–577. Even though *Grenzschicht* occurs only once in this paper, it is the term that Prandtl later preferred.

[52]A similar reasoning is found in Blasius [1908] pp. 2–3. Prandtl is not likely to have used dimensionless variables, for these only became popular in later years.

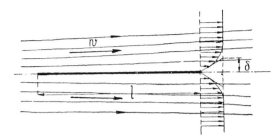

Fig. 7.9.   Flow along a flat plate. Here v is the asymptotic velocity, l the length of the plate, and is δ is the thickness of the boundary layer at the end of the plate. From Prandtl 1931b: 90.

absence of a characteristic length (such as the length of the plate), the parallel-velocity profiles at different points of the plate only differed by the y-scale. Formally, this means that, for any constant α, there are two other constants β and γ for which the substitution $u, v, x, y \to u, \gamma v, \alpha x, \beta y$ leaves the boundary-layer equations and the boundary conditions (zero velocity on the blade, u equal to U far from the plate) invariant. This is indeed the case if $\alpha = \beta^2$ and $\beta\gamma = 1$. As the solution should be unique for given boundary conditions, we have $u(x,y) = u(\alpha x, y\sqrt{\alpha})$ for any values of x, y, and α. The choice $\alpha = 1/x$ leads to $u(x, y) = u(1, y/\sqrt{x})$, in conformance with Prandtl's assertion.[53]

Prandtl then solved the resulting ordinary differential equation numerically. Integrating the stress $\mu \, \partial u/\partial y$ on both sides of the blade from the edge to the length l, he reached the resistance formula

$$R = 1.1 \, b\sqrt{\mu\rho l U^3}, \tag{7.9}$$

where b is the breadth of the blade. He thereby assumed that the boundary layer of a finite-length blade was approximately the same as the $x < l$ part of the boundary layer of an infinite plate.

Prandtl next proceeded to 'the most important result with regard to application', that is, the separation (*Ablösung*) of the fluid current from the wall in the presence of an antagonistic pressure gradient. Such a gradient typically occurs at the rear of a bluff-shaped body, where the lines of the Eulerian flow spread out, the sliding velocity diminishes, and the pressure therefore increases along the wall (through Bernoulli's law). Owing to viscous damping, Prandtl reasoned, the fluid in the transition layer may reach a point at which it does not have enough kinetic energy to surmount the pressure gradient, in which case it shoots off the wall. Prandtl drew the evolution of the velocity profile in such cases, and argued that separation occurred at the point $\partial u/\partial y = 0$, beyond which an absurd backward flow would occur if separation did not prevent it (see Fig. 7.10). He assumed the separation to result in a vortex sheet *à la* Helmholtz:

> A layer of fluid that has been set into rotation through wall friction thus pushes itself into the free fluid and there plays the same role as Helmholtz's separation layers (*Trennungschichten*) in effecting a complete reconfiguration of the motion.

---

[53]Prandtl [1905] p. 578. A similar reasoning is found in Blasius [1908] p. 5.

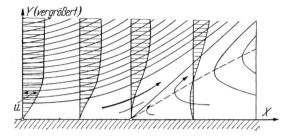

Fig. 7.10.   The evolution of a boundary
   layer in an antagonistic pressure gradient. From Prandtl [1905] p. 578.

Prandlt summed up:[54]

> The treatment of a given flow process divides itself in two mutually interacting parts:
> on one hand we have the *free fluid* that can be treated as friction-free according to
> Helmholtz's laws of vortex motion, on the other hand we have the transition layers
> on the solid boundaries, the motion of which is ruled by the free fluid, but which in
> return give to the latter its characteristic imprint.

Prandtl then showed theoretical drawings of the formation of a separation surface at the
edge of a plate and behind a cylinder (see Fig. 7.11). He emphasized the instability of these
surfaces, with the characteristic spiral unrolling identified by Helmholtz. Lastly, he described the apparatus he had used to verify (or reach?) his insights, a waterwheel-driven
water current with suspended metal dust (see Fig. 7.12), and gave the pictures of the
observed motions behind the edge of a blade and behind a cylinder (see Fig. 7.13). In
the latter case, he showed that the separation process could be prevented by pumping
off the fluid of the boundary layer though a slit on the wall of the cylinder.[55]

Comparing this communication with earlier notions of a boundary layer and a separation surface by Stokes, Thomson, Rankine, Froude, Boussinesq, and Levi-Civita, two
specificities stand out. Firstly, Prandtl was able to mathematically derive the velocity
profile within a laminar boundary layer and the resulting contribution to the resistance,
whereas his predecessors (with the exception of Boussinesq) only had qualitative knowledge of the layer. Secondly, Prandtl saw that the separation process and the departure
point of discontinuity surfaces depended on how the velocity profile of the layer evolved
along the walls, whereas previous advocates of flow separation and discontinuity surfaces
ignored viscosity and the role of viscous stress in determining the separation point.[56]

7.3.2   *Prandtl's heuristics*

According to Prandtl, a first key to his success in this and other problems was his ability to
develop an intuitive, visual understanding of the phenomena before trying to set them into
equations:

[54]Prandtl [1905] pp. 578–9.

[55]*Ibid.* pp. 580–4. Prandtl later ([1927b] pp. 768–9) gave aerodynamic illustrations of the prevention of
separation.

[56]In their correspondence of December 1898 (see earlier on p. 269), Kelvin and Stokes regarded small viscosity as
being responsible for a high-shear instability in the boundary layer. However, they did not relate the separation point
with the velocity profile in this layer. According to Prandtl, the position of the separation point does not depend on
the value of the viscosity since the condition $\partial u/\partial y = 0$ does not. Yet the separation mechanism requires a finite
value of the viscosity (see Chapter 5, pp. 214–5 for Rayleigh's discovery of a similar occurrence in 1883).

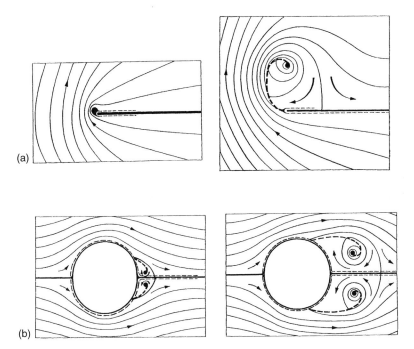

Fig. 7.11. Initial stages of the discontinuous fluid motion (a) past an edge, and (b) behind a cylinder. From Prandtl [1905] p. 579–80.

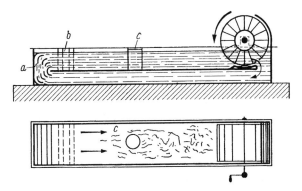

Fig. 7.12. Prandtl's apparatus for studying the flow past a solid obstacle (at c). The water is set into motion by the paddle-wheel. The four sifters at b homogenize the flow after the sharp turn at a. From Prandtl [1905] p. 581.

> Herr Heisenberg has ... alleged that I had the ability to see without calculation what solutions the equations have. In reality I do not have this ability, but I strive to form the most penetrating *intuition* [*Anschauung*] I can of the *things* that make the basis of the problem, and I try to *understand* the processes. The equations come only later, when I think I have understood the matter.

The sort of intuition he had in mind was acquired by 'special training', in the manner exemplified in his *Digest* [*Abriss*] *of the science of flow*. Instead of deriving the fundamental hydrodynamic equations and then discussing their consequences for a given hydrodynamic

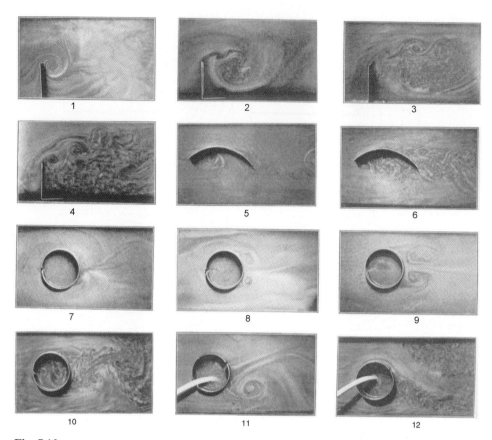

Fig. 7.13.   Pictures of the initial flow past the edges of flat and curved plates (2–6); past a cylinder, without (8–9) and with (11–12) suction through a slit. From Prandtl [1905] plate.

system, he directly applied Newton's laws of motion to slices of the tubes of flow of the system, thus combining geometrical representation and dynamical understanding of the flow. Intuition was the experience gained by working out series of concrete examples in this manner. As Prandtl remembered, 'in the examples of mechanics, I gradually got used to "see" the forces and accelerations in the equations and sketches or to "feel" them by muscular sense.' When he learned the Navier–Stokes equation, he studied examples of viscous flow in order to appreciate the relative importance of each term and thus 'to penetrate the mode of action of this equation'.[57]

There was, however, a more specific key to Prandtl's invention of boundary-layer theory:

> When the complete mathematical problem looks hopeless, it is recommended to enquire what happens when one essential parameter of the problem reaches the limit

[57]Prandtl [1948] pp. 1604–5.

zero. It is assumed that the problem is strictly soluble when this parameter is set to zero from the start and that for very small values of the parameter a simplified approximate solution is possible. Then it must still be checked whether the limiting process and the direct way lead to the same solution. Let the boundary conditions be chosen so that the answer is positive. The old saying '*Natura non facit saltus*' decides the physical soundness of the solution: in nature the parameter is arbitrarily small, but it never vanishes. Consequently, the first way [the limiting process] is the physically correct one!

From this, we may infer that Prandtl conceived the boundary layer and the separation process by requiring that the zero-viscosity limit of the viscous flow should resemble the perfect-fluid flow. The finite fluid slide on a rigid wall in the latter case suggests a thin layer of intense shear in the former. Also, Helmholtz's recourse to discontinuity surfaces (altered boundary conditions) suggests separation in slightly-viscous fluids. Reciprocally, the working out of boundary-layer dynamics informs the genesis of separation surfaces.[58]

7.3.3 Lanchester and Rayleigh

In 1907, the British automobile engineer and flight enthusiast Frederick Lanchester published his *Aerodynamics*, including a description of boundary layer and separation that was clearly independent of Prandtl's. Lanchester gave much importance to Helmholtz's surfaces of discontinuity, to the point of defining a streamlined body as a body for which motion through a fluid does not give rise to a surface of discontinuity. For non-streamlined bodies, he ascribed most of the resistance to low pressure in the dead-water region within the surface of discontinuity. Around any body within a stream of a viscous fluid, he argued, there must be a layer of dead water adhering to the surface of the body. If the viscosity is small, then this layer is extremely thin near the cutwater, but grows in the sternward direction owing to internal friction. Along a curved surface, this dead water tends to move toward the places of lower pressure. For instance, in the case of a sphere the dead water tends to accumulate near the equator (the axis being parallel to the flow). If the curvature is too high, then viscous drag is not sufficient to 'pump off' the excess of dead water, and a discontinuity surface is formed. Lanchester thus made separation depend on the competition between external pressure gradient and internal viscous stress, as Prandtl had done differently in 1904.[59]

Lanchester also investigated skin friction along a plate advancing with the velocity U through a still fluid. Following Rankine and Froude, he balanced the frictional force on the plate with the momentum increase in the boundary layer. In Prandtl's symbols, this gives

$$\mu \frac{\partial u}{\partial y} = \frac{d}{dx} \int_0^{+\infty} \rho(u - U)^2 dy. \tag{7.10}$$

Lanchester then assumed that the flow caused by the velocity αU only differed from the flow induced by the velocity U through the rescaling $u \to \alpha u$, $y \to \beta y$ of the velocity

---

[58] *Ibid.* p. 1606.

[59] Lanchester [1907] pp. 27–30. Cf. Ackroyd [1992], [1996].

profile. Since the previous balance between inertial and frictional forces must be preserved in the new flow, the relation $\alpha\beta^2 = 1$ must hold. The friction is therefore multiplied by $\alpha^{3/2}$, which means that the resistance is proportional to $U^{3/2}$. Dimensional homogeneity further requires the resistance to have the form

$$R = Cb\sqrt{\mu\rho l U^3},\tag{7.11}$$

in conformance with Prandtl's result (7.9).[60]

Besides this remarkably simple derivation of the form of the laminar resistance law for an edgewise moving plate, Lanchester explained that the $U^{3/2}$ dependence corresponded to a form of resistance intermediate between purely viscous and purely inertial. In purely viscous cases, such as Stokes's pendulum, the resistance is entirely due to the energy dissipated by the viscous stresses and is therefore proportional to the velocity. In purely inertial cases, such as eddy production at a blunt stern, the resistance corresponds to the kinetic energy of a continually-generated wake of eddies and is therefore proportional to the velocity squared. For the edgewise moving plate, both effects are combined because both heat and wake are generated. Lanchester further noted that his derivation of the $U^{3/2}$ law required the motion around the plate to be laminar. As he knew from Rankine and Froude, the flow is in fact turbulent along a ship hull. In this case, Lanchester's intuition led to a velocity exponent intermediate between 3/2 and 2, in conformance with Beaufoy's and Froude's measurements.[61]

Lanchester's derivation of the $U^{3/2}$ law intrigued Lord Rayleigh. In 1911, this champion of dimensional reasoning commented that the only changes in space and velocity scale that led to geometrically-similar motions were those for which the Reynolds number Ul/ν of the plate was left invariant. Lanchester's special rescaling assumption only made sense if the plate was so long that its length did not significantly affect the structure of the boundary layer. Rayleigh went on to derive this structure on the basis of an analogy with a problem that Stokes had long ago solved in his pendulum memoir, namely, the flow induced by an infinite plate suddenly set into constant motion in its own plane.[62]

As in the similar problem treated by Boussinesq, the only nonzero component of the induced fluid motion satisfies the equation

$$\frac{\partial u}{\partial t} = \nu\frac{\partial^2 u}{\partial y^2},\tag{7.12}$$

which has the same form as the equation for the propagation of heat. For the given boundary condition (if $t < 0$ then $u = 0$ everywhere, and if $t \geq 0$ then $u = U$ for $y = 0$ and $u = 0$ for $y = \infty$), Stokes obtained the solution by Fourier analysis as

---

[60]Lanchester [1907] pp. 50–2. As Lanchester implictly kept the ratio b/l constant, he wrote $S^{3/4}$ instead of $b\sqrt{l}$ (with $S = bl$).

[61]Lanchester [1907] pp. 70–5. Joukowski's brief discussion of laminar boundary layers ([1916] pp. 120–1) was largely erroneous, for he assumed $\delta \propto 1/U$ in order that the experimental law $R \propto U^2$ should result from the wall stress $\tau \approx \mu U/\delta$.

[62]Rayleigh [1911] pp. 39–40.

$$u = \frac{2U}{\sqrt{\pi}} \int_{y/2\sqrt{vt}}^{+\infty} e^{-\eta^2} d\eta.$$ (7.13)

The corresponding resistance per unit area is

$$\mu \frac{\partial u}{\partial y}\Big|_{y=0} = \rho U \sqrt{\frac{v}{\pi t}}.$$ (7.14)

By convolution, Stokes then determined the resistance induced by any given motion $U(t)$ of the plane. His purpose was to refine the determination of a fluid's viscosity through Coulomb's old measurements of the viscous damping of a disc oscillating in its own plane, by taking into account the fact that the fluid is at rest at the beginning of the first oscillation.[63]

Rayleigh saw in eqn (7.13) the velocity profile of a boundary layer that has developed in the time t, and he guessed that a similar profile and a similar resistance per unit area roughly applied to Lanchester's boundary layer if the time t in Stokes's problem was identified with the time x / U taken by the fluid to travel the distance x from the cutwater at velocity U. The resulting resistance for a blade of length l and width b is

$$R = b\rho \left(\frac{v}{\pi}\right)^{1/2} U^{3/2} \int_0^l x^{-1/2} \, dx = 2b\sqrt{\mu\rho l U^3},$$ (7.15)

in conformance with Lanchester's result (7.11). Rayleigh had no illusions about the practical usefulness of this result:[64]

> The fundamental condition as to the smallness of v would seem to be realized in numerous practical cases; but any one who has looked over the side of a steamer will know that the motion is not usually of the kind supposed in the theory. It would appear that the theoretical motion is subject to instabilities which prevent the motion from maintaining its simply stratified character. The resistance is then doubtless more nearly as the square of the velocity and independent of the value of v.

7.3.4 Slow reception

Neither Lanchester nor Rayleigh were aware of Prandtl's paper of 1904. Yet it did not go completely unnoticed. The towering Göttingen mathematician Felix Klein told Prandtl that his Heidelberg communication was 'the most beautiful' he had heard in the whole congress. Since the 1890s, Klein had been very active in promoting applied mathematics at Göttingen, securing private funds and recruiting competent personnel for this purpose. Since 1900, he had had an eye on Prandtl as a potential contributor to this effort. In June 1904, Prandtl accepted a call to a chair of technical physics at Göttingen. The following year, he assumed the directorship of the Technical Physics Section of the new Institute for Applied Mathematics and Mechanics. Thus, he enjoyed excellent conditions for develop-

---

[63]Stokes [1850b] pp. 130–2, 102–3.

[64]Rayleigh [1911] p. 40 (I have corrected a transposition of b and l). The more exact coefficient of Blasius [1908] is 1.33 (Prandtl's estimate was 1.1).

ing his research, having brilliant graduate students to help him develop theoretical ideas, and first-class experimental facilities to test the results.[65]

Historians of boundary-layer theory all agree that this theory remained mostly a small Göttingen affair until the 1920s. Most frequently, they hold the concision of Prandtl's paper of 1904 and the boldness of its contents responsible for this sluggish reception. This is part of the Prandtl myth. In reality, Prandtl's short paper did not have much to dazzle contemporary experts on hydrodynamics. Its two main novelties, namely, the computation of the laminar boundary-layer profile and a plausible separation mechanism, were largely irrelevant to resistance prediction in concrete cases. As Rayleigh emphasized in 1911, laminar boundary layers are rarely encountered in nature. A critical reader could also doubt that Prandtl's separation mechanism sufficed to determine the separation point in the final separated flow, the ensuing turbulent motion, and the resistance. As is now well known, the tentative separation condition $\partial u/\partial y = 0$ usually implies a failure of approximate integration procedures near this point. Even if this difficulty was solved, a more fundamental one would remain, namely that the potential gradient along the boundary layer is not a priori known, for it depends on the separated flow. Lastly, the instability of the separation surface leads to essentially unpredictable motions in the wake.[66]

In summary, Prandtl's early insights into boundary-layer theory did not bring him much closer to a practical solution of low-viscosity resistance problems. The difficulties of the determination of separated flow remain unsolved to this day. Most of the thirteen papers on boundary layers published before 1930 were mathematical studies of the laminar case under Prandtl's supervision. In the first of these, published in 1908, Heinrich Blasius skillfully integrated the boundary-layer equation through power series, for a flat plate and for a symmetric cylinder. In the latter case, he managed to approximately determine the separation point in permanent, suddenly started, and uniformly-accelerated flows.[67]

The limited value of such calculations soon became evident when testing experiments performed by Karl Hiemenz in Prandtl's laboratory led to unexpectedly violent but quite regular oscillations in the wake of the cylinder. In 1911, Prandtl's brilliant, Hungarian-born student Theodore von Kármán understood that the succession of vortices produced by the instability of the separation surface could only be stable if the vortices were arranged according to the double-alternating row of Fig. 7.14, where the distance h between the two rows is a definite fraction (0.283) of the spacing l between two successive vortices. As Rayleigh later saw, this periodic shedding of vortices explains the 'Aeolian harp' heard by sailors when strong wind blows past the shrouds of a mast. Unfortunately, this is about the only case where something simple can be said about an eddying wake.[68]

[65]Cf. Rotta [1990] p. 9 (Klein's comment, reported by Sommerfeld), Hanle [1982] chap. 3 (Klein's project), chap. 4 (Prandtl's call).

[66]For histories of boundary-layer theory, cf. Tani [1977], Dryden [1955]. For a critical assessment of separation prediction, cf. Batchelor [1967] pp. 325–9.

[67]Blasius [1908]. Cf. Tani [1977], Ackroyd et al. [2001] chap. 11.

[68]Kármán [1911]; Rayleigh [1915a]. In 1908, Henri Bénard had published a careful experimental study of what is now known as the 'Kármán vortex street'. Cf. Kármán [1954] pp. 67–72.

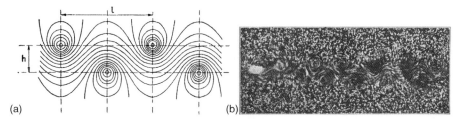

Fig. 7.14.   Lines of flow for Kármán's vortex street, (a) theoretical, and (b) experimental. From Kármán
[1911], and Prandtl [1931*b*] p. 133.

7.3.5   *Turbulent layers*

In 1913, while trying to verify the $U^{3/2}$ resistance law for a plate, Blasius found that this
laminar law ceased to be valid for a critical Reynolds number Ul/ν of about 450 000,
beyond which turbulence occurred in the boundary layer and the resistance became
proportional to $U^{1.864}$, in conformance with Beaufoy's and Froude's earlier measurements. The following year, Prandtl used the turbulent boundary layer to explain a strange
anomaly in experiments performed by Gustave Eiffel on spheres suspended in a wind
tunnel. Against any received theory, Eiffel found a sudden diminution of the resistance of
his spheres beyond a certain critical velocity. Prandtl suspected that at that point the
laminar boundary layer became unstable before the (laminar) separation point, and that
the resulting eddies 'washed away the thin wedge of quiet air behind this point', thus
retarding the separation of the flow. He succeeded in visualizing this effect with smoke in
the Göttingen wind tunnel, but found a higher critical velocity than that measured in Paris.
To explain this last anomaly, he noted that Eiffel's flow-homogenizing device caused
turbulence of the incoming air and thus induced an earlier transition of the boundary
layer from laminar to turbulent. Lastly, he confirmed this view by showing that a
turbulence-inducing wire attached around a parallel of the sphere similarly retarded the
separation of the flow (see Fig. 7.15).[69]

As Prandtl immediately saw, in the case of airships and airplanes, the boundary layer
always becomes turbulent before the laminar separation point. Consequently, the true
separation point is very close to the rear end of the flying body, the global flow is nearly
potential except in a narrow wake, and most of the resistance is frictional (unless there is
also drag-related, induced resistance). Prandtl liked to emphasize the paradoxical role of
turbulence in this felicitous cancellation of eddy resistance: 'It is precisely these turbulent
flows of low resistance around bodies that can be so closely represented by the theory of a
perfect fluid.' At the break of World War I, Prandtl worried that the boundary layer might
not be turbulent in some model experiments, which would jeopardize predictions of full-
scale resistance. Fortunately, he found this was generally not the case for the elongated
bodies that imitated zeppelins or airfoils.[70]

---

[69]Blasius [1913] pp. 25–7; Eiffel [1912]; Prandtl [1914] p. 600.

[70]Prandtl [1914] pp. 605–8, [1927] p. 773.

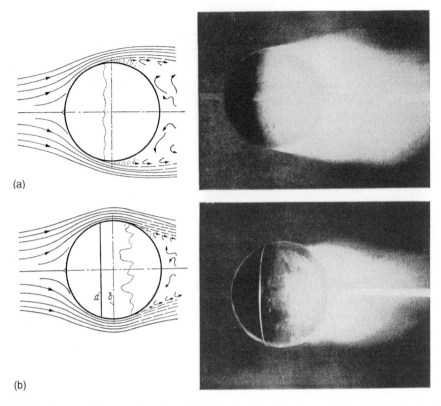

Fig. 7.15. Separated flow around a sphere: (a) with laminar boundary layer, and (b) with turbulence induced in the boundary layer through the wire *a* From Prandtl [1914] p. 605; [1926] p. 720.

7.3.6 *Instabilities*

Despite the high technical importance of turbulent boundary layers, Prandtl held back their theoretical study. His priority of the 1910s was wing theory, for which it was sufficient to know that separation only occurred at the rear edge of the wing and that the flow was laminar and potential everywhere except in the boundary layer. As the frictional resistance of the wing could be evaluated from the measured flat-surface friction, its theory could be postponed. Prandtl began his theoretical investigation of turbulence in 1921, with the onset of turbulence in Poiseuille flow and in boundary layers.[71]

At that time, the received wisdom was that Poiseuille flow was always stable under an infinitesimal perturbation, but unstable with regard to finite perturbations. Prandtl's own experiments on the critical transition of open-channel flow contradicted this view, as they showed that growing wave-like oscillations next to the walls preceded the transition to turbulence. With Oskar Tietjens's help, Prandtl examined the stability of non-viscous,

[71]Prandtl [1921*b*].

parallel, two-dimensional flow under an infinitesimal perturbation. In agreement with Rayleigh's theorem of 1880, he found that an inflection of the velocity profile led to instability. As Prandtl had known since 1904, the evolution of a boundary layer along the wall leads to an inflected profile beyond a certain point (see Fig. 7.10). At that point, the boundary layer should become unstable, as long as viscous damping does not prevent the growth of perturbations.[72]

Prandtl and Tietjens next took into account viscosity, which leads to the much more difficult problem of Kelvin and Orr (in two dimensions). In order to simplify the calculation, they replaced the continuously-curved profile with the broken profiles of Fig. 7.16. They found instability even for convex profiles, and for any value of the Reynolds number. As this was more instability than Prandtl wanted—pipe flow has to be stable for a sufficiently high viscosity—Prandtl surmised that the broken-profile idealization was not permitted. In the real, continuously-curved case, he knew from Rayleigh that, at points of the velocity profile for which the celerity of the sine-wave perturbation equals the flow velocity, a special kind of motion occurs, namely the Kelvin 'cat-eye' pattern of Fig. 7.17. Prandtl suspected a connection with the wave-like behavior he had observed as a prelude to turbulence in channel flow.[73]

In the absence of viscosity, the cat-eye motion is stationary. The most evident effect of viscosity is a damping of the whirling motion in the eyes of the pattern. Prandtl speculated that the viscous stress also induced a phase difference between the u- and v-components of the oscillations of the fluid particles, in which case the energy $\int\int (dU/dy)\rho uv\,dx\,dy$ that the unperturbed motion U conveys to the oscillatory perturbation may have a positive value which exceeds the viscous damping. Another student of Prandtl, Walter Tollmien, confirmed this intuition in 1929, thus providing one of the first proofs of the instability of

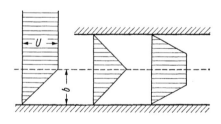

Fig. 7.16. Broken-line velocity profiles. From Prandtl [1921b] p. 691.

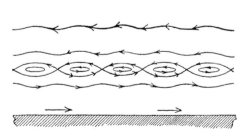

Fig. 7.17. Kelvin's cat-eye flow pattern. From Thomson [1880c] p. 187.

[72] *Ibid.* pp. 688–9.   [73] *Ibid.* pp. 689–93.

plane Poiseuille flow. Kelvin also believed the cat-eyes to cause instability, no doubt because he intuitively connected whirling motion with turbulence. He did not realize, however, the essential role that viscosity played in permitting the growth of the whirls.[74]

7.3.7 *Developed turbulence*

Prandtl's next concern was the 'developed turbulence' (*ausgebildete Turbulenz*) that occurs well after the critical point of instability has been reached. On this question he found himself in competition with Kármán, who now held the chair of mechanics and aerodynamics in the Aachen Technische Hochschule. In 1921, Kármán propounded a semi-empirical derivation of the basic properties of a turbulent boundary layer. Borrowing from Boussinesq's *Eau courantes*, he assumed that the average fluid motion obeyed an equation of the same form as the Navier–Stokes equation, but only if the ordinary viscosity was replaced by an eddy viscosity (*Turbulenzfaktor*) depending on the momentum convection caused by turbulent fluctuation. This implied that Prandtl's boundary-layer equation also held for the average motion, but only if the effective viscosity replaced the molecular viscosity.[75]

As efforts to solve this equation had been largely frustrated, even in the simpler laminar case, Kármán replaced it with the momentum equation obtained by integrating Prandtl's equation over the thickness δ of the boundary layer (taking into account the continuity equation):

$$\frac{\partial}{\partial t}\int_0^\delta \rho u\, dy + \frac{\partial}{\partial x}\int_0^\delta \rho u^2 dy - U\frac{\partial}{\partial x}\int_0^\delta \rho u\, dy = -\delta\frac{\partial P}{\partial x} - \tau_0. \qquad (7.16)$$

On the left-hand side of this equation, the first term represents the acceleration of a normal thin slice of the layer multiplied by its mass, the second term represents the difference of the momentum fluxes across the two sides of the slice, and the third term represents the momentum of the fluid that enters the tip of the slice with the asymptotic velocity U. On the right-hand side, the first term represents the impressed pressure difference on the two sides of the slice, and the second term represents the wall friction ($-\mu\, du/dy|_{y=0}$ in the laminar case). If a reasonable *Ansatz* is made on the form of the velocity profile in the boundary layer (giving u/U as a function of y/δ), then the above equation becomes a differential equation for the unknown function $\delta(x,t)$, or just $\delta(x)$ in the steady case. This mathematical problem is much easier than Prandtl's original problem.

Kármán drew his *Ansätze* for the velocity profile and wall stress of a turbulent boundary layer from pipe-retardation data. In 1913, through careful experiments performed at the Versuchsanstalt für Wasserbau und Schiffbau and through compilation of older smooth-pipe data, Blasius had found the loss of head to vary as the power 7/4 of the section-average velocity. Exploiting a private suggestion by Prandtl, Kármán used dimensional

---

[74]Prandtl [1921*b*] pp. 692–3; Tollmien [1929]. Cf. Prandtl [1930] p. 791. Prandtl borrowed the expression for the energy transfer between macro-flow and micro-perturbation from Reynolds [1894] (see Chapter 6, p. 261). Heisenberg gave another proof of this instability in 1923.

[75]Kármán [1921]. The expression *ausbebildete Turbulenz* already appeared in Prandtl [1913] p. 119.

considerations to derive the velocity profile from Blasius's law. If the (large-scale) velocity u of the water does not depend on the radius of the tube, then it can only depend on the distance y from the wall, the wall stress τ_0, the density ρ, and the kinematic viscosity ν. As the quantities $\sqrt{\tau_0/\rho}$ and ν/y both have the dimension of velocity, the only possible monomial expression for the fluid's velocity is

$$u = A \left(\frac{\tau_0}{\rho}\right)^{(n+1)/2} \left(\frac{y}{\nu}\right)^n, \tag{7.17}$$

where A is a dimensionless constant. Accordingly, the wall stress and the loss of head vary as the power $2/(n+1)$ of the average velocity in the pipe's section. Compatibility with Blasius's law then implies that $n = 1/7$. Having found satisfactory agreement with measured pipe velocity profiles, Kármán assumed the similar form

$$\frac{u}{U} = \left(\frac{y}{\delta}\right)^{1/7} \tag{7.18}$$

for the velocity profile of a turbulent boundary layer, with

$$\tau_0 = \rho \left(\frac{U}{A}\right)^{7/4} \left(\frac{\nu}{\delta}\right)^{1/4}. \tag{7.19}$$

Substituting these two expressions into the momentum equation (7.16), he found that the boundary-layer thickness δ of a flat plate (for which the impressed velocity U is a constant) varied as the power $4/5$ of the distance x from the cutwater, and that the corresponding resistance varied as the power $9/5$ of the impressed velocity U. Accordingly, the growth of a turbulent boundary layer is faster than that of a laminar one, and the resistance has nearly the same form as found by Beaufoy and Froude. Kármán found even better agreement with more recent experimental data.[76]

7.3.8 *The mixing length*

No matter how successful it was, Kármán's approach remained semi-empirical, for it borrowed the law of pipe retardation from experiment. Prandtl wanted a more fundamental theory of developed turbulence from which velocity profiles and the form of the resistance law would result without experimental input. In 1925, he announced significant progress toward this deductive goal. Starting from 'Boussinesq's formula'

$$\tau = \varepsilon \frac{du}{dy} \tag{7.20}$$

for the shear stress in a turbulent flow with the transverse, large-scale velocity gradient du/dy, he followed up Saint-Venant's idea that the effective viscosity ε resulted from momentum transfer through velocity fluctuation. He had just read a popular book by the Viennese meteorologist Wilhelm Schmidt, who subsumed the transport of momentum,

---

[76]Blasius [1913]; Kármán [1921]. Unlike Boussinesq and French hydraulicians, Kármán did not assume a finite velocity at the walls. The $y^{1/7}$ law nonetheless gives a very rapid increase of the velocity near the walls.

heat, and electricity in the atmosphere under the unified concept of 'turbulent exchange' (*Austausch*) and 'mixing' *à la* Reynolds.[77]

Knowing that ε/ρ has the dimension of length multiplied by velocity, Prandtl sought an intuitive representation of the relevant length and velocity. For this purpose, he imagined that, owing to the turbulent fluctuation, balls of fluid were constantly carried over a distance of the order l from one layer of the fluid to another, with a transverse velocity w. Identifying the resulting momentum exchange, $w\rho l\,du/dy$, with the Boussinesq stress, he obtained $\varepsilon = \rho wl$. He then made the transverse velocity w result from the collision of two balls of fluid with different u, which gives the estimate $w \approx l|du/dy|$. The resulting expression for the turbulent stress is

$$\tau = \rho l^2 \left|\frac{du}{dy}\right| \frac{du}{dy}. \tag{7.21}$$

This only improves on Boussinesq's formula if the length l is a simpler function of the flow than the coefficient ε. Prandtl showed that this was indeed the case for the 'free turbulence' occurring in the boundary layer of an air jet. With Tollmien's help, he proved that the simple *Ansatz* $l = Cx$, where x is the distance the jet has traveled from the nozzle and C is a dimensionless constant, matched observations of the layer. The case of pipes did not work so well, since Blasius's law then required that l varies as the power $6/7$ of the distance from the wall—not so simple a law.[78]

The following year, Prandtl gave a somewhat different interpretation of the length l, the one most commonly known today. He now reasoned by analogy with the notion of the mean free path in the kinetic theory of gases and on the basis of Reynolds's stress formula $\tau = \rho\overline{\tilde{u}\tilde{v}}$, where $\tilde{u}$ and $\tilde{v}$ represent the turbulent fluctuations of the velocity components. According to the fluid-ball picture, $\tilde{u} \approx \tilde{v} \approx \pm l\,du/dy$, and the stress formula (7.21) follows. Prandtl now called l the 'mixing length' (*Mischungsweg*), in conformance with Reynolds's and Schmidt's emphasis on mixing.[79]

Prandtl returned to the mixing length in his Tokyo lectures of October 1929. For flow along a smooth wall, he then noted, the simplest possible *Ansatz* is $l = \kappa y$, where y is the distance from the wall and κ is a numerical constant (l must vanish at the wall, since there is no room for fluctuation there). Within the boundary layer the stress τ is nearly independent of y, so that eqn (7.21) leads to

$$u = \frac{1}{\kappa}\sqrt{\frac{\tau}{\rho}}\,(\ln y + C). \tag{7.22}$$

Prandtl rejected this option, because it implied the absurd $u = -\infty$ for $y = 0$. In general, he went on, dimensional homogeneity requires the form $l = y\varphi(R_*)$, with $R_* = (y/\nu)\sqrt{\tau/\rho}$.

[77]Prandtl [1925] p. 716 (Boussinesq), [1927a] (Schmidt); Schmidt [1925]. Föppl ([1909] vol. 4, pp. 364–5) already emphasized *Mischbewegung*, *Platzwechsel*, and *Austausch*. Prandtl ([1913] p. 120) used *Mischbewegung* and briefly described Boussinesq's and Reynolds's approaches.

[78]Prandtl [1925]. Cf. Battimelli [1984] pp. 83–6. Darcy ([1857], see Chapter 6, p. 234), a few other French engineers, and Froude ([1869b], see earlier on p. 281) had used a similar stress formula (with an uninterpreted constant instead of ρl^2) in analogy with the quadratic form of wall friction.

[79]Prandtl [1926], [1927a].

As he had already shown in 1825, the choice $\varphi(R_*) = AR_*^{-1/7}$ leads to $u \propto y^{1/7}$ and to Blasius's law for pipe retardation. Implicitly, Prandtl confined his *Ansätze* for the mixing length to simple algebraic expressions meant to apply to the whole range of the y variable.[80]

7.3.9 *The logarithmic profile*

Kármán got rid of this prejudice in an important memoir of 1930. Instead of speculating on the form of the mixing length, he assumed that turbulent fluctuations at different locations of a fluid only differed in their temporal and spatial scales. He also implicitly assumed that these fluctuations were entirely determined by the first and second derivatives of the macroscopic velocity function $u(y)$ in plane-parallel or circular-cylindrical flows. These two assumptions together imply that the Reynolds stress τ can only depend on the characteristic length $L = u'/u''$, the characteristic time $T = 1/u'$, and the density ρ. In order to be homogenous to a pressure, it must then have the form

$$\tau = \frac{k\rho L^2}{T^2} = \frac{k\rho u'^4}{u''^2},$$

(7.23)

where k is a numerical constant. For a constant τ, this equation leads to the logarithmic profile[81]

$$u = \frac{1}{k}\sqrt{\frac{\tau}{\rho}}\,[\ln(y + C_1) + C_2].$$

(7.24)

If the wall is rough, then eddy viscosity is dominant even next to the wall, and this formula applies to arbitrary small values of the variable y. Furthermore, the characteristic length L at the bottom must be of the order of the size a of the asperities of the wall. Taking into account the vanishing of the velocity at the wall, this gives

$$u = \frac{1}{k}\sqrt{\frac{\tau}{\rho}}\ln\left(\frac{y}{a} + 1\right).$$

(7.25)

This formula holds as long as the stress τ can be regarded as constant. In a circular pipe, this stress grows linearly with the distance from the axis, as required by the balance between the pressures and stresses acting on the surface of a volume element.[82] Kármán obtained the counterparts of formulas (7.24) and (7.25) in this case, and used them to derive the retardation law.[83]

It may be noted, however, that τ remains approximately constant as long as the distance y from the wall does not exceed a small fraction, say 10^{-2}, of the radius h of the tube.

---

[80]Prandtl [1930] (translation of notes taken by a Japanese auditor).

[81]Kármán [1930] pp. 58–65. Saint-Venant ([1887b] pp. 133–4) had used a logarithmic profile, with finite slides on the walls, for the flow between two coaxial circular cylinders, one of which moves along the axis (with stress as the inverse of the distance from the axis, and a constant effective viscosity).

[82]For a fluid disc of radius r and thickness dx, this balance requires $2\pi r\tau dx = \pi r^2(-dP/dx)\,dx$.

[83]*Ibid.* pp. 65–8, 74.

Denoting the wall stress by τ_0, it may also be noted that for dimensional reasons the ratio $u/\sqrt{\tau_0/\rho}$ must be a universal function of the ratio y/h provided that y is much larger than the size a of the asperities.[84] Consequently, the average value of the former ratio over the cross-section of the pipe differs from its value for $y/h = 10^{-2}$ by a universal constant (provided that $a/h \ll 10^{-2}$). Hence the average velocity must have the form

$$\bar{u} = \frac{1}{k} \sqrt{\frac{\tau_0}{\rho}} \left[\ln\left(\frac{h}{a}\right) - K \right], \tag{7.26}$$

where k and K are two universal numerical constants. The pressure gradient, which is balanced by the integral of the wall stress τ_0 over the perimeter of the pipe, is thus proportional to the square of the velocity, in conformance with the usual assumption made by hydraulicians.

If the walls are smooth, then the velocity formula (7.24) can only hold at a sufficiently large distance from them. Near the walls the flow is controlled by the viscosity ν. For dimensional reasons, the thickness of this viscous sublayer must be of the order $\nu\sqrt{\rho/\tau}$ and the velocity at the border of this layer must be of the order $\sqrt{\tau/\rho}$. Assuming that formula (7.24) begins to apply at this border, Kármán required that

$$\frac{1}{k} \sqrt{\frac{\tau}{\rho}} \left[\ln\left(\nu\sqrt{\frac{\rho}{\tau}} + C_1\right) + C_2 \right] = \alpha \sqrt{\frac{\tau}{\rho}}, \tag{7.27}$$

where α is a numerical constant. Hence the velocity profile must have the form

$$u = \frac{1}{k} \sqrt{\frac{\tau}{\rho}} \left[\ln\left(\frac{y}{\nu\sqrt{\rho/\tau}} + \beta\right) + \gamma \right], \tag{7.28}$$

where β and γ are two numerical constants. The resulting average velocity has the form

$$\bar{u} = \frac{1}{k} \sqrt{\frac{\tau}{\rho}} \left[\ln\left(\frac{h}{\nu\sqrt{\rho/\tau}}\right) - K' \right], \tag{7.29}$$

where k and K' are two numerical constants. Kármán found excellent agreement with the latest pipe-retardation data provided by the Göttingen experimentalist Johann Nikuradse. Since this epoch-making paper, the problem of pipe retardation is reduced to the empirical determination of the two numerical constants k and K'. As Kármán later remembered, his subsequent communication at the third international congress of applied mechanics in Stockholm signaled his victory in a tacit competition with his mentor:[85]

> I came to realize that ever since I had come to Aachen my old professor and I were in a kind of world competition. The competition was gentlemanly, of course. But it was first-class rivalry nonetheless, a kind of Olympic Games, between Prandtl and me, and

---

[84]Kármán ([1930] p. 61) noted this property, but did not exploit it in the rest of his calculations. The following reasoning is found in Kármán [1932] p. 409.

[85]Kármán [1930] pp. 69–72, [1967] p. 135 (quote). Cf. Battimelli [1984] pp. 86–92.

beyond that between Göttingen and Aachen. The 'playing field' was the Congress of Applied Mechanics. Our 'ball' was the search for a universal law of turbulence.

In the report of his Tokyo lectures, presumably written after he saw Kármán's paper, Prandtl admitted the logarithmic velocity profile (7.22) that he had originally rejected because of its divergence on the wall. At a sufficient distance from the wall, he now reasoned, the mixing length must have the form Cy because, in the absence of viscosity, y is the only relevant length. Hence the profile must be logarithmic. Next to the wall the flow is laminar and the mixing length must be $\nu\sqrt{\rho/\tau}$. In 1833, Prandtl added that, in the case of the rough wall, the natural choice $l = a + Cy$ immediately leads to Kármán's velocity profile (7.25). In the case of a smooth wall, he directly replaced the roughness with the length $\nu\sqrt{\rho/\tau}$ and thus obtained a formula similar to Kármán's profile (7.28).[86]

Commenting on Kármán's 'very much noticed paper', Prandtl noted that his and Kármán's approach coincided only for a constant stress τ.[87] For a variable τ, both approaches become more arbitrary: Prandtl's does not take into account another characteristic length of the problem, which is $\tau/(d\tau/dy)$, while Kármán's overlooks derivatives of u of order higher than two. Fortunately, most applications only require knowledge of the velocity profile in regions of approximately constant τ. Prandtl's approach is then recommended, since it is the simpler one. Prandtl attributed this simplicity to his focus on the mixing length as the main parameter of turbulent momentum transport. Yet he could also have reasoned directly in terms of Boussinesq's eddy viscosity ε. The only expression of this parameter that can be built from y, ρ, and τ is $\kappa y\sqrt{\rho\tau}$, where κ is a dimensionless constant. Then the relation $\tau = \varepsilon\,du/dy$ leads to $du/dy = (1/\kappa y)\sqrt{\rho/\tau}$, from which Prandtl derived the logarithmic profile.

In subsequent years, Kármán's and Prandtl's derivations of the velocity profile of a turbulent boundary layer were improved in various manners. It was understood that the assumption of an overlap region between the turbulent layer and the laminar sublayer sufficed to establish the logarithmic form of the velocity profile, and more precise estimates of the numerical constants were given. From a practical point of view, the discovery of the logarithmic profile of turbulent boundary layers marked the successful completion of Prandtl's program for determining fluid resistance at high Reynolds numbers. Since 1814, it was clear that the resistance of well-designed airships, airfoils, and ship hulls, as well as hydraulic pipe retardation, depended on the formation of turbulent boundary layers. By 1930, the relevant wall stress could be computed directly from the logarithmic velocity profile in the hydraulic case, and indirectly via Kármán's momentum equation in the nautical and aeronautical cases. From an academic, Göttingen-centered activity, boundary-layer theory gradually evolved into a widely-known procedure for determining fluid resistance in the real world.[88]

We may now reflect on the reasons for this success. In their major advances on the fluid-resistance problem, Prandtl and his disciples relied on the nineteenth-century key concepts of discontinuity, similitude, instability, and mixing. However, they transcended the original use of these concepts in various manners. Whereas earlier users of Helmholtz's

[86]Prandtl [1931a], [1933].

[87]Prandtl [1933] p. 827.

[88]Cf. Tani [1977] pp. 102–3.

surface of discontinuity reasoned in a purely Eulerian context, Prandtl extracted part of the behavior of these surfaces from local, high-Reynolds-number approximations of the Navier–Stokes equation. Whereas previous similitude arguments by Stokes, Helmholtz, Rayleigh, and Froude were confined to the interpretation of model measurements and to the dimensional homogeneity of resistance formulas, Prandtl and Kármán brought them to bear on the internal processes of a system: they saw that in some circumstances different parts of the system only differed in scale. Whereas Rayleigh's and Kelvin's theories of parallel-flow instability had no practical import, Prandtl, Tietjens, and Tollmien showed that properly completed and applied to boundary layers they bore on crucial mechanisms of fluid resistance and retardation. Whereas Boussinesq and Reynolds remained unable to quantify the mixing process that they regarded as the essence of turbulence, Kármán's and Prandtl's insights into the similitude properties of this process led to accurate laws of pipe retardation and turbulent-boundary-layer resistance.

Prandtl's extraordinary ability at combining and extending received theoretical concepts within a coherent, productive picture did not completely solve the resistance problem, however. When it comes to separated flow, today's physicist can predict little more than Saint-Venant did in the mid-nineteenth century. Prandtl only told us how to avoid separation, so that the resistance be small and computable through the boundary-layer approximation. Fortunately, except for parachutes or braking flaps, low resistance is most frequently desired in technical applications.

7.4 Wing theory

In the 1890s, interest in flying contraptions grew tremendously, partly as a consequence of Otto Lilienthal's invention of the man-carrying glider in 1889 (see Fig. 7.18). The prospects of building a motor-powered, piloted airplane seemed high in some engineering quarters. They materialized in 1903 when Wilbur and Orville Wright flew the first machine of that kind. Theory played almost no part in this spectacular success. Analogies with flying animals, experiments with models, and broad engineering ability were all the inventors needed. Although the most learned of them, Samuel Langley, contributed important measurements of lift and drag, refuted the Newtonian $\sin^2 \theta$ dependence on the incidence angle, and even noted that this law would make artificial flight nearly impossible, he still refrained from higher hydrodynamic theory.[89]

The contemporary flight frenzy nonetheless prompted theoretical comments and reflections, ranging from flat rejection to elaborate support. Most negative was Lord Kelvin, who refused an invitation to join the Aeronautical Society of London with the comment: 'I have not the smallest molecule of faith in aerial navigation other than ballooning or of expectation of good results from any of the trials we hear of.' Lord Rayleigh was far more favorable. Commenting on Langley's inclined-plane measurements, he noted qualitative agreement with the formula he had derived in 1876 on the basis of Helmholtz's discontinuity surfaces; he tentatively ascribed the remaining quantitative disagreement to a viscosity-driven suction at the rear of the plate; and he agreed with Langley that the results justified optimism for the possibility of mechanical flight. Rayleigh also applied energy and momentum considerations to a global understanding of the conditions of

---

[89]Cf. Gibbs-Smith [1960], Anderson [1997].

Fig. 7.18.   Otto Lilienthal on a biplane glider in 1895. From *Deutsches Museum* collection.

flight, insisting on 'the vicarious principle' that 'if the bird does not fall, something else must fall' (a downward air current). However, he did not attempt any detailed theory of the flow around the wings of flying objects.[90]

Rayleigh presumably believed that his and Kirchhoff's solution of the two-dimensional inclined-plate problem offered a general explanation of the existence of lift. Indeed, discontinuity surfaces and dead water not only solved d'Alembert's paradox, but they also made the resistance perpendicular to the plate, which implies a finite lifting component when the plate is moving horizontally (see Fig. 7.19). Yet Rayleigh knew of a special case of fluid resistance in which the reaction was normal to the velocity of the moving body, that is, a pure lift without drag. In 1853, Gustav Magnus had explained the long-known deviation of spinning bullets by an induced whirlwind. The superposition of the whirling motion with that resulting from the translational motion of the ball implies different fluid velocities on the two sides of the ball, as indicated in Fig. 7.20. According to Bernoulli's law, this difference implies a pressure difference and a transverse deviation of the ball. Magnus tested his explanation with the device of Fig. 7.21.[91]

In a memoir of 1877 'On the irregular flight of a tennis ball', Rayleigh recalled Magnus's reasoning and noted that the most general irrotational solution of Euler's equation for the two-dimensional flow around a cylinder with constant asymptotic velocity had the form

$$\psi = \alpha\left(1 - \frac{a^2}{r^2}\right) r \sin\theta + \beta \ln r, \tag{7.30}$$

[90]Kelvin to Baden-Powell, 8 Dec. 1896, in Gibbs-Smith [1960] p. 35; Rayleigh [1891], [1883*c*], [1900] p. 462.

[91]Magnus [1853] pp. 5–7. In his *New principles of gunnery* (London, 1742), Benjamin Robins had described experiments on the deviation of spinning bullets. Cf. Barkla and Auchterlonie [1971]; Steele [1994].

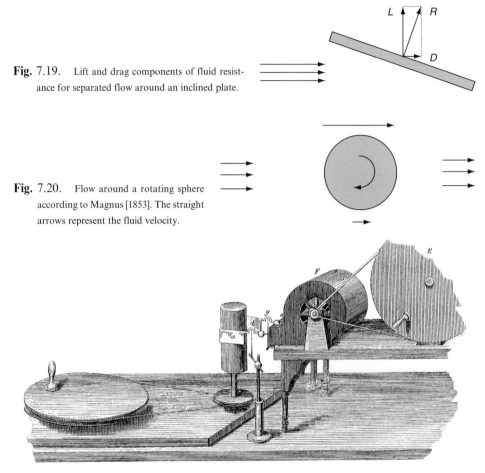

Fig. 7.19. Lift and drag components of fluid resistance for separated flow around an inclined plate.

Fig. 7.20. Flow around a rotating sphere
according to Magnus [1853]. The straight
arrows represent the fluid velocity.

Fig. 7.21. Magnus's apparatus for demonstrating the pressure difference between the two sides of a rotating
cylinder subjected to the draft from the ventilator F. The light, horizontally movable blades a and b serve to
detect the increase and decrease in pressure when the fan is turned on. From Magnus [1853] plate.

where ψ is the stream function, r is the distance from the axis of the cylinder, θ is the angle
around this axis, and a is the radius of the cylinder. The first part of this formula, already
known to Stokes, by itself satisfies the boundary conditions if α is equal to the asymptotic
velocity of the flow. The second part represents a circulation of the fluid around the axis,
with a velocity β/r at the distance r from the axis. Integrating the pressure over the surface
of the cylinder, Rayleigh found a resultant force perpendicular to the asymptotic velocity,
with the intensity $2\pi\alpha\beta$.[92]

At the very best, Rayleigh hoped this consideration to be relevant to the Magnus effect,
abstraction being made of the circulation-inducing process and of the 'unwillingness of the

---

[92]Rayleigh [1877b]. A missing factor of 2 has been corrected.

stream-lines to close in at the stern of an obstacle'. He did not dream of any application to
the problem of flight. That viscosity could possibly induce a fluid circulation around a
non-rotating, flying object was hard to conceive. For a perfect liquid, Kelvin's circulation
theorem seemed to prohibit the genesis of any circulation. In his treatise of 1895, Lamb
reproduced Rayleigh's solution of the cylinder problem as an interesting example of fluid
motion that has circulation despite being everywhere irrotational. No more than Rayleigh
did he perceive any connection with the problem of flight.[93]

In summary, Rayleigh, Lamb, and Kelvin knew too much fluid mechanics to imagine
that circulation around wings was the main cause of lift. The two men who independently
hit upon this idea lacked training in theoretical physics. One of them was an engineer, and
the other was a young mathematician.

7.4.1 *Lanchester's theory*

Frederick Lanchester was an automobile engineer and industrialist with a passion for
aeronautics. In 1892, he imagined a singular theory of what he called an 'aerofoil', that is,
the organ of sustentation of airplanes and birds. As he had 'very little acquaintance with
classical hydrodynamics', he reasoned by direct application of the laws of mechanics to
the particles of the fluid. In the first, Newtonian approximation, the fluid particles hit the
aerofoil independently of each other, which leads to a resistance proportional to the
squared sine of the inclination.[94]

In reality, Lanchester went on, the mutual interaction of the fluid particles implies that
the layers of air adjacent to the foil react on the neighboring layers, so that a stratum of air
of considerable thickness is affected (see Fig. 7.22). Then the flux of deviated particles is no
longer proportional to the sine of the inclination (as it was in Newton's reasoning) but to
the width of the stratum or 'sweep', and the resistance becomes proportional to the sine of
the inclination, in conformance with small-angle measurements. Lanchester estimated this
width from Langley's experiments with superposed planes, which showed that the sustaining power of the planes added up only when the vertical distance between them exceeded a
certain value. Substituting this value into the revised resistance formula, he obtained about

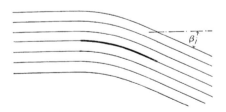

Fig. 7.22.   Provisional, constant-sweep picture of the flow past an inclined plate. From Lanchester [1907] p. 227.

[93]Rayleigh [1877*b*] p. 346; Lamb [1895] pp. 87–90.

[94]Lanchester [1894], [1907] p. 143, [1926] p. 593. On Lanchester's biography, cf. Fletcher [1996]. On his
aerodynamics, cf. Ackroyd [1992], [1996], Ackroyd *et al.* [2001] pp. 57–69. The theory of 1892 is given in
Lanchester [1907] pp. 143–62 (the manuscripts are lost).

half of the measured drag. Consequently, the flow of Fig. 7.22 could not accurately represent reality.[95]

The reason for this discrepancy, Lanchester surmised, was the dubious assumption that the air encountered by the front edge of the moving foil was at rest. In reality, air must flow from the region below the foil to the region above it in order to prevent the accumulation and rarefaction of fluid in these two regions. To make this clear, Lanchester decomposed the motion of an approximately-planar foil into a component parallel to the plane and a normal component. Then the resistance problem is the same as in the case of a falling plate subjected to a simultaneous (faster) horizontal motion. The fall of the plate, Lanchester reasoned, induces a fluid motion of the sort represented in Fig. 7.23, with an upward current or 'vortex fringe' in front of the plate. In its forward motion the plate intercepts this upward current, and thus experiences a stronger lift than it would by the sole production of a downward current.[96]

To refine his reasoning, Lanchester gave the plate infinite span and loaded it with a small weight. He assimilated the effect of the weight with the creation of an acceleration field of the form given in Fig. 7.24. The horizontal air flow with respect to the plate brings new fluid particles into this acceleration field. Their trajectory has the shape indicated in

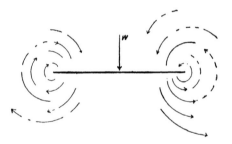

Fig. 7.23.   Vortex fringe for a plate falling through the air with the velocity *w*. From Lanchester [1907] p. 145.

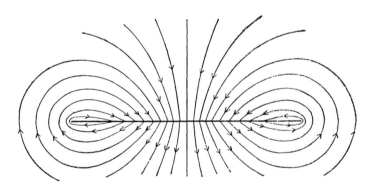

Fig. 7.24.   Acceleration field around a falling plate. From Lanchester [1907] p. 176.

[95]Lanchester [1907] (1892) pp. 144–5.   [96]*Ibid.* pp. 145–46.

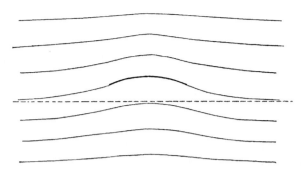

Fig. 7.25.   Horizontal flow modified by the acceleration field of Fig. 7.24. From Lanchester [1907] p. 159.

Fig. 7.25. Owing to the left–right symmetry of the acceleration field, the particles leave it with their original velocity and their original height. Therefore, the only effect of the small loading of the plate is to produce 'a supporting wave' traveling together with it. No work is needed to preserve the horizontality of the motion. In modern words, Lanchester imagined a state of fluid motion such that the lift exactly compensates the load of the plane, without any induced drag.[97]

This state of motion is only possible if the plate is given a small curvature, so that the undulating trajectories of the fluid particles do not cross the foil. Using this principle, Lanchester drew the profile marked by the thick line of Fig. 7.25. The curvature increases with the load of the plate. An aerofoil can thus produce lift with vanishing inclination, as long as this foil is curved. Lanchester regarded the observed shape of bird wings as a vindication of this theory. Lilienthal and Langley also used cambered wings in their gliders, and the former had given precise experimental proof of their superiority.[98]

Lastly, Lanchester considered the more difficult case of an aerofoil with finite span. Owing to the lateral spread of the field lines in this case, the ascending field that acts around the edges of the plane is weaker than the descending field that acts underneath and above the plane. Consequently, the fluid particles that travel through these two fields emerge with a downward velocity; there is a downward current in the wake of the foil, compensated for by two upward currents caused by the ascending field that acts alone beyond the tips of the foil. Since the accelerating field, seen from behind the foil, has a form similar to that drawn in Fig. 7.24, it must induce a whirling motion of the fluid, essentially two vortices starting from the tips of the aerofoil. The continual production of these vortices and the formation of the downward and upward currents spend energy, so that an induced drag necessarily accompanies the lift of a finite aerofoil.[99]

Lanchester expounded these ideas at the annual meeting of the Birmingham Natural History and Philosophical Society on 19 June 1894. In 1897, the Physical Society of

[97] *Ibid.* pp. 149–56. Lanchester's acceleration field has the same geometry as the velocity field of a moving plate (indeed, the motion can be regarded as being impulsively started from rest).

[98] *Ibid.* pp. 158–60. Lilienthal attributed the superiority of cambered wings to the reduced production of eddies (on his resistance measurements, cf. Anderson [1997] pp. 138–59).

[99] Lanchester [1907] pp. 156–8.

London rejected a fuller account. As Prandtl later put it, 'Lanchester's treatment is difficult to follow, since it makes a very great demand on the reader's intuitive perceptions.' Only a reader who would have known the results to be essentially correct would have bothered penetrating the car maker's odd reasoning. Lanchester must have become aware of this communication problem, since he immersed himself in Lamb's *Treatise* and sought more academically acceptable justifications of his intuition of the flow around an aerofoil.[100]

Most relevant to his thinking were Helmholtz's vortices and discontinuity surfaces, as well as Rayleigh's tennis-ball problem. Lanchester now understood that the flow he had imagined around an aerofoil belonged to the same category as Rayleigh's irrotationally-circulating flow: the same compression of the lines of flow above the flying object and rarefaction below occur in both problems. Lanchester now made circulation the essence of lift. From Helmholtz's law for the velocity induced by a linear vortex, he inferred the downward precession of the two trailing vortices of the foil. He further suggested that these vortices should be replaced by a Helmholtz vortex sheet extending behind the whole breadth of the foil, as the air skirting the upper surface of the aerofoil reaches its rear edge with a transverse velocity directed toward the axis of flight, and the air skirting the lower surface reaches the rear edge with an opposite velocity (see Fig. 7.26). Wrongly assuming that the circulation around every transverse section of the foil caused a deviation of the vortex filaments away from the axis, and taking into account the mutual twisting and the viscous diffusion of these filaments, he obtained the emblematic picture in Fig. 7.27.[101]

Lanchester published these considerations together with his earlier intuitive theory in his *Aerodynamics, constituting the first volume of a complete work on aerial flight* of 1907. The book got fair reviews in the British press, and won Lanchester an appointment to the British Advisory Committee for Aeronautics. The president of this committee, Lord Rayleigh, endorsed Lanchester's boundary-layer consideration, as was mentioned earlier. Despite these welcoming signs, Lanchester's ambition to provide guidance for aeroplane builders was largely frustrated. When the Wright brothers' machine was first flown in Europe at Le Mans in 1908, Lanchester found Wilbur Wright very ill-disposed toward theory. The pioneering constructor dryly commented that the most talkative bird

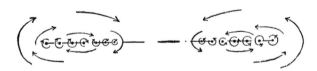

Fig. 7.26.   The vortex sheet induced by the lateral skirting of the air on the upper and lower surfaces of the aerofoil, seen from behind. From Lanchester [1907] p. 176.

[100]Lanchester [1894]; Prandtl [1927*b*] p. 753. Cf. Lanchester [1907] p. 142.

[101]Lanchester [1907] pp. 162–78. Lanchester confused the circulation around the foil with a real layer of vorticity around it. He does not seem to have understood the connection between the circulation around the foil and the vorticity of the trailing vortices.

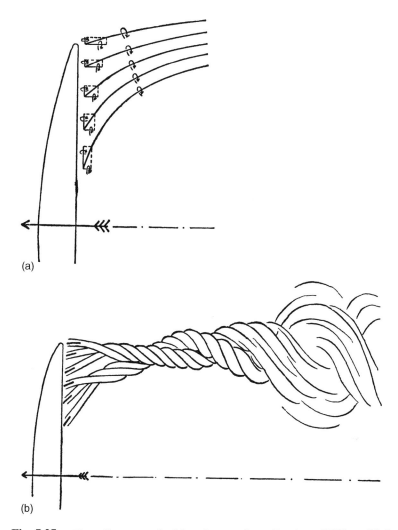

Fig. 7.27.   The trailing vortex of a flying wing according to Lanchester [1907] pp. 177–8.

(the parrot) was also a poor flier. To a letter from Lanchester in the following year, he briefly replied:[102]

> In glancing over [your paper] I note such differences in matters of information, theory, and even ideals, as to make it quite out of the question to reach common ground by more talk, as I think it will save me much time if I follow my usual plan, and let the truth make itself apparent in actual practice.

Although the British aeronautical establishment was more open to theory than the Wright brothers, it seems to have ignored Lanchester's aerofoil theory until Prandtl's

---

[102]Cf. Lanchester [1926] p. 588, Ackroyd [1992].

related theory became known after the war. The Germans were the most receptive to Lanchester's ideas. Soon after the publication of *Aerodynamics*, Prandtl's prominent colleague Carl Runge contacted Lanchester to propose a translation. He welcomed him to Göttingen in September 1908, and he arranged conversations with Prandtl, who was then busy completing the Göttingen wind channel. Although Prandtl later claimed (in Lanchester's presence) to have reached the main ideas of his wing theory before reading Lanchester, he also admitted that he and his collaborators 'were able to draw many useful ideas' from *Aerodynamics*. Kármán, who had witnessed the Göttingen encounter, suggested that Prandtl had borrowed more from the English engineer than he was conscious of.[103]

7.4.2   *Kutta's and Joukowski's theories*

In 1902, Wilhelm Kutta, an assistant in higher mathematics at the *Technische Hochschule* in Munich with an interest in Lilienthal's gliding experiments, devoted his dissertation to the flow around the simplest idealization of Lilienthal's cambered wings, namely a circular arc. His method consisted in applying a conformal transformation $z = \Phi(\zeta)$ to the incompressible flow around a circular cylinder, represented in the complex plane of the variable $z = x + iy$. As Rayleigh had shown, the most general irrotational solution to the latter problem with an asymptotic, horizontal velocity U is given by

$$\varphi + i\psi = U\left(z + \frac{a^2}{z}\right) - \frac{\Gamma}{2\pi} i \ln z, \tag{7.31}$$

where φ is the velocity potential, ψ is the stream function, a is the radius of the disc, and Γ is the cyclic period of the potential (the circulation $\oint \mathbf{v} \cdot d\mathbf{r}$). As a mathematician, Kutta had no objection against the circulatory component of this solution.[104]

Kutta applied to this flow an intricate, double-step conformal transformation that turned the circular boundary of the cylinder into a circular arc with chord parallel to the asymptotic flow. The velocity at the tips of the arc, he found out, was only finite if the circulation Γ had the specific value $2\pi h U$, where h is the maximum height of the arc. Under this condition, the flow has the shape shown in Fig. 7.28. Kutta then integrated the fluid pressure (as given by Bernoulli's law) to obtain the lift

$$L = 2\pi\rho U^2 h. \tag{7.32}$$

Comparing this theoretical result with Lilienthal's measurements, Kutta found a 25% excess that could plausibly be explained by vortex formation and a finite span.[105]

Through a consideration of energy, Kutta also related this lift to the cyclic period Γ of the potential. The work done by the lift during a (virtual) vertical displacement δy of the

---

[103]Prandtl [1927*b*] pp. 753 (quote), 776 (Lanchester remembering Göttingen); Kármán [1967] pp. 50–3. Runge, who had an English mother, was the interpreter. His and his wife Aimée's translation of *Aerodynamics* appeared in 1909.

[104]Kutta [1902*a*]. Sebastian Finsterwalder, a mathematics professor and ballooning expert at the Technische Hochschule in Munich, suggested the topic of Kutta's *Habilitationsschrift* (cf. Kutta [1910] p. 4). As was well known, the compressibility of the air can be neglected in any resistance problem for which the velocity of the air remains small compared to the celerity of sound waves.

[105]Kutta [1902*a*], [1902*b*]. Cf. Ackroyd *et al.* [2001] pp. 70–6.

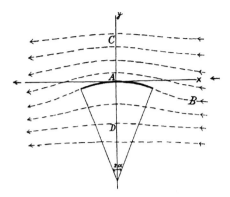

Fig. 7.28.   Kutta's flow around a circular arc. From Kutta [1902*b*] p. 133.

arc, he reasoned, should be equal to the energy produced by the annihilation of a horizontal fluid slice of the same breadth at a large positive ordinate Y and the creation of another slice at the symmetric ordinate $-Y$. This gives, for the lift L,

$$L = \lim_{Y \to +\infty} \left[-\int_{-\infty}^{+\infty} \frac{1}{2} \rho v^2(x, Y)\, dx + \int_{-\infty}^{+\infty} \frac{1}{2} \rho v^2(x, -Y)\, dx \right]. \qquad (7.33)$$

Using the asymptotic approximation

$$v^2(x, Y) \approx U^2 + \frac{\Gamma U}{2\pi} \frac{Y}{x^2 + Y^2}, \qquad (7.34)$$

and lightheartedly assuming a mutual cancellation of infinite terms, Kutta found that

$$L = \rho \Gamma U, \qquad (7.35)$$

in conformance with the result (7.32) of direct pressure integration.[106]

In summary, Kutta's mathematics led to a flow around a thin curved foil that strikingly resembled the one Lanchester predicted. Instrumental to his derivation was the condition that the velocity of the flow should remain everywhere finite, which is now called the Kutta condition. The remarkably simple formula $L = \rho \Gamma U$ is now called the Kutta–Joukowski theorem. However, Kutta did not explicitly identify Γ with the circulation of the air around the foil. Nor did he refer to Rayleigh's tennis-ball problem as the origin of formula (7.31) for the irrotational flow around a circular cylinder. In the semi-popular summary published in the *Illustrirte aeronautische Mittheilungen*, he did not give the general relation between lift and circulation. Instead, he argued that, in order to prevent the formation of vortex sheets at the extremities of the arc foil, the velocity of the air had to be tangential. This implies a higher velocity above and a lower velocity below the foil, and a lifting pressure difference by Bernoulli's law.[107]

---

[106]Kutta [1910] pp. 19–20. In this article Kutta described the reasoning as belonging to his *Habilitationschrift* [1902*a*], which I have not been able to find. Joukowski ([1910] p. 282) accepted this claim.

[107]Kutta [1902*b*]. Kutta ([1910] p. 3) credited Lanchester for the concept of wing circulation.

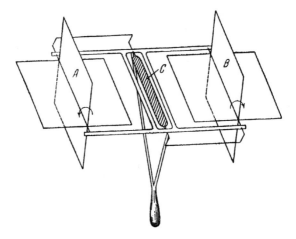

Fig. 7.29.   A tentative flying device by Joukowski. The twisted rubber band C induces the rotation of the
paddle-wheels A and B. From Joukowski [1890] p. 350.

Unlike Lanchester and Kutta, the Russian physicist Nikolai Joukowski was a highly
professional physicist of international repute, and head of the mechanics department at
Moscow University. Much of his early work was in theoretical hydrodynamics, with an
emphasis on potential flow and complex-variable methods. He published his first significant paper on aerodynamic lift in 1906, after several years of interest in the problem of
flight. From then on, he played a leading role in developing the aeronautical industry in his
country. His main contribution of 1906 was a rigorous and general derivation of the
theorem that relates circulation and lift for the two-dimensional flow around a solid
cylinder.[108]

In an address of 1890 on the theory of flight, Joukowski argued that paddle propulsion
was only possible if the fluid motion implied discontinuity surfaces, viscous stress, or
whirling motion. In the last case, he imagined and constructed the device shown in
Fig. 7.29, in which each of the rotating paddle-wheels is subjected to the upward current
induced by the rotation of the other. Although the device turned out to be too heavy to fly,
Joukowski found that the rotation of the wheels diminished its apparent weight. There is
no hint, in this communication, that whirling motion may also occur around static wings,
nor that it may imply a transverse, lifting force when the whirl progresses horizontally.
Worth noting, however, is the general idea of exploiting vortex motion for the sake of
artificial flight.[109]

Before 1906, Joukowski had read Louis Pierre Mouillard's *L'empire de l'air*, a book of
1881 familiar to several pioneers of aeronautics. By careful observation of bird flight,
Mouillard hoped to help in the successful design of gliders and 'aeroplanes'. He also
sketched a strange wing theory based on an analogy with the fall of a Bristol card (a rigid,
rectangular paper strip). The fall of the card from a horizontal position usually implies a

[108]Cf. Grigorian [1965], [1976], Strizhevskii [1957].

[109]Joukowski [1890]. I thank Yury Kolomensky, who helped me read this paper.

rotation around its axis of symmetry, as well as a deviation in the direction of the horizontal velocity component of the lower edge of the strip. By the adequate loading of the strip and the folding of its tail, Mouillard believed he could check the rotation and yet preserve the deviation from a vertical fall. In his opinion, birds flew according to this mechanism. His theoretical explanation of the rotation involved the dubious principle that the center of gravity of any falling body should be displaced by an amount proportional to the velocity of fall. He remained silent on the cause of the horizontal deviation.[110]

Although Joukowski ignored Mouillard's speculations, he credited him with the description of the 'interesting phenomena' accompanying the fall of a Bristol card. He also mentioned Wladimir Köppen's model of an aeroplane with motorized rotating wings, based on the principle that rotation prevents the fall of bodies. Joukowski justified this principle by analogy with the Magnus effect, according to which a rotating projectile is subjected to a deviating force proportional to its rotation. As he was unaware of Rayleigh's tennis-ball paper, he explained this deviation by a general theorem of his own:[111]

> If an irrotational, two-dimensional flow with asymptotic velocity [U] surrounds a closed curve [made of lines of current] on which the circulation of the velocity is [Γ], the resultant of hydrostatic pressure on this curve is perpendicular to the velocity [U] and has the value [ρΓU]. The direction of this force is obtained through a right-angle rotation of the vector [U] in the sense of negative circulation.

In his demonstration Joukowski imitated Poncelet's and Saint-Venant's recourse to momentum balance in their theories of resistance, with which he had become familiar during a formative stay in Paris. Around the closed curve made of lines of current he drew a circle of large radius (see Fig. 7.30), and required that the pressures acting on the fluid contained between the closed curve and the circle should balance the momentum increase of this fluid:

$$-\mathbf{L} - \oint P\mathbf{n} \, ds = \rho \oint \mathbf{v}(\mathbf{v} \cdot \mathbf{n} \, ds), \qquad (7.36)$$

where **L** represents the resistance (action of the fluid on the body) per unit length, and the integrals are taken over the circular trace of the fictitious cylinder (**n** being the unit normal vector). Using Bernoulli's law and retaining only first-order terms in $\mathbf{w} = \mathbf{v} - \mathbf{U}$, this gives

$$\mathbf{L} = \rho \oint [(\mathbf{U} \cdot \mathbf{w})\mathbf{n} - \mathbf{w}(\mathbf{U} \cdot \mathbf{n})] \, ds = \rho \oint \mathbf{U} \times (\mathbf{n} \times \mathbf{w}) \, ds. \qquad (7.37)$$

At large distances from the body, **w** is the velocity of a pure circulation in a direction perpendicular to **n**. Therefore, the vector **L** is directed downward when **U** is directed to the

---

[110]Mouillard [1881] pp. 210–17. Unknown to Mouillard, in 1854 James Clerk Maxwell had explained the rotation by the greater resistance of the air to the motion of the lower edge of the plane, and the deviation by periodic modulations of the net resistance and the fall velocity.

[111]Joukowski [1906a] p. 52; Köppen [1901] (falling card experiments); Moedebeck [1904] p. 179 (on Köppen's model). Joukowski tried to verify his theorem by measuring the force acting on a rotating blade in the wind tunnel of the Aerodynamic Institute of Koutchino. The director of this pioneering institute, Dimitri Riabouchinski [1909], criticized this procedure as well as Maxwell's old theory of the falling paper strip (Maxwell [1854]), which does not imply any transverse force when flow relative to a rotating blade is kept constant.

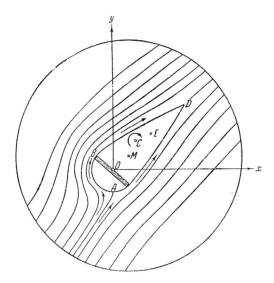

Fig. 7.30.   Joukowski's flow past a blade (shaded rectangle) rotating around the axis O*z* (perpendicular to
O*xy*). The closed curve ABD, made of two converging lines of current, separates the zone of whirling motion
from a zone of laminar, irrotational flow. From Joukowski [1906*a*].

right and the circulation is oriented trigonometrically, and the intensity *L* agrees with the
Kutta–Joukowski formula (7.35).[112]

Although Joukowski knew his theorem to apply to the case in which the closed line of
current is the frontier of an immersed solid, he only applied it to a rotating blade immersed
in a uniform stream, assuming that the flow was smooth and irrotational outside a pear-
shaped zone (ABD in Fig. 7.30) delimited by converging lines of current. In another paper
of 1906, he introduced the notion of 'bound vortices', that is, a series of virtual or real
vortex lines contained within a closed curve and able to represent the circulatory part of
the flow outside this curve. He enunciated theorems relating the force and angular
momentum resulting from the pressure on this curve to the strength of the vortex lines
and the fluid velocity on these lines. He applied these notions to the rotating blade and to
the vortex pair behind a plate immersed perpendicularly in a uniform stream. He did not
consider the case of an airfoil or wing, in which he may not yet have understood that
circulation-flow occurred.[113]

[112]Joukowski [1906*a*], [1906*b*]. Cf. Ackroyd *et al.* [2001] pp. 88–106. In the first paper, Joukowski uses the
balance of angular momentum around an arbitrary axis instead of the momentum balance. Had Joukowski
followed Saint-Venant closer, he would have used a fictitious surface of large rectangular section. In this case the
momentum variation of the enclosed fluid vanishes, and the resistance is simply given by the pressure difference on
the two horizontal walls of the fictitious cylinder.

[113]Joukowski [1906*b*]. In 1909, Joukowski obtained the two-dimensional flow around a curved plate as an
extension of Kirchhoff's flat-plate solution *with surfaces of discontinuity*: cf. Chaplygin [1911], who gives Kutta
full credit for conceiving the possibility of smooth, circulatory flow around a cambered foil. I thank my friend
Guenaddi Sezonov who helped me with the Russian (before I became aware of Anatoly Ruban's translation in
Ackroyd *et al.* [2001] pp. 88–104).

In 1910, Kutta extended his calculation of the lift of an arc of a circle to the case of an inclined flow. In the same year, Joukowski's brilliant disciple Sergei Alekseevich Chaplygin rediscovered Kutta's solution for the smooth flow around a circular arc, using the much simpler conformal transformation

$$\zeta = z - ia + \frac{\lambda^2}{z - ia},$$ (7.38)

where a and λ are real constants. Kutta and Chaplygin both noted that, in the inclined case, an infinite velocity could only be avoided at one extremity of the arc, say the rear one. In order to avoid the remaining infinite velocity and vortex-sheet formation, Kutta rounded the front edge through a complicated numerical procedure, and Chaplygin grafted a disc onto it. Joukowski then found that a horizontal shift of the origin in the z-plane magically thickened the arc-shaped foil, leaving only one sharp edge at the rear. The transformation

$$\zeta = z + \frac{\lambda^2}{z},$$ (7.39)

now called the Joukowski transformation, turns a circle of radius λ centered at the origin of the z-plane into a segment of length 4λ in the ζ-plane. A horizontal (real) shift of the origin of the circle turns the segment into a fish shape. A vertical (purely imaginary) shift of the origin turns it into a circular arc. Both shifts combined lead to a cambered fish shape (see Fig. 7.31).[114]

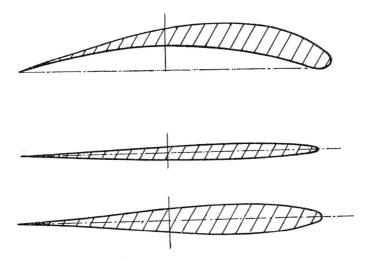

Fig. 7.31.   Joukowski's theoretical wing profiles. From Joukowski [1916] p. 105.

[114]Kutta [1910]; Chaplygin [1910]; Joukowski [1910], [1916] chap. 6. Cf. Ackroyd *et al.* [2001], chaps 12–14. For a modern account, cf. Batchelor [1967] pp. 445–9. According to Chaplygin ([1945] p. 5), his paper was already in press when Joukowski told him about Kutta's earlier work. Chaplygin ([1911] pp. 17–18) briefly mentioned the necessity of tip vortices in the case of finite span, but gave them an erroneous mustache shape.

The two-dimensional approach to wing theory culminated with these Russian findings. Joukowski's suggestion for obtaining a finite drag within this theory was, however, misconceived. He believed that two-dimensional vortex production at the front of the wing would account for observed drags, and ignored the effect of finite span, even though he had read Lanchester's book and approved the concept of trailing vortices.[115]

7.4.3   Prandtl's theory

In the years 1910–1918, Prandtl and his collaborators combined Lanchester's intuition of the motion around a three-dimensional aerofoil with the mathematical precision of two-dimensional theories. His fullest publications on this topic appeared in 1918/19, with a delay due to wartime secrecy. As Prandtl himself noted, the organization of these papers does not reflect the historical course of his thoughts. This course may, however, be inferred from Prandtl's few historical remarks, from earlier fragmentary publications, and from the logic of the subject.[116]

One of Prandtl's earliest contributions must have been the explanation of the process by which circulation is produced around a streamlined two-dimensional aerofoil. While Kutta said nothing on this process, Lanchester's falling-plate reasoning could not pass for a proper hydrodynamic demonstration. Yet for anyone versed in fluid mechanics, circulation was only admissible if its genesis could be reconciled with the theorems by Lagrange and Kelvin that seemed to forbid it. Due to his familiarity with Helmholtz's vortex sheets, Prandtl easily solved the paradox as follows.[117]

The irrotational, non-circulatory flow around the aerofoil involves infinite velocity at the rear edge. In order to keep the velocity finite, a vortex sheet must be generated at the beginning of the motion, as shown in Fig. 7.32. This process is perfectly compatible with Kelvin's theorem, which only forbids vorticity for fluid particles that have never been in contact with a wall. Neither does the theorem forbid a change in the velocity circulation around the body. On the contrary, when applied to a curve enclosing both the body and the emerging vortex sheet (the dotted line in Fig. 7.32), this theorem implies that the velocity circulation around the body should increase by an amount equal to the total vorticity of the vortex sheet. After a brief time, this circulation reaches the value for which

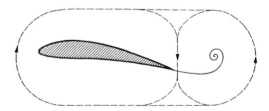

Fig. 7.32. Transient pattern of the flow around a wing, with vortex production at the trailing edge. From Prandtl [1921a] p. 464.

[115]Joukowski [1916] pp. 184–5, [1910] p. 282 (approving Lanchester).

[116]Cf. Prandtl [1918] p. 322.

[117]Prandtl's systematic use of Helmholtz's and Kelvin's vortex theorems in wing theory presupposes that the compressions of the air are negligible, which is true for widely subsonic flight.

the Kutta condition of smooth flow is satisfied, the vortex sheet production ceases, and the resulting vortex flows away.[118]

For a wing of finite span, Prandtl reasoned, circulation must exist at least around the central sections of the wing in order to make lift possible. Such circulation, however, cannot exist without permanent vortex production. This is a consequence of the theorem according to which the variation of the circulation around a loop during a continuous deformation or a displacement of this loop is equal to the number of vortex filaments cut by the loop.[119] Consider a loop that embraces a section of the wing, and move it toward one of the tips of the wing. As the circulation necessarily vanishes at the tip (since the loop shrinks to a point), it must cross vortex filaments on its way. Hence vorticity is necessarily produced near the tips. More generally, vorticity must be produced whenever the circulation varies between successive sections of the wing. According to one of Helmholtz's theorems, the generated vorticity must follow the fluid motion. Therefore, a trailing vortex sheet is formed behind the wing, with an intensity depending on the rate of variation of the circulation along the span of the wing (see Fig. 7.33).[120]

Prandtl once said that he had reached this picture while puzzling over Lanchester's trailing vortex.[121] There are significant differences, however. Whereas Lanchester reasoned in an intuitive, qualitative manner based on the 'field of force' of a falling plate, Prandtl applied Helmholtz's vortex theorems to derive a precise quantitative connection between the circulation around the wing and the trailing vortex. Prandtl had the vortex filaments follow the main flow, whereas Lanchester erroneously gave them a sideways inclination. Prandtl related the variation of the circulation and the production of vorticity to the variation of the wing's section along its span, whereas Lanchester reasoned on a constant section.

Lastly and most importantly, Prandtl was able to apply his picture of wing flow to a quantitative determination of lift and drag, whereas Lanchester's considerations remained mostly qualitative. In a first approximation, Prandtl reasoned, the lift is the sum of the lifts given by the Kutta–Joukowski theorem applied to the successive sections of the wing (as if they belonged to infinite cylinders), and there is no drag. In a second approximation, the velocity field of the trailing vortex must be taken into account. In the vicinity of a given section of the wing, this induced flow is approximately uniform and in the downward vertical direction (see Fig. 7.34). Therefore, the net flow impressed on this section has a downward inclination, and the corresponding reaction, being rotated by the same angle, now has a finite drag component and a slightly diminished lift component.[122]

Prandtl had this general picture by 1912. The mathematical implementation did not go as smoothly as he had hoped. The simplest conceivable case is that of constant

---

[118]Prandtl [1913] pp. 118–19, [1918] pp. 325–8, [1921a] pp. 463–4.

[119]This theorem results from the divergenceless character of the vorticity: the flux of the vorticity across the surface swept by the loop must be equal to the variation of its flux across a surface bounded by the loop, which by Stokes's theorem is equal to the circulation around the loop.

[120]Prandtl [1913] p. 112, [1918] pp. 324–5, [1921a] pp. 465–6.

[121]Prandtl [1948] p. 1607n.

[122]Prandtl [1918] pp. 337–8, [1921a] p. 477. On Lanchester's few quantitative attempts, cf. Ackroyd [1992].

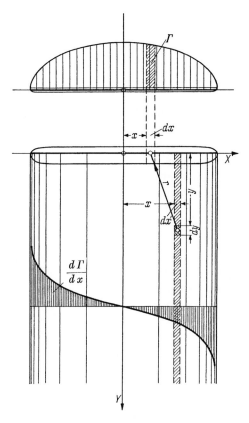

Fig. 7.33.   A trailing vortex sheet with vorticity profile dΓ/dx corresponding to the (elliptic) circulation profile Γ around a flying wing. From Prandtl [1918] p. 337.

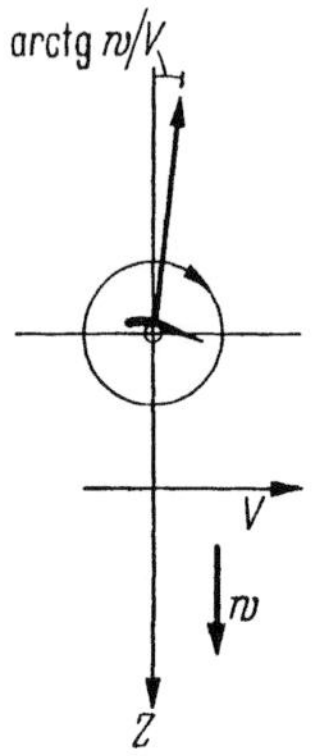

Fig. 7.34.   The Inclination of the resistance owing to the vertical induced velocity w superposed to the unperturbed, horizontal air flow V. From Prandlt [1918] p. 337.

circulation Γ along the span of the wing, for which the trailing vortex has the horseshoe shape of Fig. 7.35. The corresponding value of the velocity $w(x)$ of the induced flow at the abcissa x along the span of the wing is given by the Biot and Savart law as

$$w(x) = \frac{\Gamma}{4\pi}\left(\frac{1}{a-x} + \frac{1}{a+x}\right), \tag{7.40}$$

if $2a$ is the span of the wing. According to the reasoning outlined above, the resulting drag is

$$D = \rho \int_{-a}^{+a} \Gamma w \, \mathrm{d}x. \tag{7.41}$$

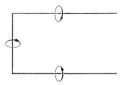

Fig. 7.35. The horseshoe vortex behind a flying wing. From Prandtl [1921*a*] p. 466.

This integral diverges logarithmically. Prandtl was thus compelled to use a variable circulation $\Gamma(x)$. By the above-mentioned theorem, a vortex filament of intensity $\Gamma'(x)\mathrm{d}x$ trails behind the element $\mathrm{d}x$ of the wing's span. The resulting induced velocity is

$$w(x) = \frac{1}{4\pi} \int_{-a}^{+a} \frac{\Gamma'(\xi)}{\xi - x} \, \mathrm{d}\xi. \tag{7.42}$$

The divergence of this integral for $\xi = x$ is easily avoided by taking its principal value in Cauchy's sense. Prandtl tried a number of simple expressions for the circulation profile $\Gamma(x)$, but kept obtaining an infinite result for the drag integral (7.41). For a while, he put this difficulty on hold, and considered the non-divergent effect of the induced velocity on other wings in the same aeroplane. In the case of a biplane or a single wing interacting with its mirror image through a solid wall, his collaborator Albert Betz published considerations of this kind and their wind-tunnel confirmation in 1912/14.[123]

As last, in November 1913, Betz and Prandtl found that the elliptic profile

$$\Gamma(x) = \Gamma_0 \sqrt{1 - \frac{x^2}{a^2}} \tag{7.43}$$

yielded the constant induced velocity $w = \Gamma_0/a$. For a horizontal velocity V of the wing, the lift

$$L = \rho V \int_{-a}^{+a} \Gamma \, \mathrm{d}x \tag{7.44}$$

takes the value

$$L = \frac{\pi}{2}\rho V \Gamma_0 a, \tag{7.45}$$

and the drag is

$$D = \frac{w}{V}L = \frac{\Gamma_0}{aV}L. \tag{7.46}$$

Remembering that the circulation Γ_0 is proportional to the velocity V (owing to Kutta's smooth-flow condition), the lift is proportional to the squared velocity of the wing and to

[123]Prandtl [1913]; [1918] p. 339, 342; Betz [1912], [1914]. Cf. Anderson [1997] p. 285.

its span. The ratio of drag to lift is independent of the wing's velocity, and diminishes with the span, as was to be expected.[124]

In sum, Prandtl and Betz accidentally discovered the elliptically-loaded wing in an attempt to avoid drag-integral divergence. They soon realized that this circulation profile was the one for which the drag was a minimum for a given lift. Lagrange's variational method leads to a simple demonstration of this fact (not Prandtl and Betz's original one). Denoting by λ the Lagrange parameter for the constraint of constant lift, the minimum drag corresponds to

$$\delta D - \lambda \delta L = \rho \int_{-a}^{+a} [\Gamma \delta w + (w - \lambda V) \, \delta \Gamma] \, dx = 0. \tag{7.47}$$

The relation (7.42) between w and Γ further leads to

$$\int_{-a}^{+a} \Gamma \, \delta w \, dx = \int_{-a}^{+a} w \, \delta \Gamma \, dx, \tag{7.48}$$

by analogy with the symmetry of mutual inductance coefficients.[125] Hence the vanishing of the integral in eqn. (7.47) for an arbitrary variation $\delta \Gamma$ requires w to be a constant. This only happens for the elliptic circulation profile.

As Prandtl noted, the corresponding pattern of the induced flow in a vertical plane containing the wing is that of a horizontal plate suddenly set into motion with the downward velocity w, for this is the only irrotational flow that satisfies the boundary conditions (see Fig. 7.23). Amazingly, this flow is exactly the one on which Lanchester based his elementary, intuitive reasoning![126]

With this treatment of the elliptically-loaded lifting line, Prandtl had in hand the basic elements of his wing theory. During the war, Betz and another outstanding collaborator, Max Munk, helped Prandtl solve the following problems.

(i) Determine the form and size of the sections of the wing that produce a given circulation $\Gamma(x)$.
(ii) Determine the circulation $\Gamma(x)$ and the corresponding drag and lift for a given shape of the wing.

The first problem is easily solved by noting that, according to the Kutta–Joukowski theory, the circulation around a given section of the wing has the form

$$\Gamma(x) = (\alpha \theta' + \beta) V l(x) \tag{7.49}$$

to first order in the effective angle of attack θ', if $l(x)$ denotes the chord of the section at the abscissa x. To first order in the induced velocity $w(x)$, the effective angle of attack differs

---

[124]Prandtl [1918] p. 342, [1921a] pp. 478–80. Hints to these results are in Betz [1914].

[125]Munk obtained this relation in 1918, cf. Prandtl [1921a] p. 489.

[126]Prandtl [1921a] pp. 464–5.

from the real angle of attack θ by the amount $-w/V$. Taking into account the relation (7.42) between Γ and w, Prandtl obtained the equation

$$\Gamma(x) + l(x)\frac{\alpha}{4\pi} \int_{-a}^{+a} \Gamma'(\xi)(\xi - x)^{-1}\mathrm{d}\xi = (\alpha\theta + \beta)Vl(x). \qquad (7.50)$$

Problem (i), that is, the determination of the chord $l(x)$ for a given circulation $\Gamma(x)$, only involves a simple integration. In contrast, the inverse problem (ii) involves an integro-differential equation that required the full skills of Betz and Munk.[127]

By the end of the war, Prandtl and his collaborators could legitimately claim a mathematical, quantitative solution of the wing problem. The only leftover task was to justify the various approximations that Prandtl had introduced at various steps of the reasoning. For this purpose, Prandtl started his memoir of 1918 with the exact, general equations of the problem. For a given wing at a given inclination, a first equation gives the velocity field as a function of the asymptotic velocity V, the trailing vortex sheet, and fictitious bound vortices that replace the boundary conditions on the wing. Reciprocally, the trailing vortex sheet depends on the velocity field through two conditions, namely, that the vortex filaments must be lines of flow for the velocity field, and their intensity must be given by the gradient of the circulation along the span of the wing. In principle, this mutual coupling should determine both the vortex sheet and the velocity field, and a three-dimensional generalization of the Kutta–Joukowski theorem then gives the force acting on the wing.[128]

In practice, various approximations must be made. Treating the circulation and the induced velocities as small quantities, Prandtl argued that, in a first (linear) approximation, the vortex sheet was parallel to the unperturbed flow V and the corresponding velocity field simply added to the unperturbed flow in the force formula. He also argued that, in the calculation of this first-order induced velocity field, the aerofoil could be replaced by a line of vorticity $\Gamma(x)$—hence the name 'lifting-line theory' now given to his wing theory. Lastly, he argued that along most of the span the motion could be regarded as being approximately two-dimensional, which makes the circulation a function of the angle of attack and the sectional form only.[129]

Although Prandtl's justifications for these assumptions lacked rigor, experiments performed during the war in the Göttingen wind tunnel vindicated them. Post-war British and American experiments further confirmed Prandtl's theory. The purely empirical methods of early aeronautics gradually made room for refined theoretical considerations. In particular, Prandtl and his group computed the effect that the walls of the tunnel had on the vortex trail of the wings, and subtracted it from raw model data in order to improve full-scale predictions. After some hesitation on the British side, by the mid-1920s this 'Prandtl correction' became a routine procedure in any wind-tunnel experiment.[130]

---

[127]Prandtl [1918] pp. 339–40, [1921a] pp. 484–7.

[128]Prandtl [1918] pp. 329–35.

[129]Ibid. pp. 335–9. Prandtl ([1918] pp. 336) used the words *tragender Faden* and *tragende Linie*.

[130]Cf. Anderson [1997] pp. 292–4, Cf. Epple [2002]. On the Prandtl correction, cf. Hashimoto [2000] pp. 231–5.

No matter how much it owes to Lanchester's intuitive mechanics, to Kutta's conformal transformations, or to Joukowski's interest in bird flight, the Göttingen wing theory may be seen as a splendid application of Helmholtz's theory of vortex motion, including discontinuity surfaces and conformal methods. As in the case of boundary-layer theory, Prandtl astutely combined and extended nineteenth-century concepts through intuitive pictures related to asymptotic approximations. Under the stimulus of the rising field of aeronautics and with the strong support of Göttingen institutions, his group put an end to the engineers' legitimate distrust of the theoretical predictions of fluid mechanics.

8

CONCLUSION

Hydrodynamics evolved considerably in the course of its application to various phenomena. So did all major theories of mathematical physics. The myths that make Newton the sole creator of mechanics, Cauchy the father of the theory of elasticity, Clausius the founder of thermodynamics, and Maxwell the unique inventor of modern electrodynamics do not resist historical analysis. These theories have changed so much since the first formulations of their fundamental principles and equations that a modern physicist who reads the mythical founders can barely recognize a kinship with present theories. This estrangement is not limited to notations and styles of presentation, but runs very deeply into the conceptual structure of the theory.

In many cases, these structural changes have occurred during attempts to apply the theory to a specific class of phenomena. For example, William Rowan Hamilton's attempt to apply mechanics to light rays led to the Hamiltonian formulation of mechanics; Charles-Eugène Delaunay's application of the same theory to the motion of the Moon yielded a new perturbation theory based on action and angle variables; Saint-Venant's application of the theory of elasticity to the flexion and torsion of prisms produced the semi-inverse method of approximation; the application of thermodynamics to mixtures and chemical reactions led to the concept of thermodynamic potential; Hendrik Lorentz's application of Maxwell's electrodynamics to certain optical phenomena led him to separate ether and matter; the application of quantum mechanics to solid-state physics engendered the theory of bands; and its application to field-mediated interactions prompted Richard Feynman's path-integral formulation.

In this small sample, four kinds of theory change are involved. In an order of increasing magnitude, they imply new methods of resolution or approximation (Saint-Venant, Delaunay), new derived concepts (thermodynamic potentials, bands), a reformulation of the foundations (Hamilton, Feynman), and the replacement of a basic principle (Lorentz). Although such innovations are most frequent during the early applications of a theory, they may occur many years later. They affect the very life of the theory, that is, the class of problems to which it is believed to be relevant, the communities that use it, the way it is taught, its conceptual hierarchy, the attached paradigms, and its relationships to other theories.

Such wide-ranging feedback effects of application are rarely acknowledged. Most commonly, applications are regarded as 'runs' of a theory, for utilitarian purposes or for transmitting implicit knowledge to students. According to Thomas Kuhn, applications contribute to the smooth, gradual expansion, and consolidation of normal science. Significant conceptual change can only result from the accumulation of major anomalies, in which case a global revolution occurs and a new paradigm emerges.[1] The above-cited

---

[1] See, e.g., Kuhn [1961].

examples of application-induced change fit neither the smooth paradigmatic phase nor the revolutionary one. They do not lead to the overthrow of the theory, yet they entail transformations of such a magnitude that the word 'application' sounds inadequate. The phenomena are not passively subjected to a rigidly established theory, but instead react upon the content and structure of the theory. They *challenge* the theory and may thus induce important adaptive transformations.

What distinguishes the history of hydrodynamics from that of other physical theories is not so much the tremendous effect of challenges from phenomenal worlds, but rather it is the slowness with which these challenges were successfully met. Nearly two centuries elapsed between the first formulation of the fundamental equations of the theory and the deductions of laws of fluid resistance in the most important case of large Reynolds numbers. In contrast, the theories of mechanics, electrodynamics, and thermodynamics were almost immediately useful in making predictions in the intended domains of application. Hydrodynamics is probably the only theory whose promises to comprehend a range of phenomena took so long to be fulfilled.

The reasons for this extraordinary delay are easily identified a posteriori. They are the infinite number of degrees of freedom and the nonlinear character of the fundamental equations, both of which present formidable obstacles to obtaining solutions in concrete cases. Moreover, instability often deprives the few known exact solutions of any physical relevance. Although unstable solutions also occur in ordinary mechanics, they do not interfere with the most common applications. In contrast, almost every theoretical description of a natural or man-made flow involves instabilities.

These difficulties have barred progress along purely mathematical lines. They have also made physical intuition a poor guide, and a source of numerous paradoxes. Hydrodynamicists therefore sought inspiration in concrete phenomena. Challenged to understand and act in real worlds, they developed a few innovative strategies. One was to modify the fundamental equations, introducing for instance Navier's viscous term, or still other terms of higher order (as a few French engineers tried to do). Another was to give up the continuity of the solutions of Euler's equation, and to study the evolution of the resulting singularities. Helmholtz pursued this approach without leaving the realm of the perfect liquid. The instability of laminar solutions was also evoked, and the resulting turbulence subjected to a statistical analysis or absorbed in the parameters of semi-empirical, effective theories of large-scale flow. Rules of similarity were used, either to predict the properties of full-scale flows from model measurements or to limit the form of resistance and retardation laws. When none of that worked, the Columbus-egg method was still available, where the hydrodynamicist could try to determine the concrete conditions under which the few flows he could predict would actually occur. This 'streamlining' strategy proved quite fruitful, because the computable flows happen to be those for which fluid resistance is a minimum.

None of these strategies sufficed to fully master the real flows for which they were intended. Prandtl's ultimate success depended on combining them within the asymptotic framework of high Reynolds numbers (quasi-inviscid fluid) and large aspect ratios (quasi-two-dimensional flow). The role of a small viscosity, Prandtl reasoned, is to produce boundary layers of high shear, and vortex sheets to which Helmholtz's theory of vortex motion may be applied in a second step. Vortex sheets are always unstable, and boundary layers often are so. These instabilities lead to turbulence. Similitude and statistical considerations allow a quantitative determination of the average effects of turbulence in cases of non-separated flow. When separation occurs, the hydrodynamicist is left with Columbus's egg, unless strong resistance is desired, in which case he can appeal to model measurements combined with similitude arguments.

Engagement with and challenges from the real worlds of flow were essential to the development of the above-mentioned strategies. The challenged theorists strove to find new solutions and to develop new methods of approximation. Experience indicated some general properties of the motion, such as the existence of boundary layers, the random character of turbulence, the sudden character of the Reynolds transition, or the formation of trailing vortices. Experimentation on ship models induced reflection on the conditions under which similitude applied. The focus on specific systems, such as Stokes's 'boxes of water' or Helmholtz's organ pipes, permitted instructive comparisons between explicit solutions of the fundamental equations and real flows. Altogether, there were many ways in which practical concerns oriented theorists in the conceptual maze of fluid dynamics.

The evolution from a paper theory to an engineering tool thus depended on transgressions of the limits between academic hydrodynamics and applied hydrodynamics. The utilitarian spirit of Victorian science, the Polytechnique ideal of a theory-based engineering, a touch of Helmholtz's eclectic genius, and the Göttingen pursuit of applied mathematics all contributed to the fruitful blurring of borders between physics and engineering. The 'sagacious geometers' who answered d'Alembert's ancient call for a solution to his resistance paradox all visited the real worlds of flow.

APPENDIX A

MODERN DISCUSSION OF D'ALEMBERT'S PARADOX

A solid body is set into motion within an infinite, homogeneous, perfect liquid and kept moving at the constant velocity $\mathbf{U}$. According to a theorem by Lagrange, the resulting flow admits a potential φ (as long as the fluid motion remains thoroughly continuous). Owing to the incompressibility of the fluid, this potential must satisfy Laplace's equation $\Delta\varphi = 0$. Consequently, at every instant it is completely determined (up to a constant) by the boundary conditions that the velocity $\mathbf{v}$ should vanish at infinity and that the normal component of $\mathbf{v} - \mathbf{U}$ should vanish on the surface of the body. The velocity field therefore follows the body in its motion, which means that the flow pattern is steady from the point of view of an observer bound to the body.

The first non-constant term in the multipolar expansion of the potential at a large distance from the body is dipolar, since a single pole would imply a divergent flux from the body, in contradiction with the incompressibility of the fluid. Hence the fluid velocity varies asymptotically as the inverse cube of the distance from the body.

The most direct way to determine the force impressed on the body by the fluid is to compute the pressure integral

$$\mathbf{R} = -\int_\sigma P \, \mathrm{d}\mathbf{S} \tag{A.1}$$

over the surface σ of the body. According to a theorem by Green, this is also equal to the integral

$$\mathbf{R} = \int_{\text{fluid}} \nabla P \mathrm{d}\tau - \int_\Sigma P \, \mathrm{d}\mathbf{S}, \tag{A.2}$$

where the second integral is taken over a spherical surface Σ surrounding the body, and the first over the volume of the fluid contained between the surfaces σ and Σ (see Fig. A.1). The surface integral tends to zero as r^{-4} when the radius r of the sphere approaches zero, because Bernoulli's law applies to the pressure P. Euler's equation gives

$$\nabla P = -\rho \frac{\partial \mathbf{v}}{\partial t} - \rho(\mathbf{v} \cdot \nabla)\mathbf{v} = \rho[(\mathbf{U} - \mathbf{v}) \cdot \nabla]\mathbf{v}, \tag{A.3}$$

or, using the incompressibility condition $\nabla \cdot \mathbf{v} = 0$,

$$\partial_i P = \rho \partial_j[(U_j - v_j)v_i]. \tag{A.4}$$

Ostrogradski's theorem then gives

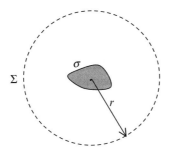

Fig. A.1.   Integration surfaces for a discussion of d'Alembert's paradox.

$$\int \nabla P \, d\tau = -\rho \int_\sigma \mathbf{v}[(\mathbf{U} - \mathbf{v}) \cdot d\mathbf{S}] + \rho \int_\Sigma \mathbf{v}[(\mathbf{U} - \mathbf{v}) \cdot d\mathbf{S}].$$ (A.5)

The integral over the surface of the body vanishes since the normal component of $\mathbf{U} - \mathbf{v}$ vanishes. The integral over the sphere Σ tends to zero as r^{-1} when its radius r tends to infinity. Consequently, the resistance vanishes.

The former reasoning amounts to applying the momentum principle to the fluid contained between the surfaces σ and Σ in a reference system bound to the body: as the flow is steady in this system, the sum of the pressures applied to the fluid on these two surfaces must be equal to the flux of the momentum tensor $\rho(v_i - U_i)(v_j - U_j)$ across them. Taking into account the incompressibility of the fluid, this flux is identical to the right-hand side of eqn (A.5).

Although it is tempting to apply the momentum principle to the whole, infinite volume of the fluid in the reference system for which the fluid is at rest at infinity, this is not possible because the total momentum of the fluid diverges logarithmically. In contrast, Borda's application of the energy principle (conservation of live force) turns out to be perfectly legitimate, because the total energy of the fluid is finite and well defined. According to Borda's simple reasoning of 1766, the work of the resistance during the motion of the body must be equal to the variation of the energy of the fluid motion, which is nil since the flow pattern is invariant. This reasoning only proves the nullity of the drag component of the resistance. Recourse to the momentum principle is necessary to prove the nullity of the lift component.[1]

The derivations given above of d'Alembert's paradox crucially depend on the infinite extent of the fluid. If there is a wall or a free surface in the vicinity of the body, then the surface Σ can no longer be rejected to infinity, and the resistance generally takes a finite value. The only exception is the case of a body moving in a direction parallel to a cylindrical wall. In the vicinity of a free surface, the body experiences a resistance even if it moves in a direction parallel to the surface, owing to the constant production of surface waves.

Another way to escape the nullity of the resistance within a perfect liquid is to reduce the dimensionality of the space. In two dimensions, the irrotational character of the motion no

---

[1] Borda [1766]. See Chapter 7, p. 267.

longer requires the velocity to asymptotically vary as the inverse cube of the distance from the body. If there is circulation around the body, then the velocity varies as the inverse of the distance and the counterpart of the Σ integral no longer tends to zero. The drag still vanishes by Borda's reasoning (the fluid-energy integral still converges). However, there is a lift proportional to the circulation according to the Kutta–Joukowski theorem (see Chapter 7, pp. 313–4). This result does not contradict the general fact that the resistance must vanish if the fluid motion remains continuous while the body is set into motion, because circulation and vorticity must both vanish under this condition.

BIBLIOGRAPHY

Bibliographic abbreviations

ACP: *Annales de chimie et de physique*

AHES: *Archive for history of exact sciences*

AP: *Annalen der Physik*

APC: *Annales des Ponts et Chaussées*

BAR: British Association for the Advancement of Science, *Report*

BB: Akademie der Wissenschaften zu Berlin, mathematisch-physikalische Klasse, *Sitzungsberichte*

BSM: Société de Mathématiques, *Bulletin*

BSP: Société Philomatique, *Bulletin*

CR: Académie des Sciences, *Comptes-rendus hebdomadaires des séances*

DSB: *Dictionary of scientific biography*, C. C. Gillispie (ed.), 16 vols (1970–1980), New York.

EM: *Exercices de mathématiques* (journal edited by Augustin Cauchy)

GN: Gesellschaft der Wissenschaften zu Göttingen, mathematisch-physikalische Klasse, *Nachrichten*

HN: Helmholtz *Nachlass*, Akademie der Wissenschaften, Berlin

HSPS: *Historical studies in the physical (and biological) sciences*

HVR: Hermann von Helmholtz, *Vorträge und Reden*, 2 vols (1896), Braunschweig.

HWA: Hermann von Helmholtz, *Wissenschafliche Abhandlungen*, 3 vols (1882, 1883, 1895), Leipzig.

JMPA: *Journal de mathématiques pures et appliquées*

JRAM: *Journal für die reine und angewandte Mathematik*

MAS: Académie (Royale) des Sciences, *Mémoires* (*de physique et de mathématiques*)

MASB: Académie Royale des Sciences et des Belles-Lettres de Berlin, *Mémoires*

MSE: Académie des Sciences de l'Institut de France, *Mémoires présentés par divers savants*

PGA: Ludwig Prandtl, *Gesammelte Abhandlungen zur Mechanik, Hydro- und Aerodynamik*, 3 vols (1961), Göttingen

PLMS: London Mathematical Society, *Proceedings*

PLPSM: Literary and Philosophical Society of Manchester, *Proceedings*

PM: *Philosophical magazine*

PRI: Royal Institution, *Proceedings*

PRS: Royal Society of London, *Proceedings*

PRSE: Royal Society of Edinburgh, *Proceedings*

PT: Royal Society of London, *Philosophical transactions*

ReP: Osborne Reynolds, *Papers on mechanical and physical subjects*, 3 vols (1900–1903), Cambridge.

RSP: Lord Rayleigh, *Scientific papers*, 6 vols (1899–1920), Cambridge.

SMPP: George Gabriel Stokes, *Mathematical and physical papers*, 5 vols (1880–1905), Cambridge.

ST: David B. Wilson (ed.), *The correspondence between Sir George Gabriel Stokes and Sir William Thomson, Baron Kelvin of Largs*, 2 vols (1990), Cambridge.

TCPS: Cambridge Philosophical Society, *Transactions*

TINA: Royal Institution of Naval Architects, *Transactions*

TMPP: William Thomson, *Mathematical and physical papers*, 6 vols (1882–1911), Cambridge.

TRSE: Royal Society of Edinburgh, *Transactions*

Bibliography of primary literature

Abbe, Cleveland [1890] 'A plea for terrestrial physics,' American Association for the Advancement of Science, *Proceedings* **39**, pp. 55–79.

Airy, George Biddell [1838] 'On the intensity of light in the neighborhood of a caustic', *TCPS* **6**, pp. 379–402.

—— [1845] 'Tides and waves', *Encyclopedia metropolitana* **5**, pp. 291–396.

Arago, François, Babinet, J., and Piobert, G.

—— [1842] (Report on Poiseuille [1844]), *CR* **15**, pp. 1167–86.

Arnol'd, Vladimir I. [1966] 'Sur un principe variationnel pour les écoulements stationnaires des liquides parfaits et ses applications aux problèmes de stabilité non linéaires', *Journal de mécanique* **5**, pp. 29–43.

Baschin, Otto [1899] 'Die Entstehung wellenähnlicher oberflächenformen', *Zeitschrift der Gesellschaft für Erdkunde zu Berlin* **34**, pp. 409–34.

Basset, Alfred B. [1888] *A treatise on hydrodynamics*, 2 vols, Cambridge.

Batchelor, George Keith [1967] *An introduction to fluid dynamics*, Cambridge.

Baumgarten, André [1847] 'Notice sur le moulinet de Woltmann destiné à mesurer les vitesses de l'eau, sur son perfectionnement et sur les expériences faites avec cet instrument', *APC* **14**, pp. 326–74.

Bazin, Henry [1865] 'Recherches expérimentales sur la propagation des ondes', *MSE* **19**, pp. 495–652.

Beaufoy, Mark [1834] *Nautical and hydraulic experiments*, London.

Becker, Henry A., and Massaro, T. A. [1968] 'Vortex evolution in a round jet', *Journal of fluid mechanics* **31**, pp. 435–448.

Bélanger, Jean-Baptiste [1828] *Essai sur la solution numérique de quelques problèmes relatifs au mouvement permanent des eaux courantes.* Paris.

—— [1841/42] *Notes sur l'hydraulique*, Paris (lithographed course for the Ecole Royale des Ponts et Chaussées).

Bélidor, Bernard Forest [1819] *Architecture hydraulique*, Claude Louis Navier (ed.), Paris.

Belt, Thomas [1859] 'An inquiry into the origin of whirlwinds', *PM* **4**, pp. 47–53.

Bernoulli, Daniel [1738] *Hydrodynamica, sive de viribus et motibus fluidorum commentarii*, Strasbourg.

—— [2002] *Die Werke von Daniel Bernoulli*, vol. **5**, Gleb K. Mikhailov (ed.), Basel.

Bernoulli, Johann [1714] 'Meditatio de natura centri oscillationis', *Opera omnia* **2**, pp. 168–86.

—— [1742] 'Hydraulica nunc primum detecta ac demonstrata directe ex fundamentis pure mechanicis. Anno 1732', *Opera omnia* **4**, pp. 387–493.

Bertrand, Joseph [1848] 'Note sur la similitude en mécanique', Ecole Polytechnique, *Journal* **32**, pp. 189–197.

—— [1868a] 'Théorème relatif au mouvement le plus général d'un fluide', *CR* **66**, pp. 1227–1230.

—— [1868b] 'Note relative à la théorie des fluides. Réponse à la communication de M. Helmholtz', *CR* **67**, pp. 267–269.

—— [1868c] 'Observations nouvelles sur un mémoire de M. Helmholtz,' *CR* **67**, pp. 469–472.

—— [1868d] 'Réponse à la note de M. Helmholtz,' *CR* **67**, pp. 773–775.

Betz, Albert [1912] 'Auftrieb und Widerstand eines Doppledeckers', *Zeitschrift für Flugtechnik und Motorluftschiffahrt* **3**, pp. 217–20; **4**, (1913), pp. 1–3.

—— [1914] 'Die gegenseitige Beeinflüssung zweier Tragflächen', *Zeitschrift für Flugtechnik und Motorluftschiffahrt* **5**, pp. 253–258.

Bezold, Wilhelm von [1888] 'Zur Thermodynamik der Atmosphäre. Zweite Mitteilung', *BB*, pp. 1189–1206.

Bidone, George [1820] 'Expériences sur le remou et sur la propagation des ondes', *Memorie della reale Accademia delle scienze di Torino* **25**, pp. 21–112.

Birkhoff, Garrett [1950] *Hydrodynamics: A study in logic, fact, and similitude*, Princeton.

Bjerknes, Vilhelm [1898] 'Über einen hydrodynamischen Fundamentalsatz und seine Anwendung besonders auf die Mechanik der Atmosphäre und des Weltmeeres', Kongliga Svenska, Vetenskaps-Akademiens, *Handlingar* **31**, pp. 3–38.

Bjerknes, Vilhelm, et al. [1910] *Dynamic meteorology and hydrography*, Washington.

—— [1933] *Physikalische Hydrodynamik mit Anwendung auf die dynamische Meteorologie*, Berlin.

Blasius, Heinrich [1908] 'Grenzschichten in Flüssigkeiten mit kleiner Reibung,' *Zeitschrift für Mathematik und Physik* **56**, pp. 1–37.

—— [1913] 'Ähnlichkeitsgesetz bei Reibungsvorgängen in Flüssigkeiten,' Verein Deutscher Ingenieure, *Mitteilungen über Forschungsarbeiten* **131**, pp. 1–40.

Boileau, Pierre [1846] 'Etudes expérimentales sur les cours d'eau,' *CR* **22**, pp. 212–6; *CR* **24**, pp. 957–960.

—— [1854] *Traité de la mesure des eaux courantes*, Paris.

Borda, Jean-Charles de [1763] 'Expériences sur la résistance des fluides,' *MAS*, pp. 356–376.

—— [1766] 'Sur l'écoulement des fluides par les orifices des vases,' *MAS*, pp. 579–607.

—— [1767] 'Expériences sur la résistance des fluides,' *MAS*, pp. 495–503.

Bossut, Charles [1786/87] *Traité théorique et expérimental d'hydrodynamique*, 3 vols, Paris.

Boussinesq, Joseph [1868] 'Mémoire sur l'influence des frottements dans les mouvements réguliers des fluides', *JMPA* **13**, pp. 377–424.

—— [1870] 'Essai théorique sur les lois trouvées expérimentalement par M. Bazin pour l'écoulement uniforme de l'eau dans les tuyaux découverts', *CR* **71**, pp. 389–393.

—— [1871*a*] 'Théorie de l'intumescence liquide appelée *onde solitaire* ou *de translation*, se propageant dans un canal rectangulaire', *CR* **72**, pp. 755–759.

—— [1871*b*] 'Sur le mouvement permanent varié de l'eau dans les tuyaux de conduite et dans les canaux découverts', *CR* **73**, pp. 34–38, 101–105.

—— [1871*c*] 'Théorie générale des mouvements qui sont propagés dans un canal rectangulaire horizontal', *CR* **73**, pp. 256–260.

—— [1872*a*] 'Théorie des ondes liquides périodiques', *MSE* **20**, pp. 509–615 (presented to the French Academy on 19 April 1869).

—— [1872*b*] 'Théorie des ondes et des remous qui se propagent le long d'un canal rectangulaire horizontal, en communiquant au liquide contenu dans ce canal des vitesses sensiblement pareilles de la surface au fond', *JMPA* **17**, pp. 55–108 (presented to the Académie des Sciences on 13 November 1871).

—— [1872*c*] 'Essai sur la théorie des eaux courantes', *CR* **75**, pp. 1011–15 (extraits du mémoire présenté à l'Académie des Sciences sur l'écoulement des eaux dans un canal prismatique).

—— [1877] 'Essai sur la théorie des eaux courantes', *MSE* **23**, pp. 1–680.

—— [1880*a*] 'Sur la manière dont les frottements entrent en jeu dans un fluide qui sort de l'état de repos, et sur leur effet pour empêcher l'existence d'une fonction des vitesses', *CR* **90**, pp. 736–739.

—— [1880*b*] 'Quelques considérations à l'appui d'une note du 29 mars, sur l'impossibilité d'admettre, en général, une fonction des vitesses dans toutes les questions de l'hydraulique où les frottements ont un rôle notable', *CR* **90**, pp. 967–969.

—— [1883] *Notice sur les travaux scientifiques de M. Joseph Boussinesq*, Lille.

—— [1890] 'Théorie du mouvement permanent qui se produit près de l'entrée évasée d'un tube fin: application à la deuxième série d'expériences de Poiseuille', *CR* **110**, pp. 1238–1244.

—— [1891] 'Sur la manière dont les vitesses, dans un tube cylindrique de section circulaire, évasé à son entrée, se distribuent depuis cette entrée jusqu'aux endroits où se trouve établi un régime uniforme', *CR* **113**, pp. 9–15.

—— [1897] *Théorie de l'écoulement tourbillonnant et tumultueux des liquides dans les lits rectilignes à grande section*, Paris.

Bresse, Jacques [1860] *Cours de mécanique appliquée. Second partie. Hydraulique*, Paris.

Brillouin, Marcel [1911] 'Les surface de glissement de Helmholtz et la résistance des fluides', *ACP* **23**, pp. 145–230.

Caflisch, Russel E. [1990] 'Analysis for the evolution of vortex sheets', UCLA Computational and Applied Mathematics, *Report* pp. 91–05, November.

Cauchy, Augustin [1818] 'Seconde note sur les fonctions réciproques', *BSP*, pp. 178–80.

—— [1823] 'Recherches sur l'équilibre et le mouvement des corps solides ou fluides, élastiques ou non élastiques', *BSP*, pp. 9–13 (extract of a memoir read on 30 September 1822).

—— [1827*a*] 'Théorie de la propagation des ondes à la surface d'un fluide pesant d'une profondeur indéfinie', *MSE* **1**, pp. 1–123.

—— [1827*b*] 'Sur la pression ou la tension dans les corps solides', *EM* **2**, pp. 42–57.

—— [1827*c*] 'Sur la condensation et la dilatation des corps solides', *EM* **2**, pp. 60–69.

—— [1827*d*] 'Sur les relations qui existent, dans l'état d'équilibre d'un corps solide ou fluide, entre les pressions ou tensions et les forces accélératrices', *EM* **2**, pp. 108–111.

—— [1828*a*] 'Sur les équations qui expriment les conditions d'équilibre ou les lois du mouvement d'un corps solide, élastique ou non élastique', *EM* **3**, pp. 160–187.

—— [1828*b*] 'De la pression ou tension dans un système de points matériels', *EM* **3**, pp. 213–236.

—— [1829] 'Sur l'équilibre et le mouvement d'un système de points matériels sollicités par des forces d'attraction ou de répulsion naturelle', *EM* **4**, pp. 129–139.

Cauchy, Augustin [1833] *Sept leçons de physique générale*, posthumously published by François Moigno, Paris, 1868 (Torino lectures).
—— [1840] 'Mémoire sur les deux espèces d'ondes planes qui peuvent se propager dans un système isotrope de points matériels', *CR* **10**, pp. 905–918.
—— [1841] 'Mémoire sur les dilatations, condensations et les rotations produites par un changement de forme dans un système de points matériels', *Exercises d'analyse et de physique mathématique* **2**, pp. 302–330.
—— [1845] 'Note relative à la pression totale supportée par une surface finie dans un corps solide ou fluide', *CR* **20**, pp. 1764–1766.
Challis, James [1833] 'Report on the present state of the analytical theory of hydrostatics and hydrodynamics', *BAR*, pp. 131–151.
Chaplygin, Sergei Alekseevich [1910] 'On the pressure exerted by a plane-parallel flow upon an obstructing body', Moscow Mathematical Society, *Mathematical abstracts* **28**, 120–66 (Russian, translated in Chaplygin [1956] pp. 1–16).
—— [1911] 'Results of theoretical considerations on the motion of aircraft', Moscow Aeronautical Society, *Bulletin* (Russian, translated in Chaplygin [1956] pp. 17–22).
—— [1956] *The selected works on wing theory of Sergei A. Chaplygin*, translated by Maurice A. Garbell, San Francisco.
Clausius, Rudolph [1849] 'Über die Veränderungen welche in den bisher gebräuchlichen Formeln für das Gleichgewicht und die Bewegung elastischer fester Körper durch neuere Beobachtungen nothwendig geworden sind', *AP* **76**, pp. 46–67.
Clebsch, Rudolph [1859] 'Über die Integration der hydrodynamischen Gleichungen', *JRAM* **56**, pp. 1–10.
Colson, DeVer [1954] 'Wave-cloud formation at Denver', *Weatherwise* **7**, pp. 34–35.
Coriolis, Gaspard [1836] 'Sur l'établissement de la formule qui donne la figure des remous, et sur la correction qu'on doit y introduire pour tenir compte des différences de vitesse dans les divers points d'une même section d'un courant', *APC*, pp. 314–35.
Couette, Maurice [1887] 'Oscillations tournantes d'un solide de révolution en contact avec un fluide visqueux', *CR* **105**, pp. 1064–1067.
—— [1888] 'Sur un nouvel appareil pour l'étude du frottement des liquides', *CR* **107**, pp. 388–390.
—— [1890*a*] 'Etudes sur le frottement des liquides', *ACP* **21**, pp. 414–424.
—— [1890*b*] 'Distinction de deux régimes dans le mouvement des fluides', *Journal de physique* **9**, pp. 414–426.
Coulomb, Charles Augustin [1800] 'Expériences destinées à déterminer la cohérence des fluides et les lois de leur résistance dans les mouvements très lents', Institut National des Sciences et des Arts, *Mémoires de sciences mathématiques et physiques* **3**, pp. 246–305.
Cournot, Antoine [1828] Review of Navier's memoir on fluid motion, *Bulletin des sciences mathématiques* **10**, pp. 11–14.
D'Alembert, Jean le Rond [1743] *Traité de dynamique*, Paris.
—— [1744] *Traité de l'équilibre et du mouvement des fluides*, Paris.
—— [1747] *Réflexions sur la cause générale des vents*, Paris.
—— [1752] *Essai d'une nouvelle théorie de la résistance des fluides*, Paris.
—— [1761] 'Remarques sur les lois du mouvement des fluides', in *Opuscules mathématiques*, vol. **1**, pp. 137–68, Paris.
—— [1768] 'Paradoxe proposé aux géomètres sur la résistance des fluides', in *Opuscules mathématiques*, vol. **5**, 34th memoir, pp. 132–8, Paris.
—— [1780] *Opuscules mathématiques*, vol. **8**. Paris.
Darcy, Henry [1856] *Les fontaines publiques de Dijon. Exposition et application des principes à suivre et des formules à employer dans les questions de distribution d'eau*, Paris.
—— [1857] *Recherches expérimentales relatives au mouvement de l'eau dans les tuyaux*, Paris.
Darcy, Henry, and Bazin, Henry [1865*a*] *Recherches hydrauliques. Première partie: Recherches expérimentales sur l'écoulement de l'eau dans les canaux découverts*, Paris.
—— [1865*b*] *Recherches hydrauliques. Deuxième partie: Recherches expérimentales relatives aux remous et à la propagation des ondes*, Paris.
Dines, William Henry [1891] 'On wind pressure upon an inclines surface', *PRS* **48**, pp. 233–257.
Drazin, Philip G., and Reid, William H. [1981] *Hydrodynamic stability*. Cambridge.
Du Buat, Pierre [1786] *Principes d'hydraulique, vérifiés par un grand nombre d'expériences faites par ordre du gouvernement*, 3 vols, Paris. Reedited in 1816.

Earnshaw, Samuel [1838] 'On fluid motion, so far as it is expressed by the equation of continuity', *TCPS* **6**, pp. 202–233.

—— [1849] 'The mathematical theory of the two great solitary waves of the first order', *TCPS* **8**, pp. 326–41 (read on 8 December 1845).

Eiffel, Gustave [1912] 'Sur la résistance des sphères dans l'air en mouvement', *CR* **155**, pp. 1597–1599.

Espy, James P. [1841] *The philosophy of storms*. Boston.

Euler, Leonhard [1740] 'Inquisitio physica in causam fluxus ac refluxus maris,' in *Pièces qui ont remporté le prix de l'Académie Royale des Sciences de Paris en 1740*, pp. 235–350, Paris.

—— [1745] *Neue Grundsätze der Artillerie, aus dem englischen des Herrn Benjamin Robins übersetzt und mit vielen Anmerkungen versehen*, Berlin (from B. Robins, *New principles of gunnery*, London, 1742).

—— [1750] 'Découverte d'un nouveau principe de mécanique', *MASB*, also in *Opera omnia*, ser. 2, vol. **5**, pp. 81–108, Lausanne, 1957.

—— [1752] 'Sur le mouvement de l'eau par des tuyaux de conduite', *MASB*, also in *Opera omnia*, ser. 2, vol. **15** (Lausanne, 1957), pp. 219–250.

—— [1755*a*] 'Principes généraux de l'état d'équilibre d'un fluide', *MASB*, also in Euler [1954], pp. 2–53.

—— [1755*b*] 'Principes généraux du mouvement des fluides', *MASB*, also in Euler [1954], pp. 54–91.

—— [1755*c*] 'Continuation des recherches sur la théorie du mouvement des fluides', *MASB*, also in Euler [1954], pp. 92–132.

—— [1756] 'Principia motus fluidorum', *Novi commentarii academiae scientiarum Petropolitanae* (written in 1752), also in Euler [1954] pp. 133–68.

—— [1760a] (written in 1751) 'Recherches sur le mouvement des rivières,' *MASB* **16**, pp. 101–18, also in Euler [1954], pp. 272–88.

—— [1760b], 'Dilucidationes de resistencia fluidorum,' *NACP* **8**, pp. 197–229, also in Euler [1954], pp. 215–243.

—— [1954] *Opera omnia*, ser. 2, vol. **XII.**, Clifford Truesdell (ed.), Lausanne.

—— [1998] *Commercium epistolium*, ser. 4A, vol. **2**, Emil Fellmann and Gleb Mikhajlov (eds), Basel.

Exner, Felix [1925] *Dynamische Meteorologie*, 2nd edn, Vienna.

Fernando, Harind J. S. [1991] 'Turbulent mixing in stratified fluids', *Annual review of fluid mechanics* **23**, pp. 455–493.

Ferrel, William [1860] *The motions of fluids and solids relative to the earth's surface*, New York (reprint from the *Mathematical monthly*).

FitzGerald, George Francis [1885] 'On a model illustrating some properties of the ether', Royal Dublin Society, *Scientific proceedings*, also in FitzGerald [1902] pp. 142–56.

—— [1889] 'On an electromagnetic interpretation of turbulent liquid motion', *Nature*, also in FitzGerald 1902: pp. 254–261.

—— [1899*a*] 'On the energy per cubic centimeter in a turbulent liquid when transmitting laminar waves', *BAR*, also in FitzGerald [1902], pp. 484–486

—— [1899*b*] 'On a hydrodynamical hypothesis as to electromagnetic actions', Royal Dublin Society, *Scientific proceedings*, also in FitzGerald [1902], pp. 472–477,

—— [1902] *The scientific writings of the late George Francis FitzGerald*. Dublin.

Flachat, Eugène [1866] *Navigation à vapeur transocéanienne: Etudes scientifiques*, 2 vols and atlas. Paris

Flamant, Alfred [1891] *Mécanique appliquée: Hydraulique*, Paris.

Föppl, August [1899] *Vorlesungen über technische Mechanik. Vierter Band: Dynamik*, Leipzig.

—— [1909–1911] *Vorlesungen über technische Mechanik*, 3rd edn, 6 vols, Leipzig.

Forchheimer, Philipp [1905] 'Hydraulik', in *Encyklopädie der mathematischen Wissenschaften mit Einschluss ihrer Anwendungen*, vol. **4.3**, pp. 324–472, Leipzig.

—— [1914] *Hydraulik*, Leipzig. Reedited in 1924 and 1930.

—— [1927] 'Wasserströmungen', in *Handbuch der Physik*, Hans Geiger and Karl Scheel (eds), vol. **7**, Berlin.

Fresnel, Augustin [1821] 'Sur le calcul des teintes que la polarisation développe dans les lames cristallines', in *Oeuvres complètes*, 3 vols (1866, 1868, 1870), vol. **1**, pp. 609–53.

Fourier, Joseph [1818] 'Note relative aux vibrations des surfaces élastiques et au mouvement des ondes', *BSP*, pp. 129–135.

Froude, Robert [1869] 'On the hydraulic internal scraping of the Torquay water-main,' *BAR*, pp. 210–211.

Froude, William [1862] 'On the rolling of ships. Appendix 2: On the dynamical structure of oscillating waves,' *TINA* 3, pp. 45–62.

—— [1868] 'Observations and suggestions on the subject of determining by experiment the resistance of ships', memorandum sent to E. J. Reed, Chief Constructor of the Navy, dated December [1868], in Froude [1957] pp. 120–7.

—— [1869*a*] 'On some difficulties in the received views of fluid friction,' *BAR*, pp. 211–214.

—— [1869*b*] 'Explanations', in Merrifield [1869] pp. 43–7.

Froude, William [1872] 'Experiments on the surface-friction experienced by a plane moving through water', *BAR*, pp. 118–124.

—— [1874*a*] 'Reports to the Lords Commissioners of the Admiralty on experiments for the determination of the frictional resistance of water on a surface, under various conditions, performed at Chelston Cross, under the authority of their Lordships,' *BAR*, pp. 249–55.

—— [1874*b*] 'On experiments with H.M.S. Greyhound,' *TINA* 15, pp. 36–59.

—— [1875] 'The theory of "streamlines" in relation to the resistance of ships,' *Nature* 13, pp. 50–2, 89–93, 130–3, 169–72.

—— [1877*a*] 'The fundamental principles of the resistance of ships,' *PRI* 8, pp. 188–213.

—— [1877*b*] 'Experiments upon the effect produced on the wave-making resistance of ships by length of parallel middle body,' *TINA* 18, pp. 77–97.

—— [1957] *The papers of William Froude*. London.

Gerstner, Franz Joseph von [1802] 'Theorie de Wellen,' Böhmische Gesellschaft der Wissenschaften, *Abhandlungen*, also in *AP* 32, (1809), pp. 412–45.

—— [1804] *Theorie der Wellen samt einer daraus abgeleiteten Theorie der Deichprofile* (Prague).

Girard, Pierre-Simon [1816] 'Mémoire sur le mouvement des fluides dans les tubes capillaires et l'influence de la température sur ce mouvement', Institut National des Sciences et des Arts, *Mémoires de sciences mathématiques et physiques* (1813–1815), pp. 249–380 (read on 30 April and 6 May 1816).

—— [1817] 'Mémoire sur l'écoulement linéaire de diverses substances liquides par des tubes capillaires de verre' *MAS*, 1 (1816), pp. 187–274 (read on 12 January 1817).

Graeff, Michel [1882] *Traité d'hydraulique*, 3 vols. Paris.

Green, George [1838] 'On the motion of waves in a variable canal of small depth and width', *TCPS,* 6, pp. 457–462.

—— [1839] 'Note on the motion of waves in canals,' *TCPS* 7, pp. 87–96.

Hadamard, Jacques [1903] *Leçons sur la propagation des ondes et sur les équations de l'hydrodynamique.* Paris.

Hadley, George [1735] 'Concerning the cause of the general trade winds,' *PT* 39, pp. 58–62.

Hagen, Gotthilf [1839] 'Über die Bewegung des Wassers in engen cylindrischen Röhren,' *AP* 46, pp. 423–442.

—— [1854] 'Über den Einfluss der Temperatur auf die Bewegung des Wassers in Röhren,' Akademie der Wissenschaften zu Berlin, *Mathematische Abhandlungen*, pp. 17–98.

Hagenbach, Eduard [1860] 'Über die Bestimmung der Zähigkeit einer Flüssigkeit durch den Ausfluss aus Röhren,' *AP* 109, pp. 385–426.

Hankel, Hermann [1861] *Zur allgemeinen Theorie der Bewegung der Flüssigkeiten*, Göttingen: Preisschrift.

Havelock, Thomas Henry [1908] 'The propagation of groups of waves in dispersive media, with applications to waves on water produced by a travelling disturbance,' *PRS* 81, pp. 398–430.

Heisenberg, Werner [1924] 'Über Stabilität und Turbulenz von Flüssigkeitsströmen,' *AP* 74, pp. 577–627.

Helmholtz, Hermann von [1852/53] 'Bericht über die theoretische Akustik betreffenden Arbeiten vom Jahre 1848 und 1849,' *Die Fortschritte der Physik* 4, pp. 124–5; 5 (1853), pp. 93–8, also in *HWA* 1, pp. 233–55.

—— [1856] 'Über Combinationstöne' *AP* 99, pp. 497–540, also in *HWA* 1, pp. 263–302.

—— [1857] 'Über die physiologischen Ursachen der musikalischen Harmonie', *HVR* 1, pp. 121–55 (Bonn lecture).

—— [1858] 'Über Integrale der hydrodynamischen Gleichungen, welche den Wirbelbewegungen entsprechen,' *JRAM* 55, pp. 25–55, also in *HWA* 1, pp. 101–34.

—— [1859] 'Theorie der Luftschwingungen in Röhren mit offenen Enden,' *JRAM* 57, pp. 1–72, also in *HWA* 1, pp. 303–382.

—— [1863*a*] *Die Lehre von den Tonempfindungen als physiologische Grundlage für die Theorie der Musik.* Braunschweig.

—— [1863b] 'Über den Einfluss der Reibung in der Luft auf die Schallbewegung,' Naturhistorisch-medizinischer Verein zu Heidelberg, *Verhandlungen* 3, pp. 16–20, also in *HWA* 1, pp. 382–387.

—— [1865] 'Eis und Gletcher' (Frankfurt and Heidelberg, 1865), in *Populäre wissenschaftliche Vorträge. Erster Heft*, pp. 93–134, Braunschweig.

—— [1867] 'On the integrals of the hydrodynamical equations, which express vortex-motion' *PM* 33, pp. 485–512. (P. G. Tait's translation of Helmholtz [1858]).

—— [1868a] 'Sur le mouvement le plus général d'un fluide. Réponse à une communication précédente de M. S. Bertrand,' *CR* 67, pp. 221–225, also in *HWA* 1, pp. 134–139.

—— [1868b] 'Sur le mouvement des fluides. Deuxième réponse à M. S. Bertrand, *CR* 67, pp. 754–57, also in *HWA* 1, pp. 140–44.

—— [1868c] 'Réponse à la note de Monsieur Bertrand du 19 octobre,' *CR* 67, pp. 1034–1035, also in *HWA* 1, pp. 145.

—— [1868d] 'Über diskontinuirliche Flüssigkeitsbewegungen,' *BB*, pp. 215–228, also in *HWA* 1, pp. 146–157.

—— [1869] 'Zur Theorie der stationären Stöme in reibenden Flüssigkeiten,' Naturhistorisch-medizinischer Verein zu Heidelberg, *Verhandlungen* 5, pp. 1–7, also in *HWA* 1, pp. 223–230.

—— [1873] 'Über ein Theorem, geometrisch ähnliche Bewegungen flüssiger Körper betreffend, nebst Anwendung auf das Problem, Luftballons zu lenken,' Königliche Akademie der Wissenschaften zu Berlin, *Monatsberichte*, pp. 501–514, also in *HWA* 1, pp. 158–171.

—— [1876] 'Wirbelstürme und Gewitter' (Hamburg, 1875), *Deutsche Rundschau* 6, also in *HVR* 2, pp. 139–63.

—— [1877] *Die Lehre von den Tonempfindungen als physiologische Grundlage für die Theorie der Musik*, 4th edn, Braunschweig.

—— [1886] 'Wolken- und Gewitterbildung,' Physikalische Gesellschaft zu Berlin, *Verhandlungen* 5, pp. 96–97, also in *HWA* 3, pp. 287–288.

—— [1888] 'Über atmospherische Bewegungen I,' *BB*, also in *HWA* 3, pp. 289–308.

—— [1889] 'Über atmospherische Bewegungen. Zweite Mittheilung,' *BB*, pp. 761–780, also in *HWA* 3, pp. 309–332.

—— [1890] 'Die Energie der Wogen und des Windes', *HWA* 3, pp. 333–355.

—— [1891] 'Errinerungen,' *HVR* 2, pp. 1–21.

Helmholtz, Hermann von, and Piotrowski, Gustav von [1860] 'Über Reibung von tropfbarer Flüssigkeiten,' Akademie der Wissenschaften zu Wien, Mathematisch-naturwissenschaftliche Classe, *Sitzungsberichte*, also in *HWA* 1, pp. 172–222.

Hicks, William M. [1881–82] 'Report on recent progress in hydrodynamics', *BAR* (1881), pp. 57–88; *BAR* (1882), pp. 39–70.

Humphreys, Andrew, and Abbot, Henry [1868] 'Report on the physics and hydraulics of the Mississippi river,' *Indian engineer* 5, pp. 165–194.

Jacobson, Heinrich [1860] 'Beiträge zur Hämodynamik,' *Archiv für Anatomie, Physiologie und wissenschaftliche Medicin*, pp. 80–112.

Joukowski (Zhukovskii), Nikolai [1890] 'The theory of flight', *Report of the eighth congress of the Russian naturalists and physicians* (Russian), also in *Collected works*, vol. 4 (1949), pp. 344–51.

—— [1906a] 'De la chute dans l'air de corps légers de forme allongée, animés d'un mouvement rotatoire,' Institut Aérodynamique de Koutchino, *Bulletin* 1, pp. 51–65.

—— [1906b] 'On adjoint vortices', Obshchestvo liubitelei estestvoznaniia, antropologii I etnografii, Moskva, *Izviestiia* 13, pp. 12–25 (Russian).

—— [1910] 'Über die Konturen der Tragflächen der Drachenflieger,' *Zeitschrift für Flugtechnik und Motorluftschiffahrt* 1, pp. 281–284.

—— [1916] *Aérodynamique*, Paris (lectures at the Imperial Technical School in Moscow).

Kármán, Theodore von

—— [1911] 'Über den Mechanismus des Widerstandes, den ein bewegter Körper in einer Flüssigkeit erfährt,' *GN* pp. 509–517; *GN* (1912), pp. 547–556.

—— [1921] 'Über laminare und turbulente Reibung,' *Zeitschrift für angewandte Mathematik und Mechanik* 1, pp. 233–252.

—— [1930] 'Mechanische Ähnlichkeit und Turbulenz,' *GN*, pp. 58–76.

—— [1932] 'Theorie des Reibungswiderstands,' in *Buchwerk der Konferenz über hydromechanische Probleme des Schiffantriebs, Hamburg*, also in *Collected works*, vol. 2, pp. 394–414 (1956), London.

—— [1967] *The wind and beyond*, Boston.

Kelland, Philip [1840] 'On the theory of waves,' *TRSE* **14**, pp. 497–545.

—— [1844] 'On the theory of waves. Part 2,' *TRSE* **15**, pp. 101–144.

Kinsman, Blair [1965] *Wind waves: Their generation and propagation on the ocean surface*, Englewood Cliffs.

Kirchhoff, Gustav [1850] 'Über das Gleichgewicht und die Bewegung einer elastischen Scheibe,' *JRAM*, also in Kirchhoff [1882] pp. 237–79.

—— [1869] 'Zur Theorie freier Flüssigkeitsstrahlen,' *JRAM* **70**, pp. 289–298, also in Kirchhoff [1882] pp. 416–27.

—— [1876] *Vorlesungen über Mechanik*, Leipzig.

Kirchhoff, Gustav [1882] *Gesammelte Abhandlungen*, Leipzig.

Kirsten, Christa, *et al.* (eds) [1986] *Dokumente einer Freundschaft: Briefwechsel zwischen Hermann von Helmholtz und Emil du Bois-Reymond*, Berlin.

Klein, Felix [1910] 'Über die Bildung von Wirbeln in reibungslosen Flüssigkeiten,' *Zeitschrift für Mathematik und Physik* **58**, pp. 259–262.

Knibbs, George H. [1897] 'On the steady flow of water in uniform pipes and channels,' Royal Society of New South Wales, *Journal and proceedings* **31**, pp. 314–355.

Köppen, Wladimir [1901] 'Beiträge zur Mechanik des Fluges und schwebenden Falles,' *Illustrirte aeronautische Mittheilungen* **4**, pp. 149–159.

Korteweg, Diederik, and de Vries, G. [1895] 'On the change of form of long waves advancing in a rectangular canal, and on a new type of long stationary waves,' *PM* **39**, pp. 422–443.

Koženy, Josef [1920] *Die Wasserführung der Flüsse*. Leipzig.

Krümmel, Otto [1911] *Handbuch der Ozeanographie*. Stuttgart.

Kutta, Wilhelm [1902a] *Habilitationsschrift*, University of Munich.

—— [1902b] 'Auftriebskräfte in strömenden Flüssigkeiten,' *Illustrirte aeronautische Mittheilungen* **6**, pp. 133–135.

—— [1910] 'Über eine mit den Grundlagen des Flugproblems in Beziehung stehende zweidimensionale Strömung', Königlich Bayerische Akademie der Wissenschaften, Mathematisch-physikalische Klasse, *Sitzungsberichte* **40**, 2. Abhandlung (58pp.).

Lagrange, Joseph Louis [1761] 'Application de la méthode exposée dans le mémoire précédent à la solution de différens problèmes de dynamique,' *Miscellanea taurinensia*, also in *Oeuvres*, vol. **1**, (1869), pp. 365–468.

—— [1779] 'Sur la construction des cartes géographiques,' *MASB*, also in *Oeuvres*, vol. **4** (1869), pp. 637–692.

—— [1781] 'Mémoire sur la théorie du mouvement des fluides,' *MASB*, also in *Oeuvres*, vol. **4**, (1869), pp. 695–750.

—— [1788] *Traité de mécanique analitique*, Paris.

—— [1811/15] *Traité de mécanique analytique*, 2nd edn, also in *Oeuvres*, vols. **11–12**.

—— [1867–1892] *Oeuvres*, 14 vols, Paris.

Lamb, Horace [1879] *A treatise on the mathematical theory of the motion of fluids*, Cambridge.

—— [1895] *Hydrodynamics*, Cambridge.

—— [1904] 'On deep-water waves,' London Mathematical Society, *Proceedings* **2**, pp. 371–401.

—— [1916] 'On wave-patterns due to a travelling disturbance,' *PM* **31**, pp. 539–548.

—— [1932] *Hydrodynamics*, 6th edn, New York: Dover.

Lanchester, Frederick [1894] 'The soaring of birds and the possibilities of mechanical flight' (19 June 1894), briefly reported in Birmingham Natural History and Philosophical Society, *Reports of meetings*.

—— [1907] *Aerodynamics, constituting the first volume of a complete work on aerial flight*, London.

—— [1908] *Aerodonetics, constituting the second volume of a complete work on aerial flight*, London.

—— [1926] 'Sustentation in flight', Royal Aeronautical Society, *Journal* **30**, pp. 587–606. (Wilbur Wright memorial lecture, 27 May 1926).

Langley, Samuel [1891] *Experiments in aerodynamics*. Washington.

Laplace, Pierre Simon de [1775/76] 'Recherches sur plusieurs points du système du monde', *MAS* (1775), pp. 52–182; *MAS* (1776), pp. 177–267, pp. 525–52, also in Laplace [1878–1912], vol. **9**, pp. 171–310.

—— [1776] 'Sur les ondes,' *MAS*, pp. 542–552, also in Laplace [1878–1912], vol. **9**, pp. 301–10.

—— [1878–1912], *Oeuvres complètes*, 14 vols, Paris.

Larmor, Joseph (ed.) [1907] *Memoirs and scientific correspondence of the late Sir George Gabriel Stokes*, 2 vols. Cambridge.

Le Conte, John [1858] 'On the influence of musical sounds on the flame of a jet of coal-gas, *PM* **15**, pp. 235–239.

Levi-Civita, Tullio [1901] 'Sulla resistenzia dei mezzi fluidi,' Accademia dei Lincei, *Rendiconti* **10**, pp. 3–9.

—— [1907] 'Sie e leggi di resistenza,' Circolo mathematico di Palermo, *Rendiconti* **23**, pp. 1–37.

—— [1925] 'Détermination rigoureuse des ondes permanentes d'ampleur finie,' *Mathematische Annalen* **93**, pp. 264–314.

Lighthill, Michael James [1957] 'River waves', First symposium on naval hydrodynamics, *Proceedings*, NAS/NRC publication **515**, pp. 17–44, also in *Collected works*, 4 vols. (1997), vol. **3**, Oxford.

—— [1978] *Waves in fluids* (Cambridge), pp. 269–279.

Lin, Chia Chiao [1966] *The theory of hydrodynamic stability*, 2nd edn. Cambridge.

Lorenz, Edward [1967] *The nature and theory of the general circulation of the atmosphere*, Geneva.

Love, Augustus E. H. [1901] 'Hydrodynamik: Theoretische Ausführungen', in *Encyklopädie der mathematischen Wissenschaften mit Einschluss ihrer Anwendungen*, vol. **4.3**, pp. 86–149.

Luvini, Giovanni [1888] 'Les cyclones et les trombes,' *La lumière électrique* **30**, pp. 368–372.

Magnus, Gustav [1853] 'Über die Abweichung der Geschosse, und: Über eine auffallende Erscheinung bei rotirenden Körpern', *AP* **88**, pp. 1–29.

Margules, Max [1906] 'Über Temperaturschichtung in stationär bewegter und ruhender Luft', *Meteorologische Zeitschrift*, Hannband, pp. 243–54.

Mariotte, Edme [1686] *Traité du mouvement des eaux et des autres corps fluides*, Paris.

Mathieu, Emile [1863] 'Sur le mouvement des liquides dans les tubes de très petit diamètre', *CR* **57**, pp. 320–324.

Maxwell, James Clerk [1854] 'On a particular case of the descent of a heavy body in a resisting medium,' Cambridge and Dublin Mathematical Society, *Journal*, also in Maxwell, *Scientific papers*, 2 vols (1890), vol. **2**, pp. 115–18, Cambridge.

—— [1990] *The scientific letters and papers of James Clerk Maxwell*, Peter Harman (ed.), 3 vols (1990–2003), Cambridge.

Mayer, Alfred [1878] 'A note on experiments on floating magnets', *Nature* **17**, pp. 487–488.

Merrifield, Charles Watkins [1869] 'Report of a committee, consisting of Mr. C. W. Merrifield, F.R.S., Mr. G. P. Bidder, Captain Douglas Galton, F.R.S., Mr. F. Galton, F.R.S., Professor Rankine, F.R.S., and Mr. W. Froude, appointed to report on the state of existing knowledge on the stability, propulsion, and sea-going qualities of ships, and as to the application which it may be desirable to make to Her Majesty's Government on these subjects', *BAR*, pp. 20–47.

—— [1870] 'The experiments recently proposed on the resistance of ships', *TINA* **11**, pp. 80–93.

Moedebeck, Hermann [1904] *Taschenbuch zum praktischen Gebrauch für Flugtechniker und Luftschiffer*, Berlin.

Moigno, François [1868a] *Leçons de mécanique analytique, rédigées principalement d'après les méthodes d'Augustin Cauchy et étendues aux travaux les plus récents* (Paris, 1868).

—— [1868b] (Untitled) *Les mondes* **17**, pp. 531, 577–8, 620–3.

Moseley, Henry [1830] *A treatise on hydrostatics and hydrodynamics*, Cambridge.

Mouillard, Louis Pierre [1881] *L'empire de l'air; essai d'ornithologie appliquée à l'aviation*, Paris.

Navier, Claude Louis [1820] 'Sur la flexion des plans élastiques', lithograph in the archive of the Ecole Nationale des Ponts et Chaussées (read on 14 August 1820 at the Academy).

—— [1822] 'Sur les lois du mouvement des fluides, en ayant égard à l'adhésion des molécules', *ACP* **19**, (1821) (in fact 1822), pp. 244–260 (read on 18 March 1822).

—— [1823a] Extract of Navier [1820], *BSP*, pp. 95–102.

—— [1823b] Extract of Navier [1827], *BSP*, pp. 177–181.

—— [1823c] 'Mémoire sur les lois du mouvement des fluides', *MAS* **6**, (published 1827), pp. 389–440 (read on 18 March and 16 December 1822).

—— [1823d] *Rapport à Monsieur Becquey, Directeur Général des Ponts et Chaussées et des Mines; et Mémoire sur les ponts suspendus*, Paris.

—— [1827] 'Mémoire sur les lois de l'équilibre et du mouvement des corps élastiques', *MAS* **7**, pp. 375–394 (read on 14 May 1821).

—— [1828a] 'Note relative à l'article … [Poisson [1829a]]', *ACP* **38**, pp. 304–314.

—— [1828b] 'Remarque … [about Poisson's reply (Poisson [1828a])]', *ACP* **39**, pp. 145–151.

—— [1829a] Letter to Arago, with a closing note by Arago, *ACP* **40**, pp. 99–107.

—— [1829b] 'Note relative à la question de l'équilibre et du mouvement des corps solides élastiques', *BSM* **11**, pp. 243–253.

——— [1830] *Rapport à Monsieur Becquey, Directeur Général des Ponts et Chaussées et des Mines; et Mémoire sur les ponts suspendus. Edition augmentée d'une notice sur le pont des Invalides*, Paris.

——— [1838] *Résumé des leçons données à l'Ecole des Ponts et Chaussées sur l'application de la mécanique à l'établissement des constructions et des machines. Deuxième partie: Leçons sur le mouvement et la résistance des fluides, la conduite et la distribution des eaux*, Paris.

——— [1864] *Résumé des leçons données à l'Ecole des Ponts et Chaussées sur l'application de la mécanique à l'établissement des constructions et des machines. Première section: De la Résistance des corps solides*, 3rd edn, Paris (annotated by Saint-Venant).

Newton, Isaac [1687] *Philosophiae naturalis principia mathematica*, London.

Newton, Isaac [1713] *Philosophiae naturalis principia mathematica*, 2nd edn, Cambridge.

Orr, William [1907] 'The stability or instability of the steady motions of a perfect liquid and of a viscous fluid. Part II: A viscous fluid,' Royal Irish Academy, *Proceedings* **27**, pp. 69–138.

Partiot, Henri [1858] 'Mémoire sur le mascaret', *CR* **47**, pp. 651–654 (extract).

——— [1871] 'Mémoire sur les marées fluviales', *CR* **73**, pp. 91–95 (extract).

Poiseuille, Jean-Louis [1844] 'Recherches expérimentales sur le mouvement des liquides dans les tubes de très petit diamètre', *MSE* **9**, pp. 433–543 (read on 14 December 1840, 28 December 1840, and 11 January 1841).

Poisson, Siméon Denis [1814] 'Mémoire sur les surfaces élastiques', Institut National des Sciences et des Arts, *Mémoires de sciences mathématiques et physiques* **9**, pp. 167–226.

——— [1816] 'Mémoire sur la théorie des ondes', *MAS* **1**, pp. 71–186 (read on 2 October and 18 December 1815, published in 1818).

——— [1817*a*] 'Mémoire sur la théorie des ondes,' *BSP*, pp. 85–89.

——— [1817*b*] 'Mémoire sur la théorie des ondes', *ACP* **5**, pp. 122–142.

——— [1828*a*] 'Réponse ... [to Navier's [1828*a*]]', *ACP* **38**, pp. 435–440.

——— [1828*b*] Letter to Arago, *ACP* **39**, pp. 204–211.

——— [1829*a*] 'Mémoire sur l'équilibre et le mouvement des corps élastiques', *MAS* **8**, pp. 357–570 (read on 14 April 1828).

——— [1829*b*] 'Note sur le problème des ondes', *MAS* **8**, pp. 571–580.

——— [1831*a*] 'Mémoire sur les équations générales de l'équilibre et du mouvement des corps solides élastiques et des fluides', Ecole Polytechnique, *Journal* **20**, pp. 1–174 (read on 12 October 1829).

——— [1831*b*] *Nouvelle théorie de l'action capillaire*, Paris.

——— [1832] 'Mémoire sur les mouvements simultanés d'un pendule et de l'air environnant', *MAS* **11**, pp. 521–582.

Poncelet, Jean Victor [1831] 'Sur quelques phénomènes produits à la surface libre des fluides, en repos ou en mouvement, par la présence des corps solides qui y sont plus ou moins plongés, et spécialement sur les ondulations et les rides permanentes qui en résultent', *ACP* **46**, pp. 5–25.

——— [1836] *Cours de mécanique appliquée aux machines*, Paris (lithographed course for the Ecole des Ponts et Chaussées, from lectures delivered in 1828).

——— [1839] *Introduction à la mécanique industrielle, physique ou expérimentale*, 2nd edn, Paris.

——— [1845] *Traité de mécanique appliquée aux machines. II*, Liège.

Prandtl, Ludwig [1905] 'Über Flüssigkeitsbewegung bei sehr kleiner Reibung', in III. Internationaler Mathematiker-Kongress in Heidelberg vom 8. bis 13. August 1904, *Verhandlungen*, A. Krazer (ed.), pp. 484–91, Leipzig, also in *PGA* **2**, pp. 575–84.

——— [1913] 'Flüssigkeitsbewegung', in *Handwörterbuch der Naturwissenschaften* **4**, pp. 101–40, Jena.

——— [1914] 'Der Luftwiderstand von Kugeln', *GN*, also in *PGA* **2**, pp. 597–608.

——— [1918] 'Tragflügeltheorie. I. Mitteilung', *GN*, also in *PGA* **1**, pp. 322–345.

——— [1919] 'Tragflügeltheorie. II. Mitteilung', *GN*, also in *PGA* **1**, pp. 346–372.

——— [1921*a*] 'Applications of modern hydrodynamics to aeronautics', National Advisory Committee for Aeronautics, *Technical report*, also in *PGA* **1**, pp. 433–515.

——— [1921*b*] 'Bemerkungen zur Entstehung der Turbulenz', *Zeitschrift für angewandte Mathematik und Mechanik*, also in *PGA* **2**: pp. 687–696.

——— [1925] 'Bericht über Untersuchungen zur ausbebildeten Turbulenz', *Zeitschrift für angewandte Mathematik und Mechanik*, also in *PGA* **2**: pp. 714–718.

——— [1926] 'Bericht über neuere Turbulenzforschung', in *Hydraulische Probleme*, Berlin, also in *PGA* **2**, pp. 719–30.

—— [1927a] 'Über die ausgebildete Turbulenz', in II. Internationaler Kongress Technischer Mechanik (1926), *Verhandlungen*, also in *PGA* **2**, pp. 736–51.

—— [1927b] 'The generation of vortices in fluids of small viscosity', Royal Aeronautical Society, *Proceedings* (Wilbur Wright memorial lecture, 16 May 1927), also in *PGA* **2**, pp. 752–777.

—— [1930] 'Turbulenz und ihre Entstehung', Aeronautical Research Institute, Tokyo Imperial University, *Journal* (in Japanese) (notes from lecture given on 21 October 1929 at the Imperial University of Tokyo), translated in *PGA* **2**, pp. 787–797.

—— [1931a] 'On the role of turbulence in technical hydrodynamics,' World engineering congress in Kyoto, *Proceedings*, also in *PGA* **2**, pp. 798–811.

—— [1931b] *Abriss der Strömunglehre*, Braunschweig.

—— [1933] 'Neuere Ergebnisse der Turbulenzforschung,' Kaiser-Wilhelm-Institut für Strömungslehre, *Mitteilungen*, also in *PGA* **2**, pp. 819–845.

—— [1948] 'Mein Weg zu hydrodynamischen Theorien', *Physikalische Blätter*, also in *PGA* **3**, pp. 1604–1608.

—— [1949] *Führer durch die Strömungslehre*, Braunschweig.

—— [1961] *Gesammelte Abhandlungen zur Mechanik, Hydro- und Aerodynamik*, 3 vols, Göttingen.

Prony, Gaspard de, Bossut, Charles, and Coulomb, Charles Augustin

—— [1799] Report on Venturi [1797], Institut de France, Académie des Sciences, *Procès-verbaux des séances* **1**, (1795–1799), pp. 271–272.

Prony, Gaspard de, and Girard, Pierre-Simon [1834] 'Rapport [on Vicat [1833]]', *APC*, 1st semester, pp. 293–304.

Rankine, William [1858] 'Resistance of ships', *PM* **16**, pp. 238–239 (letter to the editors of 26 August 1858).

—— [1862] 'Abstract of an investigation of the resistance of ships,' Franklin Institute, *Journal* **73**, pp. 22–29.

—— [1863] 'On the exact form of waves near the surface of deep waters', *PT* **153**, pp. 127–138.

—— [1864] 'On the computation of the probable engine-power and speed of propelled ships', *TINA* **5**, pp. 316–331.

—— [1865] 'On plane water-lines in two dimensions', *PT* **154**, pp. 369–391.

—— [1868a] *A manual of applied mechanics*, 4th edn, London.

—— [1868b] 'Elementary demonstration of principles relating to stream-lines', *The engineer*, also in *Miscellaneous papers* (1891), pp. 522–9, London.

—— [1870] 'On stream-line surfaces', *TINA* **11**, pp. 175–181.

—— [1871] 'On the mathematical theory of streamlines, especially those with four foci and upwards', *PT* **161**, pp. 267–303.

Rayleigh, Lord (William Strutt) [1876a] 'On waves', *PM,* also in *RSP* **1**, pp. 251–271.

—— [1876b] 'On the resistance of fluids', *PM* **11**, pp. 430–441.

—— [1876c] 'Notes on hydrodynamics', *PM* **11**, pp. 441–447.

—— [1877a] 'On progressive waves', *PLMS*, also in *RSP* **1**, pp. 322–327.

—— [1877b] 'On the irregular flight of a tennis-ball', *Messenger of mathematics*, also in *RSP* **1**, pp. 344–346.

—— [1877/78] *Theory of sound*, 2 vols. (London).

—— [1879] 'On the instability of jets', *PLMS*, also in *RSP* **1**, pp. 361–371.

—— [1880] 'On the stability, or instability, of certain fluid motions', *PLMS*, also in *RSP* **1**, pp. 474–487.

—— [1881] 'On the velocity of light', *Nature*, also in *RSP* **1**, pp. 537–540.

—— [1883a] 'On the circulation of air observed in Kundt's tubes, and on some allied acoustical problems', *PT*, also in *RSP* **2**, pp. 239–257.

—— [1883b] 'The form of standing waves on the surface of running water', *PLMS*, also in *RSP* **2**, pp. 258–267.

—— [1883c] 'The soaring of birds', *Nature*, also in *RSP* **2**, pp. 194–197.

—— [1884a] 'Acoustical observations, V', *PM*, also in *RSP* **2**, pp. 268–275.

—— [1884b] Presidential address, *BAR*, also in *RSP* **2**, pp. 332–354.

—— [1887] 'On the stability or instability of certain fluid motions, II', *PLMS*, also in *RSP* **3**, pp. 17–23.

—— [1891] 'Experiments in aërodynamics', *Nature*, also in *RSP* **3**, pp. 491–495.

—— [1892] 'On the question of the stability of the flow of fluids', *PM*, also in *RSP* **3**, pp. 575–584.

—— [1894–1896] *Theory of sound*, 2nd edn, 2 vols. (London).

—— [1895] 'On the stability or instability of certain fluid motions, III', *PLMS*, also in *RSP* **4**, pp. 203–208.

—— [1900] 'The mechanical principles of flight', *PLPSM*, also in *RSP* **4**, pp. 462–479.

—— [1904] 'Fluid friction on even surfaces', *PM* (note to a paper by Albert Zahm), also in *RSP* **5**, pp. 196–197.

—— [1909] 'Note as to the application of the principle of dynamical similarity,' Advisory Committee for Aeronautics, *Report* (1909–10), also in *RSP* **5**, pp. 532–533.

—— [1911] 'On the motion of solid bodies through viscous liquid', *PM*, also in *RSP* **6**, pp. 29–40.

—— [1914] 'Fluid motion', *PRI*, also in *RSP* **6**, pp. 237–249.

—— [1915*a*] 'Aeolian tones', *PM*, also in *RSP* **6**, pp. 315–325.

—— [1915*b*] 'The principle of similitude', *Nature*, also in *RSP* **6**, pp. 300–305.

—— [1918] 'A proposed hydraulic experiment', *PM*, also in *RSP* **6**, pp. 552–553.

Reech, Ferdinand [1844] *Mémoire sur les machines à vapeur et leur application à la navigation présenté à l'Académie Royale des sciences.* Paris.

—— [1852] *Cours de mécanique d'après la nature généralement flexible et élastique des corps*, Paris.

—— [1869] 'Sur la théorie des ondes liquides périodiques', *CR* **68**, pp. 1099–1101.

Reusch, Friedrich Eduard [1860] 'Über Ringbildung in Flüssigkeiten', *AP* **110**, pp. 309–316.

Reye, Theodor [1864] 'Über vertikale Luftströme in der Atmosphäre', *Zeitschrift für Mathematik und Physik* **9**, pp. 250–276.

—— [1872] *Die Wirbelstürme, Tornados und Wettersäulen in der Erdatmosphäre mit Berücksichtigung der Stürme in der Sonnen-Atmosphäre*, Hannover.

Reynolds, Osborne [1873] 'The causes of the racing of screw steamers investigated theoretically and by experiment,' Institution of Naval Architects, *Transactions*, also in *ReP* **1**, pp. 51–58.

—— [1874*a*] 'The cause of the racing of the engines of screw steamers investigated theoretically and by experiment', *PLPSM*, also in *ReP* **1**, pp. 51–58.

—— [1874*b*] 'On the forces caused by evaporation from, and condensation at, a surface', *PRS*, also in *ReP* **1**, pp. 67–74.

—— [1874*c*] 'On the extent and action of the heating surface of steam boilers', *PLPSM*, also in *ReP* **1**, pp. 81–85.

—— [1875] 'On the action of rain to calm the sea', *PLPSM*, also in *ReP* **1**, pp. 86–88.

—— [1876] 'On the forces caused by the communication of heat between a surface and a gas; and on a new photometer', *PT*, also in *ReP* **1**, pp. 170–182.

—— [1877*a*] 'On vortex motion', *PRI*, also in *ReP* **1**, pp. 184–198.

—— [1877*b*] 'On the rate of progression of groups of waves and the rate at which energy is transmitted by waves', *Nature*, also in *ReP* **1**, pp. 198–203.

—— [1878] 'On the internal cohesion of liquids and the suspension of a column of mercury to a height more than double that of the barometer', *PLPSM*, also in *ReP* **1**, pp. 231–243.

—— [1879] 'On certain dimensional properties of matter in the gaseous state', *PT*, also in *ReP* **2**, pp. 257–390.

—— [1883] 'An experimental investigation of the circumstances which determine whether the motion of water shall be direct or sinuous, and of the law of resistance in parallel channels', *PT* **174**, pp. 935–982, also in *ReP* **2**, pp. 51–105.

—— [1884] 'On the two manners of motion of water', *PRI*, also in *ReP* **2**, pp. 153–162.

—— [1886] 'On the theory of lubrication and its application to Mr Beauchamp Tower's experiments, including an experimental determination of the viscosity of olive oil', *PT*, also in *RSP* **1**, pp. 228–310.

—— [1893] 'Study of fluid motion by the method of coloured bands', *PRI* also in *ReP* **2**, pp. 524–534.

—— [1895] 'On the dynamical theory of incompressible viscous fluids and the determination of the criterion', *PT* also in *ReP* **2**, pp. 535–577.

Riabouchinsky, Dimitri [1909] 'Recherches sur la pression que supportent des plaques rectangulaires en rotation dans un courant aérien', Institut Aérodynamique de Koutchino, *Bulletin* **3**, pp. 30–45.

Riemann, Bernhard [1860] 'Ein Beitrag zu den Untersuchungen über die Bewegung eines flüssigen gleichartigen Ellipsoides', Königliche Gesellschaft der Wissenschaften zu Göttingen, *Abhandlungen* **9**, pp. 3–36.

Rogers, William [1858] 'On the formation of rotary rings by air and liquids under certain conditions of discharge', *American journal of science and arts* **26**, pp. 246–258.

Runge, Carl, and Prandtl, Ludwig [1906] 'Das Institut für angewandte Mathematik und Mechanik', in *Die physikalische Institute der Universität Göttingen*, pp. 95–111, Leipzig.

Russell, John Scott [1834] 'Notice of the reduction of an anomalous fact in hydrodynamics, and of a new law of the resistance to the motion of floating bodies', *BAR*, pp. 531–544.

—— [1835*a*] 'Experimental researches into the laws of motion of floating bodies', *BAR*, pp. 16–17.

—— [1835*b*] 'On the solid of least resistance', *BAR*, pp. 107–108.

—— [1837*a*] 'On the mechanism of waves, in relation to the improvement of navigation', *BAR*, pp. 130–131.

—— [1837*b*] 'Notice of experimental researches into the law of certain hydrodynamical phenomena that have not hitherto been reduced into conformity with known laws', *PRSE* **1** (1832–1844), pp. 160–161.

—— [1837*c*] 'Report of the committee on waves, appointed by the B.A. at Bristol in 1836', *BAR*, pp. 417–496.

—— [1838] 'On the terrestrial mechanism of the tides', *BAR*, pp. 179–181.

—— [1839] 'Experimental researches into the laws of certain hydrodynamical phenomena that accompany the motion of floating bodies, and have not previously been reduced into conformity with the known laws of the resistance of fluids', *TRSE* **14**, pp. 47–109.

—— [1841] 'Report of the Committee on the form of vessels', *BAR*, pp. 325.

—— [1842*a*] 'Report of a Committee on the form of ships', *BAR*, pp. 104–105.

—— [1842*b*] 'On the abnormal tides of the Frith of Forth', *BAR*, pp. 115–116.

—— [1843*a*] 'Report of a series of observations on the tides of the Frith of Forth and the East Coast of Scotland', *BAR*, pp. 110–111.

—— [1843*b*] 'Notice of a report of the Committee on the form of ships', *BAR*, pp. 112–115.

—— [1845] *Report on waves*, London, also published in *BAR*, (1844).

—— [1852] 'On wave-line ships and yachts', *PRI*, pp. 115–119.

—— [1854] 'On the progress of naval architecture and steam navigation, including a notice of the large ship of the Eastern Navigation Company', *BAR*, pp. 160–161.

—— [1857] 'Mechanical structure of the "Great Eastern" steam ship', *BAR*, pp. 195–198.

—— [1865] *The modern system of naval architecture*, 3 *in folio* vols, London.

—— [1885] *The wave of translation in the oceans of water, air, and ether*, London (posthumous memoirs originally submitted to the Royal Society).

Sabine, Edward [1829] 'On the reduction to a vacuum of the vibrations of an invariable pendulum', *PT*, pp. 207–239.

Saint-Venant, Adhémar Barré de [1834] 'Mémoire sur la dynamique des fluides', MS, Archives de l'Académie des Sciences, *Pochette de séance* for 14 April 1834.

—— [1834/35] 'Mémoire sur les eaux courantes considérées dans un lit de figure variable', MS, Archives de l'Ecole Polytechnique, Fond Saint-Venant, carton 21.

—— [1837] *Leçons de mécanique appliquée faites par intérim par M. de Saint-Venant, Ingénieur des Ponts et Chaussées, de 1837 à 1838* (lithographed course), Paris

—— [1838] 'Mémoire sur le calcul des effets des machines à vapeur, contenant des équations générales de l'écoulement permanent ou périodique des fluides, en tenant compte de leurs dilatations et de leurs changements de température, et sans supposer qu'ils se meuvent par tranches parallèles, ni par filets indépendants', *CR* **6**, pp. 45–47.

—— [1843*a*] 'Mémoire sur le calcul de la résistance et de la flexion des pièces solides à simple et à double courbure; en prenant simultanément en considération les différents efforts auxquels elles peuvent être soumises', *CR* **17**, pp. 942–954, pp. 1020–1031.

—— [1843*b*] 'Sur la définition de la pression dans les corps fluides ou solides en repos ou en mouvement', *BSP*, pp. 134–138.

—— [1843*c*] 'Note à joindre au mémoire sur la dynamique des fluides, présenté le 14 avril 1834', *CR* **17**, pp. 1240–1243.

—— [1844] 'Mémoire sur la question de savoir s'il existe des masses continues, et sur la nature probable des dernières particules des corps', *BSP*, pp. 3–15.

—— [1846*a*] 'Mémoire sur la perte de force vive d'un fluide, aux endroits où sa section d'écoulement augmente brusquement ou rapidement', *CR* **23**, pp. 147–153.

—— [1846*b*] 'Solution d'un paradoxe proposé par d'Alembert aux géomètres', *BSP* pp. 25–29, pp. 72–78, pp. 120–121 (read on 7 March 1846).

—— [1851*a*] 'Mémoire sur des formules et des tables nouvelles pour la solution des problèmes relatifs aux eaux courantes', *Annales des mines* **20**, pp. 183–357.

—— [1851*b*] 'Sur la prise en considération de la force centrifuge dans le calcul du mouvement des eaux courantes et sur la distinction des torrents et des rivières', in Saint-Venant [1887*b*] pp. 244–70.

—— [1855] 'Mémoire sur la torsion des prismes, avec des considérations sur leur flexion, ainsi que sur l'équilibre intérieur des solides élastiques en général, et des formules pratiques pour le calcul de leur résistance à divers efforts s'exerçant simultanément', *MSE* **14**, pp. 233–560.

—— [1864*c*] *Notice sur les travaux et titres scientifiques de M. de Saint-Venant*. Paris.

—— [1868*a*] Lectures 21 and 22 of Moigno [1868*a*].

—— [1868*b*] 'Note complémentaire sur le problème du mouvement que peuvent prendre divers points d'un solide ductile ou d'un liquide contenu dans un vase, pendant son écoulement par un orifice extérieur', *CR* **67**, p. 279.

—— [1869] 'Rapport sur un mémoire de M. Lévy …' *CR* **68**, pp. 582–594.

Saint-Venant, Adhémar Barré de [1870] 'Démonstration élémentaire de la formule de propagation d'une onde ou d'une intumescence dans un canal prismatique; et remarque sur les propagations du son et de la lumière, sur les *ressauts*, ainsi que sur la distinction entre des *rivières* et des *torrents*', *CR* **71**, pp. 186–195.

—— [1871*a*] 'Théorie du mouvement non permanent des eaux, avec application aux crues des rivières et à l'introduction des marées dans leur lit', *CR* **73**, pp. 147–154, 237–240.

—— [1871*b*] 'Sur la houle et le clapotis', *CR* **73**, pp. 521–528, 589–593.

—— [1872] 'Sur l'hydrodynamique des cours d'eau', *CR* **74**, pp. 570–577, 649–657, 693–701, 770–774.

—— [1873] 'Rapport sur un mémoire de M. Boussinesq présenté le 28 Octobre 1872 [Boussinesq [1877]]', *CR* **76**, pp. 924–943, also in Boussinesq [1877] pp. I–XXII.

—— [1876] 'Note soumise à M. Chasles … à l'appui de la candidature de M. Boussinesq [to a Sorbonne chair]', in Saint-Venant to Boussinesq, 22 April 1876, Bibliothèque de l'Institut.

—— [1880] 'Analyse succincte des travaux de M. Boussinesq', recommendation to the Academy of Sciences, MS (May 1880), 23 pages, in the Boussinesq file of the Académie des Sciences.

—— [1885] 'Mouvements des molécules de l'onde dite solitaire', *CR* **101**, pp. 1101–1105, 1215–1218, 1445–1447.

—— [1887*a*] Annotated translation of Gerstner [1802], *APC* **13**, pp. 31–86.

—— [1887*b*] *Résistance des fluides: Considérations historiques, physiques et pratiques relatives au problème de l'action dynamique mutuelle d'un fluide et d'un solide, dans l'état de permanence supposé acquis par leurs mouvements*, Paris, also in *MAS* **44**, (1888).

—— [1887*c*] 'Des diverses manières de poser les équations du mouvement varié des eaux courantes', *APC* **13**, pp. 148–228.

—— [1888] 'De la houle et du clapotis', *APC* **15**, pp. 705–773 (published posthumously by A. Flamant).

Schmidt, Wilhelm [1925] *Der Massenaustausch in freier Luft und verwandte Erscheinungen*, Hamburg.

Schneebeli, Heinrich [1874] 'Zur Theorie der Orgelpfeifen', *AP* **153**, pp. 301–305.

Sommerfeld, Arnold [1908] 'Ein Beitrag zur hydrodynamischen Erklärung der turbulenten Flüssigkeitsbewegungen', Congresso internazionale dei matematici, *atti* (Rome), also in *Gesammelte Schriften* vol. **1**, (1968), pp. 599–607, Braunschweig.

—— [1949] *Mechanik der deformierbaren Medien*, 2nd edn, Leipzig.

Stokes, George Gabriel [1842] 'On the steady motion of incompressible fluids', *TCPS* also in *SMPP* **1**, pp. 1–16.

—— [1843] 'On some cases of fluid motion', *TCPS* also in *SMPP* **1**, pp. 17–68.

—— [1846*a*] 'Report on recent researches on hydrodynamics', *BAR* also in *SMPP* **1**, pp. 157–187.

—— [1846*b*] 'Supplement to a memoir on some cases of fluid motion', *TCPS* also in *SMPP* **1**, pp. 188–196.

—— [1846*c*] 'On the constitution of the luminiferous ether, viewed with reference to the phenomenon of the aberration of light', *PM* also in *SMPP* **1**, pp. 153–156.

—— [1847*a*] 'On the theory of oscillatory waves', Cambridge Philosophical Society, *Transactions* also in *SMPP* **1**, pp. 197–225.

—— [1847*b*] 'On the critical values of the sums of periodic series', in *SMPP* **1**, pp. 236–313

—— [1848*a*] 'Notes on hydrodynamics. III. On the dynamical equations', *SMPP* **2**, pp. 1–7.

—— [1848*b*] 'On the constitution of the luminiferous ether', *PM* also in *SMPP* **2**, pp. 8–13.

—— [1848*c*] 'Notes on hydrodynamics. IV. Demonstration of a fundamental theorem', *SMPP* **2**, pp. 36–50.

—— [1849*a*] 'On the theory of the internal friction of fluids in motion, and of the equilibrium and motion of elastic solids', *TCPS* (read in 1845), also in *SMPP* **1**, pp. 75–129.

—— [1849*b*] 'Notes on hydrodynamics. IV. On waves', Cambridge and Dublin Mathematical Society, *Journal*. also in *SMPP* **2**, pp. 220–250.

—— [1850*a*] 'On the numerical calculation of a class of definite integrals and infinite series', *TCPS*, also in *SMPP* **2**, pp. 329–357.

—— [1850*b*] 'On the effect of the internal friction of fluids on the motion of pendulums', *TCPS* (read on 9 December), 1850 also in *SMPP* **3**, pp. 1–141

—— [1876] Smith prize examination papers for 2 February 1876, in *SMPP* **5**, pp. 362.

—— [1880*a*] Appendix A to Stokes [1847*a*] on 'Gerstner's wave', *SMPP* **1**, pp. 219–225.

—— [1880*b*] Appendix B to Stokes [1847*a*] on 'the highest irrotational oscillatory wave', *SMPP* **1**, pp. 226–229.

—— [1880*c*] 'Supplement to a paper on the theory of, oscillatory waves', *SMPP* **1**, pp. 314–326.

—— [1880–1905] *Mathematical and physical papers*, 5 vols, Cambridge.

—— [1891] 'Note on the solitary wave', *PM* also in *SMPP* **5**, pp. 160–162.

—— [1907] *Memoir and scientific correspondence of the late Sir George Gabriel Stokes*, 2 vols, ed. J. Larmor, Cambridge.

Tait, Peter Guthrie [1876] *Lectures on some recent advances in physical sciences*, 2nd edn, London.

Thomson, James [1878] 'On the flow of water in uniform *régime* in rivers and other open channels', *PRS* **28**, pp. 114–126.

Thomson, William (Lord Kelvin) [1849] 'Notes on hydrodynamics. On the *vis-viva* of a liquid in motion', Cambridge and Dublin Mathematical Society, *Journal*, also in *TMPP* **1**, pp. 107–112.

—— [1865] 'On the convective equilibrium of temperature in the atmosphere', *PLPSM*, also in *TMPP* **3**, pp. 255–260

—— [1867] 'On vortex atoms', *PM* **34**, pp. 15–24, also in *TMPP* **4**, pp. 1–12.

—— [1869] 'On vortex motion', *TRSE* **25**, pp. 217–260, also in *TMPP* **4**, pp. 13–66.

—— [1871*a*] 'The influence of wind and capillarity on waves in water supposed frictionless', letter to Tait, 16 Aug. 1871, *PM*, also in *TMPP* **4**, pp. 76–79.

—— [1871*b*] Letter to Tait, 23 Aug. 1871, *PM*, also in *TMPP* **4**, pp. 79–80.

—— [1871*c*] 'Ripples and waves', *Nature*, also in *TMPP* **4**, pp. 86–92.

—— [1871*d*] 'Hydrokinetic solutions and observations', *PM*, also in *TMPP* **4**, pp. 69–92 (contains [1871*a*] and [1871*b*]).

—— [1876] 'Vortex statics', *PRSE*, also in *TMPP* **4**, pp. 115–128.

—— [1878] 'Floating magnets', *Nature*, also in *TMPP* **4**, pp. 135–140.

—— [1880*a*] 'Vibrations of a columnar vortex', *PRSE*, also in *TMPP* **4**, pp. 152–165.

—— [1880*b*] 'Maximum and minimum energy of vortex motion', *BAR*, also in *TMPP* **4**, pp. 172–183.

—— [1880*c*] 'On a disturbing infinity in Lord Rayleigh's solution for waves in a plane vortex stratum', *Nature*, also in *TMPP* **4**, pp. 186–187.

—— [1886] 'On stationary waves in flowing water', *PM*, also in *TMPP* **4**, pp. 270–302.

—— [1887*a*] 'On the formation of coreless vortices by the motion of a solid through an inviscid incompressible fluid', *PRS*, also in *TMPP* **4**, pp. 149–151.

—— [1887*b*] 'On the stability of steady and periodic fluid motion', *PM*, also in *TMPP* **4**, pp. 166–172.

—— [1887*c*] 'Rectilinear motion of viscous fluid between two parallel planes', *PM*, also in *TMPP* **4**, pp. 321–330.

—— [1887*d*] 'Broad river flow down an inclined plane bed', *PM*, also in *TMPP* **4**, pp. 330–337.

—— [1887*e*] 'On the propagation of laminar motion through a turbulently moving inviscid fluid', *BAR*, also in *TMPP* **4**, pp. 308–320.

—— [1887*f*] 'On ship waves', Institution of Mechanical Engineers, *Minutes of proceedings* pp. 409–34 [lecture delivered at the "Conversazione" in the Science and Art Museum, Edinburgh, on 3 Aug 1887]. Incomplete version in *Popular lectures and addresses*, 3 vols (1891), vol. **3**, pp. 450–500, London.

—— [1887*g*] 'On the waves produced by a single impulse in water of any depth, or in a dispersive medium', *PM*, also in *TMPP* **4**, pp. 303–306.

—— [1889] 'On the stability and small oscillations of a perfect liquid full of nearly straight coreless vortices', Royal Irish Academy, *Proceedings*, also in *TMPP* **4**, pp. 202–204.

—— [1894] 'On the doctrine of discontinuity of fluid motion, in connection with the resistance against a solid moving through a fluid', *Nature*, also in *TMPP* **4**, pp. 215–230.

—— [1905] 'Deep water ship-waves', *PM*, also in *TMPP* **4**, pp. 368–393.

—— [1906] 'Deep sea ship-waves', *PM*, also in *TMPP* **4**, pp. 394–418.

—— [1907] 'Investigation of deep-sea waves of three classes: (1) from a single displacement; (2) from a group of equal and similar displacement; (3) by a periodically varying surface-pressure', *PM*, also in *TMPP* **4**, pp. 419–456.

Thomson, William, and Tait, Peter Guthrie [1867] *Treatise on natural philosophy* Oxford.

Tollmien, Walter [1929] 'Über die Enstehung der Turbulenz', *GN*, pp. 21–44.

—— [1947] 'Asymptotische Integration der Störungsdifferentialgleichung ebener laminarer Strömungen bei hohen Reynoldschen Zahlen', *Zeitschrift für angewandte Mathematik und Mechanik* **25**, pp. 33–50; **27**, pp. 70–83.

Tredgold, Thomas (ed.) [1826] *Tracts on hydraulics*, London.

Tyndall, John [1867] 'On the action of sonorous vibrations on gaseous and liquid jets', *PM* **33**, pp. 375–391.

Vaulthier, Pierre [1836] 'De la théorie du mouvement permanent des eaux courantes et de ses applications à la solution de plusieurs problèmes d'hydraulique', *APC*, pp. 241–313.

Venturi, Giovanni Battista [1797] *Recherches expérimentales sur le principe de communication latérale dans les fluides*, Paris.

Vicat, Louis [1833] 'Recherches expérimentales sur les phénomènes physiques qui précèdent et accompagnent la rupture ou l'affaissement d'une certaine classe de solides', *APC*, pp. 201–268.

—— [1834] 'Observations sur le rapport [Prony and Girard [1834]]', *APC*, 1st semester, pp. 305–312.

Weber, Heinrich, and Weber, Wilhelm [1825] *Wellenlehre auf Experimente gegründet, oder über die Wellen tropfbarer Flüssigkeiten mit Anwendung auf die Schall- und Lichtwellen*, Leipzig.

Weisbach, Julius [1863] *Lehrbuch der Ingenieur- und Maschinen-Mechanik*, 4th edn, 3 vols, Braunschweig.

Wertheim, Guillaume [1848] 'Mémoire sur la vitesse du son dans les liquides', *CR* **27**, pp. 150–152; *ACP* **23**, pp. 434–475.

—— [1851] 'Mémoire sur les vibrations sonores de l'air', *ACP* **33**, pp. 385–432.

Whewell, William [1833] 'Essay toward a first approximation to a map of cotidal lines', *PT*, pp. 147–236.

—— [1834] 'On the empirical laws of the tides in the Port of London, with some reflections on the theory', *PT*, pp. 15–45.

Wien, Wilhelm [1900] *Lehrbuch der Hydrodynamik*, Leipzig.

Wilson, David B. (ed.) [1990] *The correspondence between Sir George Gabriel Stokes and Sir William Thomson, Baron Kelvin of Largs*, 2 vols, Cambridge.

Woltman [*sic*.], Reinhardt [1791–1799] *Beiträge zur hydraulischen Architectur*, Berlin.

Yamada, H., and Matsui, T. [1978] 'Preliminary study of mutual slip-through of a pair of vortices, *Physics of fluids* **21**, pp. 292–294.

Young, Thomas [1813] 'A theory of tides, including the consideration of resistance', *Nicholson journal*, also in Young [1855], vol. **2**, pp. 262–390.

—— [1823] 'Tides', *Encyclopaedia Britannica*, supt., also in Young [1855], vol. **2**, pp. 296–355.

—— [1855] *Miscellaneous works*, 3 vols. (London).

Zamminer, Friedrich [1856] 'Über die Schwingungsbewegung der Luft in Röhren', *AP* **97**, pp. 173–212.

Bibliography of secondary literature

Abell, Westcott Stile [1933] 'William Froude. His life and work', Devonshire Association for the Advancement of Science, Literature and Arts, *Transactions* **65**, pp. 43–76

Ackeret, Jakob [1957] 'Vorrede', in L. Euler, *Opera omnia*, ser. 2, vol. **15**, pp. VII–LX, Lausanne.

Ackroyd, John A. D. [1992] 'The 31st Lanchester lecture: Lanchester the man', *The aeronautical journal* **96**, pp. 119–140.

—— [1996] 'Lanchester's *Aerodynamics*', in *The Lanchester Legacy: A trilogy of Lanchester works*, J. Fletcher (ed.), vol. **3**: *A celebration of genius*, pp. 61–98.

Ackroyd, John A.D, Axcell, Brian P., and Ruban, Anatoly I. [2001] *Early developments of modern aerodynamics*, Oxford.

Allen, Jack [1970] 'The life and work of Osborne Reynolds', in McDowell and Jackson [1970] pp. 1–82.

Anderson, John D. [1997] *A history of aerodynamics and its impact on flying machines*, Cambridge.

Arnold, David H. [1983] 'The *mécanique physique* of Siméon Denis Poisson: The evolution and isolation in France of his approach to physical theory', *Archive for history of exact sciences* **28**, pp. 243–367; **29**, pp. 37–94.

Barkla, Hugh, and Auchterlonie, L. J. [1971] 'The Magnus or Robins effect on rotating spheres,' *Journal of Fluid Mechanics* **47**, pp. 437–47.

Batchelor, George Keith [1996] *The life and legacy of G.I. Taylor*, Cambridge.

Battimelli, Giovanni [1984] 'The mathematician and the engineer: Statistical theories of turbulence in the 1920s', *Rivista di storia della scienza* **1**, pp. 73–93.

—— [1990] 'Taylor, Geoffrey Ingram' *DSB* **18**, pp. 896–898.

Belhoste, Bruno [1991] *Augustin-Louis Cauchy: A biography*, New York.

—— [1994] 'Un modèle à l'épreuve. L'Ecole Polytechnique de 1794 au second Empire', in B. Belhoste, A. Dahan, and A. Picon, *La formation polytechnicienne. 1794–1994*, pp. 9–30, Paris.

—— [1997] 'Navier, Saint-Venant et la création de la mécanique des fluides visqueux', *APC* **82**, pp. 4–9.

—— [2003] *La formation d'une technocratie: L'Ecole polytechnique et ses élèves de la Révolution au Second Empire*, Paris.

Benvenuto, Eduardo [1991] *An introduction to the history of structural mechanics*, 2 vols, New York.

—— [1998] 'Adhémar Barré de Saint-Venant: The man, the scientist, the engineer', Accademia Nazionale dei Lincei, *Atti dei convegni Lincei* **140**, pp. 7–34.

Bernhardt, Karl Heinz [1973] 'Der Beitrag Hermann von Helmholtz' zur Physik der Atmosphäre', Humboldt-Universität zu Berlin, *Wissenschaftliche Zeitschrift*, Mathematisch-naturwissenschaftliche Reihe **22**, pp. 331–340.

Billingham, John, and King, A. C. [2000] *Wave motion* (Cambridge).

Blaquière, Constant [1931] *Vie de Joseph Boussinesq*, Béziers.

Blay, Michel [1985] 'Varignon et le statut de la loi de Torricelli', *Archives internationales d'histoire des sciences* **35**, pp. 330–345.

—— [1992] *La naissance de la Mécanique analytique: La science du mouvement au tournant des XVIIe et XVIIIe siècles*, Paris.

Boussinesq, Joseph, and Flamant, Alfred [1886] *Notice sur la vie et les travaux de Barré de Saint-Venant*, Paris.

Brown, David K. [1992] 'William Froude and "The way of a ship in the sea"', The Devonshire Association for the Advancement of Science, Literature and the Arts, *Report and transactions* **124**, pp. 207–231.

Brown, Glenn, Garbrecht, Jürgen, and Hager, Willi (eds.) [2003] *Henry P. G. Darcy and other pioneers in hydraulics: Contributions in celebration of the 200th birthday of Henry Philibert Gaspard Darcy*, Reston.

Brush, Stephen, and Landsberg, Helmut [1985] *The history of geophysics and meteorology. An annotated bibliography*, New York.

Bullough, Robin, [1988] 'The wave "par excellence", the solitary, progressive great wave of equilibrium of the fluid—an early history of the solitary wave', in *Solitons* to M. Lakshmanan (ed.), pp. 7–42, New York.

Bullough, Robin and Caudrey, P. J. [1995] 'Solitons and the Korteweg-de Vries equations: Integrable systems in 1834–1995', *Acta applicandae mathematicae* **39**, pp. 193–228.

Cahan, David (ed.) [1993] *Hermann von Helmholtz and the foundations of nineteenth-century science*, Berkeley.

Calero, Julián Simón [1996] *La génesis de la mecánica de los fluidos (1640–1780)*, Madrid.

Cannone, Marco, and Friedlander, Susan [2003] 'Navier: Blow up and collapse', American Mathematical Society, *Notices* **50**(1), pp. 7–13.

Cartwright, David Edgar [1999] *Tides: A scientific history*, Cambridge.

Casey, James [1992] 'The principle of rigidification', *AHES* **43**, pp. 329–383.

Clanet, Christophe, and Villermaux, Emmanuel [2002] 'Life of a smooth liquid jet', *Journal of fluid mechanics* **462**, pp. 307–340.

Costabel, Pierre [1983] *La signification d'un débat sur trente ans (1728–1758): La question des forces vives*, Paris.

Craik, Alex [2004] 'The origins of water wave theory', *Annual reviews of fluid mechanics* **36**, pp. 1–28.

—— [2005] 'George Gabriel Stokes on water wave theory', *Annual reviews of fluid mechanics* **37**, pp. 23–42.

Crosland, Maurice [1967] *The Society of Arcueil*, Cambridge.

Dahan, Amy [1989a] 'La propagation des ondes en eau profonde et ses développements mathématiques (Poisson, Cauchy 1815–1825)', in *The history of modern mathematics*, David E. Rowe and John McCleary (eds), vol. **2**: *Institutions and applications*, pp. 128–68, Boston.

—— [1989b] 'La notion de pression: de la métaphysique aux diverses mathématisations', *Revue d'histoire des sciences et des techniques* **42**, pp. 79–108.

—— [1992] *Mathématisations: Augustin Cauchy et l'Ecole française*, Paris.

Darrigol, Olivier [1998] 'From organ pipes to atmospheric motions: Helmholtz on fluid mechanics', *HSPS* **29**, pp. 1–51.

—— [2000] *Electrodynamics from Ampère to Einstein*, Paris.

—— [2001] 'God, waterwheels, and molecules: Saint-Venant's anticipation of energy conservation', *HSPS* **31**, pp. 285–353.

—— [2002a] 'Between hydrodynamics and elasticity theory: The first five births of the Navier–Stokes equation', *AHES* **56**, pp. 95–150.

—— [2002b] 'Turbulence in 19th-century hydrodynamics', *HSPS* **32**, pp. 207–262.

—— [2002c] 'Stability and instability in nineteenth-century fluid mechanics', *Revue d'histoire des mathématiques* **8**, pp. 5–65.

—— [2003] 'The spirited horse, the engineer, and the mathematician: Water waves in nineteenth-century hydrodynamics', *AHES* **58**, pp. 21–95.

Deacon, Margaret [1971] *Scientists and the sea, 1650–1900: A study of marine science*, London.

Dobson, Geoffrey J. [1999] 'Newton's errors with the rotational motion of fluids', *AHES* **54**, pp. 243–254.

Douysset, Elie (Undated) 'Un grand savant. Un grand chrétien', undated brochure, in the Boussinesq file of the Académie des Sciences.

Dryden, Hugh, [1955] 'Fifty years of boundary-layer theory and experiments', *Science* **121**, pp. 375–379.

Dugas, René [1950] *Histoire de la mécanique*, Paris.

Eckert, Michael [2002] 'Euler and the fountains of Sanssouci', *AHES* **56**, pp. 451–468.

—— [2005] *The dawn of fluid dynamics: A discipline between science and technology*, Berlin.

Ellis, Alexander J. [1885] *On the sensations of tone as a physiological basis for the theory of music*, New York (comments to his translation of Helmholtz [1877]).

Emmerson, George [1977] *John Scott Russell: A great Victorian engineer and naval architect* (London).

Epple, Moritz [1998] 'Topology, matter, and space, I: Topological notions in nineteenth-century natural philosophy', *AHES* **52**, pp. 297–392.

—— [2002] 'Präzision versus Exaktheit: Konfligierende Ideale der angewandten mathematischen Forschung. Das Beispiel der Tragflügeltheorie,' *Berichte zur Wissenschaftsgeschichte* **25**, pp. 171–93.

Everitt, C. W. Francis [1974] 'Maxwell, James Clerk', *DSB* **9**, pp. 198–320.

Farge, Marie, and Guyon, Etienne [1999] 'A philosophical and historical journey through mixing and fully-developed turbulence', in *Mixing: Chaos and turbulence*, ed. Hugues Chate *et al*, pp. 11–36, New York.

Fletcher, John, [1996] 'Frederick William Lanchester', in *The Lanchester legacy: A trilogy of Lanchester works*, J. Fletcher (ed.), vol. **3**: *A celebration of genius*, pp. 4–22.

Fox, Robert [1971] *The caloric theory of gases: From Lavoisier to Regnault*, Oxford.

—— [1974] 'The rise and fall of Laplacian physics', *HSPS* **4**, pp. 89–136.

Fraser, Craig [1985] 'D'Alembert's principle: The original formulation and application in Jean d'Alembert's "*Traité de dynamique*" (1743)', *Centaurus* **28**, pp. 31–61, 145–159.

Friedman, Robert [1989] *Appropriating the weather: Vilhelm Bjerknes and the construction of a modern meteorology*, Ithaca.

Galison, Peter [1997] *Image and logic: A material culture of microphysics*, Chicago.

Garber, Elisabeth [1976] 'Thermodynamics and meteorology (1850–1900)', *Annals of science* **33**, pp. 51–65.

Garbrecht, Günther (ed.) [1987] *Hydraulics and hydraulic research: A historical review*. Rotterdam.

Gibbs-Smith, Charles [1960] *The aeroplane: An historical survey of its origins and developments*, London.

Gibson, Alfred Herbert [1946] *Osborne Reynolds and his work in hydraulics and hydrodynamics*, London.

Gillmor, Charles Stewart [1971] *Coulomb and the evolution of physics and engineering in eighteenth-century France*, Princeton.

Grattan-Guinness, Ivor [1984] 'Work for the workers: Advances in engineering mechanics and instruction in France, 1800–1830', *Annals of science* **41**, pp. 1–33.

—— [1990] *Convolutions in French mathematics, 1800–1840*, 3 vols, Basel.

Grigorian, Ashot T. [1965] 'Die Entwicklung der Hydrodynamik und Aerodynamik in den Arbeiten von N. J. Shukowski und S. A. Tschaplygin', *NTM* **5**, pp. 39–62.

—— [1971] 'Chaplygin, Sergei Alekseevich', *DSB* **3**, pp. 194–196.

—— [1976] 'Zhukovsky, Nikolay Egorovich', *DSB* **14**, pp. 619–622.

Grimberg, Gérard [1998] *D'Alembert et les équations aux dérivées partielles en hydrodynamique*, Thèse, Université Paris 7.

Hankins, Thomas [1968] 'Introduction' to english translation of d'Alembert [1743], pp. ix–xxxvi, New York.

Hanle, Paul [1982] *Bringing aerodynamics to America*, Cambridge.

Harman, Peter [1982] *Energy, force, and matter. The conceptual development of nineteenth century physics*, Cambridge.

Hashimoto, Takehiko [2000] 'The wind tunnel and the emergence of aeronautic research in Britain', in *Atmospheric flight in the twentieth century*, Alex Roland and Peter Galison (eds), pp. 223–40, Dordrecht.

Hatfield, Gary [1993] 'Helmholtz and classicism: The science of aesthetics and the aesthetics of science', in Cahan [1993] pp. 522–58.

Heidelberger, Michael [2006] 'Applying models in fluid dynamics,' *International studies in the philosophy of science* **20**, pp. 49–67.

Heilbron, John L. [1993] *Weighing imponderables and other quantitative science around 1800*, Berkeley.

Hiebert, Elfrida, and Hiebert, Erwin [1994] 'Musical thought and practice: Links to Helmholtz's *Tonempfindungen*', in Krüger [1994] pp. 295–314.

Hörz, Herbert [1997] 'Helmholtz und die Meteorologie: Bemerkungen zu Briefen von Meteorologen', in *Geomagnetism and aeronomy with special historical case studies*, ed. W. Schröder, *IAGA newsletters* **29**, pp. 201–24.

Hunt, Bruce [1991] *The Maxwellians*, Ithaca.

Kármán, Theodore von [1954] *Aerodynamics: Selected topics in the light of their historical development*, Ithaca (page numbers refer to the first McGraw-Hill paperback edition, 1963).

—— [1967] *The wind and beyond*, Boston.

Khrgian, Aleksandr Khristoforovich [1970] *Meteorology. A historical survey*, 2nd edn., vol. **1**, Jerusalem.

Knott, Cargill Gilston [1911] *Life and scientific work of Peter Guthrie Tait*, Cambridge.

Knudsen, Ole [1971] 'From Lord Kelvin's notebook: Ether speculations', *Centaurus* **16**, pp. 41–53.

Koenigsberger, Leo [1902] *Hermann von Helmholtz*, 3 vols, Braunschweig.

Kragh, Helge [2002] 'The vortex atom. A Victorian theory of everything', *Centaurus* **44**, pp. 32–114.

Kranakis, Eda [1997] *Constructing a bridge: An exploration of engineering culture, design and research in nineteenth-century France and America*, Cambridge, MA.

Krüger, Lorenz (ed.) [1994] *Universalgenie Helmholtz. Rückblick nach hundert Jahren*, Berlin.

Kuhn, Thomas [1961] 'The function of measurement in modern physical science', *Isis* **52**, pp. 161–193.

Kutzbach, Gisela [1979] *The thermal theory of cyclones: A history of meteorological thought in the nineteenth century*, Lancaster.

Lamb, Horace [1895] *Hydrodynamics* (Cambridge).

—— [1913] 'Osborne Reynolds, 1842–1912', *PRS* **A88**, pp. xv–xxi.

—— [1932] *Hydrodynamics*, 6th edn, New York: Dover.

Larmor, Joseph (ed.) [1907] *Memoirs and scientific correspondence of the late Sir George Gabriel Stokes*, 2 vols, Cambridge.

Le Tourneur, Stéphane [1954] 'Boussinesq, Joseph', in *Dictionnaire de biographie française* **7**, p. 31.

Levi, Enzo [1995] *The science of water: The foundations of modern hydraulics*, New York.

Lienhard, John [1975] 'Prandtl, Ludwig', *DSB* **11**, pp. 123–125.

Lighthill, Michael James [1968] 'Turbulence', in *Osborne Reynolds and engineering science today*, ed. D. M. McDowell and J. D. Jackson, pp. 83–146, New York.

—— [1995] 'Fluid mechanics,' in *Twentieth century physics*, ed. Laurie Brown, Abraham Pais, and Brian Pippard, pp. 795–912, New York.

Lindsay, Robert Bruce [1976] 'Strutt, John William, third baron Rayleigh', *DSB* **13**, pp. 100–107.

Liveing, George Downing [1907] 'Appreciation' [of G. G. Stokes], in Larmor [1907] vol. **1**, pp. 91–7.

McDonald, James E. [1963a] 'Early developments in the theory of the saturated adiabatic process', American Meteorological Society, *Bulletin* **44**, pp. 203–211.

—— [1963b] 'James Espy and the beginnings of cloud thermodynamics', American Meteorological Society, *Bulletin* **44**, pp. 634–641.

McDowell, Donald Malcom, and Jackson, J. D. (eds) [1970] *Osborne Reynolds and engineering science today*, New York.

McKeon, Robert M. [1974] 'Navier, Claude-Louis-Marie-Henri', *DSB* **10**, pp. 2–5.

Meleshko, Vaycheslav, and Aref, Hassan [2007], 'A bibliography of vortex dynamics 1858–1956,' *Advances in applied mechanics* **41**, pp. 197–292.

Melucci, Chiara [1996] *Scienza, spiritualità, visione politica in A. J. C. Barré de Saint-Venant: Contributi teorici e applicativi nella dinamica dei fluidi e nella scienza del miglioramento del territorio*, Doctoral dissertation, Università degli studi di Genova.

Mertz, John Theodore [1965] *A history of European thought in the nineteenth century*, New York.

Mikhailov, Gleb K. [1999] 'The origins of hydraulics and hydrodynamics in the work of the Petersburg academicians of the 18th century,' *Fluid dynamics* **34**, pp. 787–799.

—— [2002] Introduction to *Die Werke von Daniel Bernoulli*, ed. Gleb K. Mikhailov, vol. **5**, pp. 17–86, Basel.

Miles, John W. [1981] 'The Korteweg–de Vries equation: A historical essay', *Journal of fluid mechanics* **106**, pp. 131–147.

Nemenyi, Paul Felix [1962] 'The main concepts and ideas of fluid mechanics in their historical development,' *AHES* **2**, pp. 52–86.

Newman, John Nicholas [1977] *Marine hydrodynamics*, Cambridge, MA.

Olesko, Kathryn [1991] *Physics as a calling: Discipline and practice in the Königsberg seminar for physics*, Ithaca.

Parkinson, E. M. [1976] 'Stokes, George Gabriel', *DSB* **13**, pp. 74–9.

Passeron, Irène [1995] *Clairaut et la figure de la terre au XVIIIe siècle*, Thèse, Université Paris 7.

Pedersen, Kirsti [1975] 'Poiseuille, Jean Louis Marie', *DSB* **11**, pp. 62–64.

Picard, Emile [1933] *La vie et l'oeuvre de Joseph Boussinesq*, Paris.

Picon, Antoine [1992] *L'invention de l'ingénieur moderne: L'Ecole des Ponts et Chaussées, 1747–1851*, Paris.

Prony, Gaspard [1864] Biography of Navier, in Navier [1864] pp. xxxix–li.

Ravetz, Jerome [1961] 'The representation of physical quantities in eighteenth-century mathematical physics', *Isis* **52**, pp. 7–20.

Redondi, Pietro [1997] 'Along the water: The genius and the theory', in *The civil engineering of canals and railways before 1850*, ed. Mike Chrimes, pp. 143–176, Aldershot.

Reingold, Nathan [1975] 'Sabine, Edward', *DSB* **12**, pp. 49–53.

Roche, John J. [1998] *The mathematics of measurement: A critical history*, London.

Rott, Nikolaus [1992] 'Lord Rayleigh and hydrodynamic similarity', *Physics of fluids* A **4**, pp. 2595–2600.

Rotta, Julius [1990] *Die aerodynamische Versuchsanstalt in Göttingen, ein Werk Ludwig Prandtls*, Göttingen.

Rouse, Hunter, and Ince, Simon [1957] *History of hydraulics*, Ann Arbor.

Rowlinson, John Shipley [1971] 'The theory of glaciers', Royal Society of London, *Notes and records* **26**, pp. 189–204.

Russell, Robert Christopher Hamlin, and MacMillan, Donald Henry [1952] *Waves and tides*, London.

Saint-Venant, Adhémar Barré de [1864a] Commented bibliography in Navier [1864] pp. lv–lxxxiii.

Saint-Venant [1864b] 'Historique abrégé des recherches sur la résistance et sur l'élasticité des corps solides', in Navier [1864] pp. xc–cccxi.

—— [1864c] *Notice sur les travaux et titres scientifiques de M. de Saint-Venant*, Paris.

—— [1866] *Notice sur la vie et les ouvrages de Pierre-Louis-Georges, Comte Du Buat*, Lille.

—— [1887a] Annotated translation of Gerstner [1802], *APC* **13**, pp. 31–86.

—— [1887b] *Résistance des fluides: Considérations historiques, physiques et pratiques relatives au problème de l'action dynamique mutuelle d'un fluide et d'un solide, dans l'état de permanence supposé acquis par leurs mouvements*, Paris.

—— [1887c] 'Des diverses manières de poser les équations du mouvement varié des eaux courantes', *APC* **13**, pp. 148–228.

—— [1888] 'De la houle et du clapotis', *APC* **15**, pp. 705–773 (published posthumously by A. Flamant).

Schaffer, Simon [1995] 'Where experiments end: Tabletop trials in Victorian astronomy', in *Scientific practice: Theories and stories of doing physics*, J. Buchwald (ed.), pp. 257–99, Chicago.

Schiller, Ludwig [1933] 'Anmerkungen', in G. Hagen, J. L. Poiseuille, and E. Hagenbach, *Drei Klassiker der Strömungslehre*, pp. 81–97, Leipzig.

Schneider-Carius, Karl [1955] *Wetterkunde-Wetterforschung. Geschichte ihrer Probleme und Erkenntnisse in Dokumenten aus drei Jahrtausenden*, Frieburg.

Séris, Jean-Pierre [1987] *Machine et communication: Du théâtre des machines à la mécanique industrielle*, Paris.

Silliman, Robert H. [1963] 'William Thomson: Smoke rings and nineteenth-century atomism', *Isis* **54**, pp. 461–474.

Silver, Robert Simpson [1970] 'Reynolds flux concept in heat and mass transfer', in McDowell and Jackson [1970] pp. 176–89.

Smith, Crosbie [1998] *The science of energy: A cultural history of energy physics in Victorian Britain*, Chicago.

Smith, Crosbie, and Wise, Norton [1989] *Energy and empire: A biographical study of Lord Kelvin*, Cambridge.

Smith, George E. [1998] 'Newton's study of fluid mechanics', *International journal of engineering science* **36**, pp. 1377–1390.

Steele, Brett [1994] 'Muskets and pendulums: Benjamin Robins, Leonhard Euler, and the ballistics revolution,' *Technology and culture* **35**, pp. 348–82.

Stein, Howard [1981] ' "Subtler forms of matter" in the period following Maxwell', in *Conceptions of ether: Studies in the history of ether theories, 1740–1900*, ed. Geoffrey Cantor and M. J. S. Hodge, pp. 309–40, New York.

Strizhevskii, Semen Iakovlevich [1957] *Nikolai Zhukovsky, founder of aeronautics*, Moscow.

Tani, Itiro [1977] 'History of boundary-layer theory', *Annual review of fluid mechanics* **9**, pp. 87–111.

Thirriot, Claude [1987] 'Pouvoir politique et recherche hydraulique en France au XVIIᵉ et au XVIIIᵉ siècles', in Garbrecht [1987] pp. 103–14.

Thompson, Silvanus [1910] *The life of William Thomson, Baron Kelvin of Largs*, 2 vols. London.

Timoshenko, Stephen [1953] *History of strength of materials*, New York.

Todhunter, Isaac [1873] *A history of the mathematical theories of attraction and the figure of the earth*, London.

Todhunter, Isaac, and Pearson, Karl [1886, 1893] *A history of the theory of elasticity*, 2 vols, Cambridge.

Tokaty, Gregori A. [1971] *A history and philosophy of fluid mechanics*, Don Mills.

Truesdell, Clifford [1954] 'Rational fluid mechanics, 1657–1765', in L. Euler, *Opera Omnia*, ser. 2, vol. **12**, pp. IX–CXXV, Lausanne.

—— [1954b] *The kinematics of vorticity*, Bloomington.

—— [1955] 'Editor's introduction', in L. Euler, *Opera Omnia*, ser. 2, vol. **13**, VII–CV, Lausanne. Includes 'Rational fluid mechanics 1765–1786', pp. LXXII–CII.

—— [1960] *The rational mechanics of flexible or elastic bodies, 1638–1788*, published as L. Euler, *Opera Omnia*, ser. 2, vol. **11**, part 2, Zürich.

—— [1968] 'The creation and unfolding of the concept of stress', in *Essays in the history of mechanics*, pp. 184–238, Berlin.

Turner, Steven [1977] 'The Ohm–Seebeck dispute, Hermann von Helmholtz, and the origin of physiological acoustics', *British journal for the history of science* **10**, pp. 1–24.

Vilain, Christiane [2000] 'La question du "centre d'oscillation" de 1660 à 1690; de 1703 à 1743', *Physis* **37**, pp. 21–51, 439–466.

Villermaux, Emmanuel, and Clanet, Christophe [2002] 'Life of a flapping liquid sheet', *Journal of fluid mechanics* **462**, pp. 341–363.

Vogel, Stephan [1993] 'Sensation of tone, perception of sound, and empiricism', in Cahan [1993] pp. 259–87.

Warwick, Andrew [2003] *Masters of theory: Cambridge and the rise of mathematical physics*, Chicago.

Wenger, Robert [1922] 'Helmholtz als Meteorologe', *Die Naturwissenschaften* **10**, pp. 198–202.

Wilson, David B. [1987] *Kelvin and Stokes: A comparative study in Victorian physics* (Bristol).

—— [1990] Introduction to *The correspondence between Sir George Gabriel Stokes and Sir William Thomson, Baron Kelvin of Largs*, 2 vols (Cambridge), xv–xlvi.

Wise, Norton [1981] 'The flow analogy to electricity and magnetism—Part I: William Thomson's reformulation of action at a distance, *AHES* **25**, pp. 19–70.

Wolf, Charles [1889] Historical introduction to *Collection de mémoires relatifs à la physique*, vols **4–5**: *Mémoires sur le pendule* (1889, 1891), vol. **4**, pp. I–XLII, Paris.

Wright, Thomas [1983] *Ship hydrodynamics 1770–1880*, PhD dissertation, Science Museum, South Kensington, London.

—— [1992] 'Scale models, similitude and dimensions: Aspects of mid-nineteenth century engineering science', *Annals of science* **49**, pp. 233–254.

Yamalidou, Maria [1998] 'Molecular ideas in hydrodynamics', *Annals of science* **55**, pp. 369–400.

INDEX

Abbe, Cleveland, 167
Abbot, Henry, 231
Academy prizes: Berlin, 16, 19, 24, 26; Paris, 37, 43, 60; Göttingen, 153; Metz, 222
Ackroyd, John, x
Acoustics, ix_n, 146–7, 158–60, 161, 166, 180, 208
Aechylus, 202n
Airy, George Biddel, 53, 64–9, 96
Ampère, André Marie, 132
Anderson, John, x
Atmospheric motion. *See* Meteorology, Winds, Cyclones

Batchelor, George Keith, 47n
Backwater, 222–4, 226–7
Baily, Francis, 139
Balzac, Honoré de, 126n
Basset, Alfred, ix_n, 269
Battimelli, Giovanni, x
Baumgarten, André, 233
Bazin, Henry, 76, 81–3, 232–5
Beaufoy, Mark, 267, 271, 273, 278, 279n, 280, 290, 297
Belhoste, Bruno, ix
Bélanger, Jean Baptiste, 222–6
Bélidor, Bernard Forest de, 109–10
Belt, Thomas, 171
Bénard, Henri, 292
Bernoulli, Daniel: *Hydrodynamica*, 4; on efflux, 5–7; his law, 7–9, 11, 24; on d'Alembert, 15–6; on tides, 60; on organ pipes, 147; on whirling motion, 229–30; on separation, 283
Bernoulli, Johann: on live force, 5; *Hydraulica*, 9; on the compound pendulum, 9–10; on internal pressure, 11
Berthollet, Claude-Louis, 111
Bertrand, Joseph, 156–8, 234, 278
Bessel, Friedrich, 136–7, 140
Betz, Albert, 319–20
Bezold, Wilhelm von, 174
Bidone, Giorgio, 45, 225

Birkhoff, Garrett, 218
Bjerknes, Jacob, 177n
Bjerknes, Vilhelm, 177
Blaserna, Pietro, 171
Blasius, Heinrich, 284n, 285n, 292–6
Blood (and sap) circulation, 7n, 108, 145, 158
Boileau, Pierre, 133, 229, 231, 239–40
Borda, Jean-Charles de, 110, 134n, 164, 185, 230–1, 267, 327
Bossut, Charles, 104, 110, 221, 267, 273
Bouguer, Pierre, 273
Boundary conditions: for Euler's equation, 26; for the Navier-Stokes equation, 116–7, 139, 144, 211
Boundary layers: anticipated by Thomson, 205–7, 269; turbulent layers described by Rankine, 275–6; by Froude, 280–1; by Prandtl, 293; Boussinesq's theory of laminar layers, 272–3; Prandtl's, 283–9, 291–2; Lanchester's, 289–90; Rayleigh's, 290–1; Blasius's, 292; instability of laminar layer, 295; Kármán's theory of turbulent layers, 296–7, 299–300; Prandtl's, 297–9, 301–2
Bourgois, Siméon, 279n
Boussinesq, Joseph: background, 233–4; on solitary waves, 76–8; his and KdV's equation, 79–81; on river/torrents, 81–3, 238; on Poiseuille flow, 234–5; on meanders, 234; on effective viscosity, 235, 295–6, 300; on permanent, open-channel flow, 235–9; on boundary layers, 271–3
Brémontier, Théodore, 82n
Bresse, Jacques, 272
Brillouin, Marcel, 207
British Association, 47–8, 51, 56, 69, 274, 277, 279
Brunel, Isambard, 277
Bullough, Robin, ix

Canals, 48, 103n, 104n, 106, 221–2, 232. *See also* Open channels
Capillarity, 55–6, 59, 87–91, 106

Carnot, Lazare, 109

Cartwright, David Edgar, x

Cauchy, Augustin: on Lagrange's theorem, 43–4; on waves, 43–5; on stress and strain, 119–20, 137, 324; molecular theory of elasticity, 122–4; on viscous fluids, 125–6; on the instantaneous rotation of mass elements, 137, 149n; on complex variables, 163

Cavitation, 199–201, 246

Cavaillé Coll, Aristide, 166n

Chézy, Antoine, 104n, 222

Chladni, Ernst, 46, 112, 215

Celerity, 82

Challis, James, 47

Chaplygin, Sergei Alekseevich, 315

Circulation, 154, 305, 308, 310–1, 312–6

Clairaut, Alexis Claude, 20

Clebsch, Rudolph, 153

Clausius, Rudolph, 125n, 323

Complex plane (for two-dimensional flow), 21–2, 162–5. See also Conformal transformations

Compressible fluids, ix_n, 4, 110n, 316n. See also Acoustics

Condorcet, Marquis de, 221

Conformal transformations, 162–5, 310, 315. See also Complex plane

Continuity equation, 1, 17, 19, 21, 24–5, 29

Convective acceleration: for J. Bernoulli, 10–1; for d'Alembert, 10n, 20; for Euler, 24

Coriolis, Garpard, 65, 109, 171n, 225, 227, 237–9

Couette, Maurice, 212

Coulomb, Charles Augustin, 105, 110, 119, 128, 139, 222, 244

Couplet, Claude, 104, 232n

Cournot, Antoine, 101, 118–9

Craik, Alex, ix

Crookes, William, 248

Cyclones, 170–2

D'Alembert, Jean le Rond: principle of dynamics, 11–4, 278; on efflux, 14–5; on winds, 16–19; on fluid resistance, 19–23, 103, 221; his paradox, 22; on complex numbers, 21–2, 162–3

D'Alembert's paradox: d'Alembert's formulation, 22, 103, 267; Euler's anticipation, 103, 267; Saint-Venant's derivation, 134; Borda's derivation, 134n, 267; Froude's, 282; Rayleigh's solution, 165; modern discussion, 326–8

Darcy, Henry, 232–3, 275, 280n

Darwin, George, 213

Da Vinci, Leonardo, 105, 229

Delaunay, Charles-Eugène, 323

De Vries, Gustav, 84

Dimensional considerations: Fourier's, 37; Poisson's, 37; Reynolds's, 248–9; 257. See also Similarity conditions, Scale models

Dines, William Henry, 201, 269

Discontinuity surfaces: introduced by Stokes, 184–6; defined by Helmholtz, 152, 160–3; instability, 161–3; and conformal mapping, 163–5; in organ pipes, 165–6; and atmospheric motion, 172–8; debated by Stokes and Thomson, 197–207; according to Prandtl, 285–8; and resistance, 267–70; in wing theory, 303, 308, 316–7

Dove, Heinrich, 167, 170–2, 174

Drag (induced), 307, 317–21. See also Resistance

Du Buat, Pierre, 104–5, 110, 139, 221–2, 232, 267, 271

Duhamel, Jean Marie Constant, 184

Dupin, Charles, 118

Dupuis de Lôme, Stanislas, 279n

Earnshaw, Samuel, 70, 174

Ecole des Ponts et Chaussées, 109, 111, 119, 124, 129, 222

Ecole Polytechnique, 109, 119, 122, 129, 135, 141, 222, 234

Efflux: for D. Bernoulli, 5–7; for J. Bernoulli, 10–1; for D'Alembert, 14–5; for Lagrange, 35

Eiffel, Gustave, 293

Elasticity, theory of: before Navier, 111–2; Navier's, 112–5; Cauchy's, 119–21; Poisson's, 122–5; Saint-Venant's, 129–31; Stokes's, 138–9

Ellis, Alexander, 166n

Electromagnetism, 151–3, 190–1

Energetic arguments: 131, 134, 152, 158, 180–1, 190–1, 193, 216–7, 225–7, 261–2, 302–3

Ether: Stokes's views, 187; with rotating motes, 191; as a vortex sponge, 194, 196–7, 240–3
Espy, James, 167n, 170–1
Exner, Felix, 177
Euler, Leonhard: on internal pressure, 23; on hydraulics, 23n; his equations, 24–6; relations with d'Alembert, 25–6; on tides, 60; on fluid resistance, 103, 110, 267; on organ pipes, 147; on cavitation, 246n
Euler's equation: 1, 24–5
Eytelwein, Johann, 140

Faraday, Michael, 215
Ferrel, William, 170n, 171
Feynman, Richard, 323
Finsterwalder, Sebastian, 310n
FitzGerald, George Francis, 194–5, 242–3
Flachat, Eugène, 279n
Flamant, Alfred, 239
Fontaine, Alexis, 19
Föppl, August, 269, 283, 298n
Forbes, James, 167
Forchheimer, Philipp, 239
Fourier, Jean Baptiste Joseph: dimensions, 37; integrals, 34–5, 38, 41–2, 100; series, 117, 147, 212–3; on heat, 121, 135
Fresnel, Augustin, 46, 119n, 125n, 135
Froude, William: background, 277; on waves, 75, 85–6, 278, 282; on scaling rules, 278; ship models, 279–80, 282; towing tank, 280; on skin friction, 278, 280–2, 297
Froude number, 278

Gauthey, Emiland, 109
Germain, Sophie, 112
Gerstner, Franz Joseph von, 72–3, 275n
Girard, Pierre-Simon, 106–8, 116–8, 128, 139–40, 143, 243
Goethe, Wolfgang, 166
Graeff, Michel, 239
Graham, Thomas, 234, 248
Green, George, 73, 125, 148
Greenhill, Alfred, 192n
Grimberg, Gérard, ix, 2n
Group velocity, 41, 85–7

Hadley, George, 169–70
Hagen, Gotthilf, 140–1, 244–5

Hagenbach, Eduard, 143
Hamilton, William Rowan, 155, 323
Hällström, Gustav, 147
Hankel, Hermann, 153
Hanle, Paul, x
Hann, Julius, 167, 174n
Havelock, Thomas, 97–8
Heisenberg, Werner, 216, 296n
Helmholtz, Hermann: background, 146; on viscous flow, 144, 158–9; on acoustics, 146–8; on organ pipes, 148; on vortex motion, 148–58; on discontinuity surfaces, 159–66, 174–7, 285–6; on glaciers, 167; on the foehn, 167–8; on trade winds, 169–70, 172–4; on tropical storms, 171–2; anticipating front theory, 177–8; on waves, 88, 178–82; on balloon steering, 257–8
Hicks, William, 158, 192n
Hiemenz, Karl, 292
Hinch, John, 192n
Hopkins, William, 136
Houston, William (his paradox), 48–51, 68–9, 91–3
Humphreys, Andrew, 231
Hydraulic jumps, 224–5, 238
Hydraulics: early, 4n; versus hydrodynamics, vi, 4, 103–4, 221–2; Bossut's, 104, 221; Du Buat's, 104–5, 221–2; Bélidor's, 109–10; Saint-Venant's, 131, 229–31; Boileau's, 133; Eytelwein's, 140; Hagen's, 140; Bélanger's, 221–6; Coriolis's, 226–7; Darcy's and Bazin's, 232–3; Boussinesq's, 233–9; Graeff's 239; Forchheimer's, 239; Flamant's, 239. *See also* Open channels, Pipes
Huygens, Christian, 4–5, 265

Ince, Simon, ix
Instability: Kelvin-Helmholtz, 161–3, 166, 172–3, 190; of atmospheric motion, 168; of water surface under wind, 179–81, 188–90; of vortices, 194–7; of vortex sponge, 196–7; of water jets, 208; of smoke-jets, 208–10; wall-induced, 211, 213, 215–6, 295–6; of pipe flow, 140–1, 210–1, 244–5, 249–52; of laminar boundary layer, 293–5. *See also* Stability
Irrotational flow. *See* Potential flow

Jacobson, Heinrich, 143
Joukowski, Nikolai, 290n, 311, 312–6
Joule, James, 230, 246, 259

Kármán, Theodore von: on history, x; on energetic instability, 217; his vortex street, 292–3; his momentum equation, 296; on turbulent boundary layers, 296–7, 299–301
KdV equation, 78n, 81n, 84–5
Kelland, Philip, 69–70, 76
Kelvin, Lord. See Thomson, William
Kirchhoff, Gustav, 19, 148, 149n, 153, 156, 164–5, 268, 303, 315n
Klein, Felix, 207, 291
Korteweg, Diederik Johannes, 84–5
Koženy, Josef, 239
Köppen, Wladimir, 313
Krümmel, Otto, 182
Kuhn, Thomas, 323
Kutta, Wilhelm, 310–1, 315
Kutzbach, Gisela, 177

Lagrange, Joseph Louis de: on history, 2n; surface condition, 26; theorem for potential flow, 26–7, 43–4, 187, 197, 207; on narrow flow, 27; analytical mechanics, 28; on waves, 35–7; on elasticity, 112; on organ pipes, 147; on conformal mapping, 163
Lagrangian picture, 29–30, 33, 43, 65, 153
Lamb, Horace: treatise, ix$_n$, 144, 182, 269, 305; on waves, 42, 96–7
Laminar flow, 219, 239
Lanchester, Frederick: on boundary layers, 289–90; on wing theory, 305–10, 316–7
Langley, Samuel, 269, 302, 305, 307
Laplace, Pierre-Simon de: theory of tides, 32, 61–2, 65; on waves, 33–5, 37; molecular physics, 106, 110–1, 122
Le Conte, John, 208
Legendre, Adrien Marie, 130
Leibniz, Gottfried Wilhelm, 5
Lesbros, Joseph Aimé, 88
Levi-Civita, Tullio, 76, 207n, 269–70
Lévy, Maurice, 232n
Lilienthal, Otto, 302–3, 307, 310
Lift, 303, 305–8, 310–1, 313–5, 317, 319–21
Lin, Chia Chiao, 216

Live forces, principle of, 4–5
Loomis, Elias, 172n
Lorentz, Hendrik Antoon, 217, 323
Lubbock, John, 56
Luvini, Giovanni, 178

MacCullagh, James, 243
MacLaurin, Colin, 20n, 60
Magnus, Gustav, 303–5
Mariotte, Edme, 103–4, 185, 264
Mathieu, Emile, 143
Margules, Max, 177, 212n
Maxwell, James Clerk: on lines of force, 141, 155–7; on kinetic-theory, 144, 246, 248–9; whirling model of magnetic field, 220; on falling paper strips, 313; electromagnetic theory, 323
Mayer, Alfred, 192
Melucci, Chiara, x
Merrifield, Charles, 279–80
Meteorology: empirical nature, 166–7; for Helmholtz, 168–78. See also Winds, Cyclones
Mixing, 166, 174, 259, 297–8
Mixing length, 232n, 281n
Moebius, Ferdinand, 73
Miles, John, ix
Models. See Scale models
Moigno, François, 157
Mohn, Henrik, 172n
Molecular physics: Laplace's, 106, 110–1; Navier's, 111, 124–5; Cauchy's, 121; Poisson's, 122–4; Saint-Venant's, 129–31
Mouillard, Louis Pierre, 312–3
Munk, Max, 320–1

Napier, James, 274, 277
Navier, Claude Louis: background, 109–11; on energy, 109, 225; on elasticity, 111–5; on viscous fluids, 115–9; conflict with Poisson, 124–5; Pont des Invalides, 126–8; linear/nonlinear flow, 219, 229; on backwaters, 226
Navier-Stokes equation: derived by Navier, 115–6; by Cauchy, 121; by Poisson, 125–6; by Saint-Venant, 132–3; by Stokes, 137–8; by Helmholtz, 144; by Maxwell, 148; general validity, 101, 144, 249

Newton, Isaac: theory of tides, 18, 60; on waves, 33, 37; on fluid resistance, 47, 265–6, 273, 302; on fluid friction, 105, 231; on similarity, 278–9; founder of mechanics, 323

Navigation. *See* Ship propellers, Ship resistance, Ship rolling, Ship waves, Tides, Waves.

Neumann, Franz, 143

Nikuradse, Johann, 300

Notations, xvii, 2, 19

Ohm, Georg Simon, 147

Open channels: river tides, 65–6, 81–2; Saint-Venant's theory, 132–3; Boussinesq's, 233–9; Boileau's measurements, 133; Darcy's and Bazin's, 233–4; uniform permanent flow, 104, 221–2; gradually varying flow, 222–4, 236–7, 236–8; rapidly varying flow, 238; hydraulic jumps, 224–5. *See also* Rivers, Canals

Orr, William, 214, 217

Oseen, Carl Wilhelm, 187

Organ pipes. *See* Acoustics

Partiot, Henri, 82

Pendulum oscillations, 136–7, 139–40

Picon, Antoine, ix

Piotrowski, Gustav von, 144

Pipe retardation: for Du Buat, 104; Prony's law 104; for Girard, 106–8; for Stokes, 139; Eytelwein's law, 140; Hagen's experiments, 140–1; Poiseuille's, 141–3; laminar theory, 142–4; Darcy's measurements, 232; Reynolds', 254–6; fractional-exponent laws, 256; Weisbach's law, 175; Froude's involvement, 280; Blasius's law, 296, 299; Kármán's theory, 299–300. *See also* Hydraulics.

Pitot, Henri de, 232

Plateau, Joseph, 208

Poiseuille, Jean-Louis, 116, 141–4

Poisson, Siméon Denis: on waves, 37–42; on elasticity, 112, 122–4; on effective mass, 137; on organ pipes, 148

Poncelet, Jean Victor, 88, 109, 134, 226, 230, 270–1, 313

Potential flow, ix$_n$, 20–1, 26–7, 149, 276–7. *See also* Velocity potential

Prandtl, Ludwig: background, 283; Göttingen career, 291–2; laminar boundary layers, 283–9, 291–2; on instabilities, 294–5; on developed turbulence, 296; turbulent boundary layers, 293–4, 301; mixing length, 232n, 281n, 297–9; on wing theory, 316–22

Pressure: velocity dependence, 7–9, 11, 24; internal, 1, 8n, 11, 15, 23; molecular definition, 122–3, 130; negative, 162. *See also* Stress system

Prony, Gaspard de, 104, 126, 128, 140, 222

Rankine, William John Macquorn, 75, 191, 274–7

Rayleigh, Lord (John William Strutt): background, 208; on solitary waves, 83; on group velocity 87; on the fishing line, 88–91; on fluid resistance (plate), 164–5, 201, 268, 303; on the Kelvin-Helmholtz instability, 190; on sound, 208; on the stability of parallel flow, 208–10, 214; on wall-induced instability, 215; on boundary layers, 290–1; on the Aeolian harp, 292; on aeronautics, 302–3; on the Magnus effect, 303–5, 308, 310

Reech, Ferdinand, 75, 278–9

Reed, Edward, 278n, 279

Resistance: for d'Alembert, 19–22; for Euler, 103, 110; for Saint-Venant, 134–5; for Stokes, 137, 139–40, 140; Rayleigh's theory, 164–5, 201–2. *See also* Ship resistance, D'Alembert's paradox, Houston, Pendulum, Wing theory, Drag, Lift

Reusch, Friedrich Eduard, 155n

Reye, Theodor, 168–71

Reynolds, Osborne: background, 245–6; on group velocity, 86–7; Reynolds number, 210, 249; critical number, 210, 249; on the turbulent transition, 210–1, 211–2, 250–4; on two kinds of instability, 211, 252; Reynolds stress, 240, 261, 298; on vortices, 246–7; on cavitation, 246–7; on the radiometer, 248; on transpiration, 248; on pipe discharge, 254–6; on scales, 260–1; on the kinetic theory of turbulence, 260–2

Riabouchinsky, Dimitry, 313

Riemann, Bernhard, 153, 163–4

Rivers: tides in, 56, 60, 65–6, 81–2; versus torrents, 82, 227–8, 238; navigability, 222, 224. *See also* Open channels

Robison, John, 52

Rogers, William, 155n

Rouse, Hunter, ix

Routh, Edward, 208

Runge, Carl, 310

Russell, John Scott: on Houston's paradox, 48; on solitary waves, 48–51; on ship resistance, 51, 274; on waves in general, 51–9, 72–3; on tides, 56, 60; on group velocity, 85

Sabine, Edward, 136, 139

Saint-Venant, Adhémar Barré de: on history, ix; on river/torrents, 82, 227–8; on the impossibility of a continuous solid, 125; on Navier's bridge, 126–8; on elasticity, 129–31, 323; on effective viscosity, 131–3, 229–31; on fluid resistance, 134–5, 270–1; tumultuous/regular flow, 219; Boussinesq's mentor, 233–4, 238; on rapidly varying flow, 238; on pipe retardation, 256

Savart, Félix, 46, 208, 215

Savary, Félix, 132

Scale (turbulent versus laminar), 219, 229, 235, 260–2; scale invariance, 299. *See also* Dimensional considerations, Scale models

Scale models, 257–8, 277–80, 282–3, 293

Schklarewsky, Alexis, 158

Schmidt, Wilhelm, 297–8

Schneebeli, Heinrich, 166n

Seebeck, Thomas, 147

Separation (of boundary layer): anticipated by Stokes, 206–7, 269; predicted by Prandtl, 285–6; retarded by turbulence, 293–4

Ship propellers, 199

Ship resistance: Older views 273; Russell's, 51, 274; Rankine's, 274–7; Froude's, 277–82; wave component, 51, 274, 282; skin friction, 273–6, 280–1; the three causes, 275, 282; streamlining, 276–7. *See also* Houston, Scale models, Froude

Ship rolling, 73, 75

Ship waves, 51, 91–100

Silbermann, Andreas, 158–9

Similarity (or similitude) **conditions**: Stokes-Reynolds, 257; Helmholtz's, 173–4, 178, 257–8; Froude's, 278; Reech's 278–9

Simpson, Thomas, 130

Smith, Crosbie, x

Smith, Hermann, 166n

Solitary waves: Russell's discovery, 48–51; discussed by Airy, 69; by Stokes, 70–1, 76; by Thomson, 76; by Boussinesq, 76–8; by Rayleigh, 83

Solidification, principle of, 20n

Sommerfeld, Arnold, 164, 249

Stability: of vortex rings, 192; of columnar vortices, 192–4; of potential flow, 197–207; Rayleigh's criterion, 208–10, 214; of Couette flow, 212; of Poiseuille flow, 212–4; energetic approach, 216–7. *See also* Instability

Stokes, George Gabriel: background, 135–6; on water waves, 69–76; on group velocity, 86; on the pendulum, 136–7, 139–40; on the Navier-Stokes equation, 137–8; on elasticity, 138–9; on discontinuity surfaces, 184–6, 197–207; on instability of divergent flow, 184–7; on ether drag, 187; on instability of vortices, 195; debate with Thomson, 197–207; on viscous dissipation, 259–60; on separation, 206–7, 269

Stream function, 22n, 83, 163–4

Streamlining, 276–7

Stress system: defined by Cauchy, 120; by Saint-Venant, 130; for fluids, 126, 133, 138; Reynolds stress, 240, 261, 298; Maxwell's definition, 260; Prandtl's *Ansatz*, 297–9; Kármán's, 299. *See also* Navier-Stokes equation, Elasticity

Tait, Peter Guthrie, 155–6, 191, 193

Taylor, Geoffrey, 262

Thermodynamics, 167–8, 171, 174

Thomson, James, 239–40

Thomson, Joseph John, 192n

Thomson, William (Lord Kelvin): on solitary waves, 76; on capillarity waves, 87–8; on wind waves, 178, 179n, 188–90; on ship waves, 91–100; on stationary phase, 96–7; on vortex motion, 155–7;

circulation theorem, 155, 305, 316; on atmospheric temperature, 168; "mike" theorem, 190–1; on the vortex atom, 191–7; on the vortex sponge, 194–5, 196–7, 240–2; against discontinuity surfaces 197–207; on the stability of parallel flow, 212–4, 295; naming turbulence, 219; on meanders, 234; on turbulent rigidity, 240–2; on a boundary layer, 205–7, 269; against artificial flight, 302

Tides: Newton's theory, 18, 60; equilibrium theory, 60–1; Laplace's theory, 32, 61–2; Russell's, 56, 60; Young's, 62–3; Airy's, 63–8; Whewell's approach, 63–4; Boussinesq on river tides, 238

Tietjens, Oskar, 294–5

Tokaty, Gregori, ix

Tolmien, Walter, 216, 295, 298

Torricelli, Evangelista, 7

Truesdell, Clifford, ix, 2n, 15n

Turbulence: defined, 219; transition to, 210–1, 244–5; 249–262; and effective viscosity, 229–31, 235; in the ether, 242; in boundary layers, 275–6, 280–1

Turgot, Anne Robert Jacques, 221

Tyndall, John, 161, 167, 208, 258

Vaulthier, Pierre, 226

Velocity potential, 22n, 24, 26–7, 163–4. See also Potential flow

Vena contracta, 7n, 107, 164, 265n

Venturi, Giovanni Battista, 105–6, 229–30

Vicat, Louis, 128–9

Vince, Samuel, 267–8

Virtual velocities, principle of, 13, 28, 112–3

Viscosity: early notions, 104–8; molecular, 116; as temporary solidity, 121, 125–6; as shear stress, 132–3, 138; effective (eddy), 133, 229–31, 235, 296–8, 301; vanishingly small, 198–9, 215–6, 284, 289. See also Navier-Stokes equation

Vortex atom. See Vortex rings

Vortex rings, 153, 156, 190–7.

Vortex sheets. See Discontinuity surfaces

Vortex sponge, 194, 196–7, 240–3

Vortex street, 292–3

Vortices: Newton's consideration, 105; drawn by Vinci and by Venturi, 106; Helmholtz's theory, 149–52; for Reye, 168–71; tornados, 171; coreless, 195, 199–200; visualized by Reynolds, 246–8; behind a wing, 307–8, 317–8; bound vortices, 314, 321. See also Vortex rings, Discontinuity surfaces, Vortex sponge, Vorticity equation

Vorticity equation, 20, 24, 44, 149, 155, 209

Waves: Laplace's theory, 33–5; Lagrange's, 35–7; Poisson's, 37–42; Cauchy's, 43–5; Airy's, 63–9; Stokes's, 69–76; Gerstner's, 73–5; Rayleigh's, 88–91; Thomson's, 87–8, 91–100; Helmholtz's, 178–82; nonlinear, 66–7, 70–85, 179–81; breaking, 60, 68, 75, 82, 180; under wind, 179–82, 188–90; atmospheric waves, 178–9; Bidone's experiments, 45; the Webers', 45–6; Russell's, 48–59. See also Solitary waves, Tides, Ship resistance, Ship rolling, Ship waves, Capillarity, Group velocity, Rivers

Weber, Ernst Heinrich, 45–6, 73

Weber, Wilhelm, 45–6, 73, 147

Weierstrass, Karl, 164

Weisbach, Julius, 275

Wertheim, Guillaume, 147–8

Whewell, William, 56, 63–4

Wien, Wilhelm, vi, 180n, 181

Winds: d'Alembert's theory, 19–22; foehn, 167–8; trade-winds, 169–70, 172–4; Aeolian harp, 292. See also Waves (under wind)

Wing theory: Rayleigh's, 302–3; Lanchester's, 305–10; Kutta's, 310–1; Chaplygin's and Joukowski's, 315–6; Prandtl's, 316–22. See also Drag, Lift

Wise, Norton, x

Woltman, Reinhardt, 233

Wright, Orville, 302

Wright, Thomas, x

Wright, Wilbur, 302, 303–4

Young, Thomas, 46, 62–3, 119

Zamminer, Friedrich, 148, 158